x/B

D1391053

Digital Principles
and
Logic Design

Digital Principles
and
Logic Design

A. SAHA
N. MANNA

INFINITY SCIENCE PRESS LLC
Hingham, Massachusetts
New Delhi

Publisher: David F. Pallai

INFINITY SCIENCE PRESS LLC
11 Leavitt Street
Hingham, MA 02043
Tel. 877-266-5796 (toll free)
Fax 781-740-1677
info@infinitysciencepress.com
www.infinitysciencepress.com

This book is printed on acid-free paper.

A. Saha and N. Manna. *Digital Principles and Logic Design.*
ISBN: 978-1-934015-03-2

The publisher recognizes and respects all marks used by companies, manufacturers, and developers as a means to distinguish their products. All brand names and product names mentioned in this book are trademarks or service marks of their respective companies. Any omission or misuse (of any kind) of service marks or trademarks, etc. is not an attempt to infringe on the property of others.

Library of Congress Cataloging-in-Publication Data

Saha, A. (Arjit)
  Digital principles and logic design / A. Saha and N. Manna.
      p. cm.
  Includes bibliographical references and index.
  ISBN 978-1-934015-03-2 (hardcover with cd-rom : alk. paper)
  1. Electric circuits--Design and construction. 2. Digital electronics. 3. Logic design. I. Manna, N. (Nilotpal) II. Title.
  TK454.S3135 2007
  621.319'2--dc22
                          2007013970

07 8 9 5 4 3 2 1

Our titles are available for adoption, license or bulk purchase by institutions, corporations, etc. For additional information, please contact the Customer Service Dept. at 877-266-5796 (toll free in US).

Requests for replacement of a defective CD-ROM must be accompanied by the original disc, your mailing address, telephone number, date of purchase and purchase price. Please state the nature of the problem, and send the information to INFINITY SCIENCE PRESS, 11 Leavitt Street, Hingham, MA 02043.

The sole obligation of INFINITY SCIENCE PRESS to the purchaser is to replace the disc, based on defective materials or faulty workmanship, but not based on the operation or functionality of the product.

Dedication
To our parents
who have shown us
the light of the world.

CONTENTS

PREFACE

With the advancement of technology, digital logic systems became inevitable and became the integral part of digital circuit design. Digital logic is concerned with the interconnection of digital components and modules, and is a term used to denote the design and analysis of digital systems. Recent technology advancements have led to enhanced usage of digital systems in all disciplines of engineering and have also created the need of in-depth knowledge about digital circuits among the students as well as the instructors. It has been felt that a single textbook dealing with the basic concepts of digital technology with design aspects and applications is the standard requirement. This book is designed to fulfill such a requirement by presenting the basic concepts used in the design and analysis of digital systems, and also providing various methods and techniques suitable for a variety of digital system design applications.

This book is suitable for an introductory course of digital principles with emphasis on logic design as well as for more advanced courses. The contents of this book are chosen and illustrated in such a way that there does not need to be any special background knowledge on the part of the reader.

The philosophy underlying the material presented in this book is to describe the classical methods of design technique. The classical method has been predominant in the past for describing the operation of digital circuits. With the advent of integrated circuits, and especially the introduction of microprocessors, microcontrollers, microcomputers and various LSI components, the classical method seems to be far removed from practical applications. Although the classical method of describing complex digital systems is not directly applicable, the basic concepts of Boolean algebra, combinational logic, and sequential logic procedures are still important for understanding the internal construction of many digital functions. The philosophy of this book is to provide a strong foundation of basic principles through the classical approach before engaging in practical design approach and the use of computer-aided tools. Once the basic concepts are mastered, the utilization of practical design technique and design software become meaningful and allow the students to use them more effectively.

The book is divided into 11 chapters. Each chapter begins with the introduction and ends with review questions and problems. Chapter 1 presents various binary systems suitable for representation of information in digital systems and illustrates binary arithmetic. Chapter 2 describes various codes, conversion, and their utilization in digital systems.

Chapter 3 provides the basic postulates and theorems related to Boolean algebra. The various logic operations and the correlation between the Boolean expression and its implementation with logic gates are illustrated. The various methods of minimization and simplification of Boolean expressions, Karnaugh maps, tabulation method, etc. are explained in Chapter 4. Design and analysis procedures for combinational circuits are provided in Chapter 5. This chapter also deals with the MSI components. Design and implementation of combinational circuits with MSI blocks like adders, decoders, and multiplexers are explained with examples. Chapter 6 introduces LSI components—the read-only memory (ROM) and various programmable logic devices (PLD), and demonstrates design and implementation of complex digital circuits with them.

Chapter 7 starts with the introduction of various types of flip-flops and demonstrates the design and implementation of sequential logic networks explaining state table, state diagram, state equations, etc. in detail. Chapter 8 deals with various types of registers and sequence generators. Chapter 9 illustrates synchronous and asynchronous types of counters, and design and application of them in detail.

Chapter 10 discusses various methods of digital-to-analog conversion (DAC) as well as analog-to-digital conversion (ADC) techniques. Chapter 11 deals with the various logic families and their characteristics and parameters with respect to propagation delay, noise margin, power dissipation, power requirements, fan out, etc. Appendices have been provided at the end of the book as ready reference for 74-series and 4000-series integrated circuit functions and their pinout configurations.

Clear diagrams and numerous examples have been provided for all the topics, and simple language has been used throughout the book to facilitate understanding of the concepts and to enable the readers to design digital circuits efficiently.

The authors express their thanks to their respective wives and children for their continuous support and enormous patience during the preparation of this book.

The authors welcome any suggestions and corrections for the improvement of the book.

—AUTHORS

1 DATA AND NUMBER SYSTEMS

1.1 INTRODUCTION

One of the first things we have to know is that electronics can be broadly classified into two groups, *viz.* analog electronics and digital electronics. Analog electronics deals with things that are continuous in nature and digital electronics deals with things that are discrete in nature. But they are very much interlinked. For example, if we consider a bucket of water, then it is analog in terms of the content *i.e.*, water, but it is discrete in terms of the container, *i.e.*, bucket. Now though in nature most things are analog, still we very often require digital concepts. It is because it has some specific advantages over analog, which we will discuss in due course of time.

Many of us are accustomed with the working of electronic amplifiers. Generally they are used to amplify electronic signals. Now these signals usually have a continuous value and hence can take up any value within a given range, and are known as *analog signals*. The electronic circuits which are used to process such signals are called *analog circuits* and the circuits based on such operation are called analog *systems*.

On the other side, in a computer, the input is given with the help of the switches. Then this is converted into electronic signals, which have two distinct discrete levels or values. One of them is called HIGH level whereas the other is called LOW level. The signal must always be in either of the two levels. As long as the signal is within a prespecified range of HIGH and LOW, the actual value of the signal is not that important. Such signals are called *digital signals* and the circuit within the device is called a *digital circuit*. The system based on such a concept is an example of a *digital system*.

Since Claude Shannon systemized and adapted the theoretical work of George Boole in 1938, digital techniques saw a tremendous growth. Together with developments in semiconductor technology, and with the progress in digital technology, a revolution in digital electronics happened when the microprocessor was introduced in 1971 by Intel Corporation of America. At present, digital technology has progressed much from the era of vacuum tube circuits to integrated circuits. Digital circuits find applications in computers, telephony, radar navigation, data processing, and many other applications. The general properties of number systems, methods of their interconversions, and arithmetic operations are discussed in this chapter.

1.2 NUMBER SYSTEMS

There are several number systems which we normally use, such as decimal, binary, octal, hexadecimal, etc. Amongst them we are most familiar with the decimal number system. These systems are classified according to the values of the base of the number system. The number system having the value of the base as 10 is called a decimal number system, whereas that with a base of 2 is called a binary number system. Likewise, the number systems having base 8 and 16 are called octal and hexadecimal number systems respectively.

With a decimal system we have 10 different digits, which are 0, 1, 2, 3, 4, 5, 6, 7, 8, and 9. But a binary system has only 2 different digits—0 and 1. Hence, a binary number cannot have any digit other than 0 or 1. So to deal with a binary number system is quite easier than a decimal system. Now, in a digital world, we can think in binary nature, *e.g.*, a light can be either off or on. There is no state in between these two. So we generally use the binary system when we deal with the digital world. Here comes the utility of a binary system. We can express everything in the world with the help of only two digits *i.e.*, 0 and 1. For example, if we want to express 25_{10} in binary we may write 11001_2. The right most digit in a number system is called the 'Least Significant Bit' (LSB) or 'Least Significant Digit' (LSD). And the left most digit in a number system is called the 'Most Significant Bit' (MSB) or 'Most Significant Digit' (MSD). Now normally when we deal with different number systems we specify the base as the subscript to make it clear which number system is being used.

In an octal number system there are 8 digits—0, 1, 2, 3, 4, 5, 6, and 7. Hence, any octal number cannot have any digit greater than 7. Similarly, a hexadecimal number system has 16 digits—0 to 9— and the rest of the six digits are specified by letter symbols as A, B, C, D, E, and F. Here A, B, C, D, E, and F represent decimal 10, 11, 12, 13, 14, and 15 respectively. Octal and hexadecimal codes are useful to write assembly level language.

In general, we can express any number in any base or radix "X." Any number with base X, having n digits to the left and m digits to the right of the decimal point, can be expressed as:

$$a_n X^{n-1} + a_{n-1} X^{n-2} + a_{n-2} X^{n-3} + ... + a_2 X^1 + a_1 X^0 + b_1 X^{-1} + b_2 X^{-2} + ... + b_m X^{-m}$$

where a_n is the digit in the nth position. The coefficient a_n is termed as the MSD or Most Significant Digit and b_m is termed as the LSD or the Least Significant Digit.

1.3 CONVERSION BETWEEN NUMBER SYSTEMS

It is often required to convert a number in a particular number system to any other number system, *e.g.*, it may be required to convert a decimal number to binary or octal or hexadecimal. The reverse is also true, *i.e.*, a binary number may be converted into decimal and so on. The methods of interconversions are now discussed.

1.3.1 Decimal-to-binary Conversion

Now to convert a number in decimal to a number in binary we have to divide the decimal number by 2 repeatedly, until the quotient of zero is obtained. This method of repeated division by 2 is called the 'double-dabble' method. The remainders are noted down for each of the division steps. Then the column of the remainder is read in reverse order *i.e.*, from bottom to top order. We try to show the method with an example shown in Example 1.1.

Example 1.1. *Convert 26_{10} into a binary number.*

Solution.

Division	Quotient	Generated remainder
$\dfrac{26}{2}$	13	0
$\dfrac{13}{2}$	6	1
$\dfrac{6}{2}$	3	0
$\dfrac{3}{2}$	1	1
$\dfrac{1}{2}$	0	1

Hence the converted binary number is 11010_2.

1.3.2 Decimal-to-octal Conversion

Similarly, to convert a number in decimal to a number in octal we have to divide the decimal number by 8 repeatedly, until the quotient of zero is obtained. This method of repeated division by 8 is called 'octal-dabble.' The remainders are noted down for each of the division steps. Then the column of the remainder is read from bottom to top order, just as in the case of the double-dabble method. We try to illustrate the method with an example shown in Example 1.2.

Example 1.2. *Convert 426_{10} into an octal number.*

Solution.

Division	Quotient	Generated remainder
$\dfrac{426}{8}$	53	2
$\dfrac{53}{8}$	6	5
$\dfrac{6}{8}$	0	6

Hence the converted octal number is 652_8.

1.3.3 Decimal-to-hexadecimal Conversion

The same steps are repeated to convert a number in decimal to a number in hexadecimal. Only here we have to divide the decimal number by 16 repeatedly, until the quotient of zero is obtained. This method of repeated division by 16 is called 'hex-dabble.' The remainders are noted down for each of the division steps. Then the column of the remainder is read from bottom to top order as in the two previous cases. We try to discuss the method with an example shown in Example 1.3.

Example 1.3. *Convert 348_{10} into a hexadecimal number.*

Solution.

Division	Quotient	Generated remainder
$\dfrac{348}{16}$	21	12
$\dfrac{21}{16}$	1	5
$\dfrac{1}{16}$	0	1

Hence the converted hexadecimal number is $15C_{16}$.

1.3.4 Binary-to-decimal Conversion

Now we discuss the reverse method, *i.e.*, the method of conversion of binary, octal, or hexadecimal numbers to decimal numbers. Now we have to keep in mind that each of the binary, octal, or hexadecimal number system is a positional number system, *i.e.*, each of the digits in the number systems discussed above has a positional weight as in the case of the decimal system. We illustrate the process with the help of examples.

Example 1.4. *Convert 10110_2 into a decimal number.*

Solution.   The binary number given is   **1 0 1 1 0**

Positional weights   *4 3 2 1 0*

The positional weights for each of the digits are written in italics below each digit. Hence the decimal equivalent number is given as:

$$1 \times 2^4 + 0 \times 2^3 + 1 \times 2^2 + 1 \times 2^1 + 0 \times 2^0$$
$$= 16 + 0 + 4 + 2 + 0$$
$$= 22_{10}.$$

Hence we find that here, for the sake of conversion, we have to multiply each bit with its positional weights depending on the base of the number system.

1.3.5 Octal-to-decimal Conversion

Example 1.5. *Convert 3462_8 into a decimal number.*

Solution.   The octal number given is   **3 4 6 2**

Positional weights   *3 2 1 0*

The positional weights for each of the digits are written in italics below each digit. Hence the decimal equivalent number is given as:

$$3 \times 8^3 + 4 \times 8^2 + 6 \times 8^1 + 2 \times 8^0$$
$$= 1536 + 256 + 48 + 2$$
$$= 1842_{10}.$$

1.3.6 Hexadecimal-to-decimal Conversion

Example 1.6. *Convert $42AD_{16}$ into a decimal number.*

Solution. The hexadecimal number given is   **4 2 A D**

           Positional weights   *3 2 1 0*

The positional weights for each of the digits are written in italics below each digit. Hence the decimal equivalent number is given as:

$$4 \times 16^3 + 2 \times 16^2 + 10 \times 16^1 + 13 \times 16^0$$
$$= 16384 + 512 + 160 + 13$$
$$= 17069_{10}.$$

1.3.7 Fractional Conversion

So far we have dealt with the conversion of integer numbers only. Now if the number contains the fractional part we have to deal in a different way when converting the number from a different number system (*i.e.*, binary, octal, or hexadecimal) to a decimal number system or vice versa. We illustrate this with examples.

Example 1.7. *Convert 1010.011_2 into a decimal number.*

Solution. The binary number given is   **1 0 1 0. 0 1 1**

           Positional weights   *3 2 1 0 -1-2-3*

The positional weights for each of the digits are written in italics below each digit. Hence the decimal equivalent number is given as:

$$1 \times 2^3 + 0 \times 2^2 + 1 \times 2^1 + 0 \times 2^0 + 0 \times 2^{-1} + 1 \times 2^{-2} + 1 \times 2^{-3}$$
$$= 8 + 0 + 2 + 0 + 0 + 0.25 + 0.125$$
$$= 10.375_{10}.$$

Example 1.8. *Convert 362.35_8 into a decimal number.*

Solution. The octal number given is   **3 6 2. 3 5**

           Positional weights   *2 1 0 -1-2*

The positional weights for each of the digits are written in italics below each digit. Hence the decimal equivalent number is given as:

$$3 \times 8^2 + 6 \times 8^1 + 2 \times 8^0 + 3 \times 8^{-1} + 5 \times 8^{-2}$$
$$= 192 + 48 + 2 + 0.375 + 0.078125$$
$$= 242.453125_{10}.$$

Example 1.9. *Convert $42A.12_{16}$ into a decimal number.*

Solution. The hexadecimal number given is   **4 2 A. 1 2**

           Positional weights   *2 1 0 -1-2*

The positional weights for each of the digits are written in italics below each digit. Hence the decimal equivalent number is given as:

$$4 \times 16^2 + 2 \times 16^1 + 10 \times 16^0 + 1 \times 16^{-1} + 1 \times 16^{-2}$$
$$= 1024 + 32 + 10 + 0.0625 + 0.00390625$$
$$= 1066.06640625_{10}.$$

Example 1.10. *Convert 25.625_{10} into a binary number.*

Solution. Division   Quotient   Generated remainder

$$\frac{25}{2}$$ 12   1

$$\frac{12}{2}$$ 6   0

$$\frac{6}{2}$$ 3   0

$$\frac{3}{2}$$ 1   1

$$\frac{1}{2}$$ 0   1

Therefore, $(25)_{10} = (11001)_2$

Fractional Part

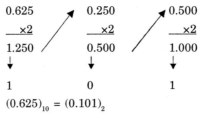

1   0   1

i.e., $(0.625)_{10} = (0.101)_2$

Therefore, $(25.625)_{10} = (11001.101)_2$

Example 1.11. *Convert 34.525_{10} into an octal number.*

Solution. Division   Quotient   Generated remainder

$$\frac{34}{8}$$ 4   2

$$\frac{4}{8}$$ 0   4

Therefore, $(34)_{10} = (42)_8$

Fractional Part

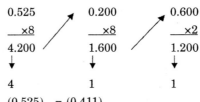

4   1   1

i.e., $(0.525)_{10} = (0.411)_8$

Therefore, $(34.525)_{10} = (42.411)_8$

Example 1.12. *Convert 92.85_{10} into a hexadecimal number.*

Solution. Division   Quotient   Generated remainder

$$\frac{92}{16}$$ 5   12

$$\frac{5}{16} \qquad\qquad 0 \qquad\qquad\qquad 5$$

Therefore, $(92)_{10} = (5C)_{16}$

Fractional Part

$$0.85 \qquad\qquad 0.60$$
$$\underline{\times 16} \qquad\qquad \underline{\times 16}$$
$$13.60 \qquad\qquad 9.60$$
$$\downarrow \qquad\qquad\qquad \downarrow$$
$$13 \qquad\qquad\quad 9$$

i.e., $(0.85)_{10} = (0.D9)_{16}$

Therefore, $(92.85)_{10} = (5C.D9)_{16}$

1.3.8 Conversion from a Binary to Octal Number and Vice Versa

We know that the maximum digit in an octal number system is 7, which can be represented as 111_2 in a binary system. Hence, starting from the LSB, we group three digits at a time and replace them by the decimal equivalent of those groups and we get the final octal number.

Example 1.13. *Convert 101101010_2 into an equivalent octal number.*

Solution.
The binary number given is	101101010

Starting with LSB and grouping 3 bits	101 101 010
Octal equivalent	5 5 2

Hence the octal equivalent number is $(552)_8$.

Example 1.14. *Convert 1011110_2 into an equivalent octal number.*

Solution.
The binary number given is	1011110
Starting with LSB and grouping 3 bits	001 011 110
Octal equivalent	1 3 6

Hence the octal equivalent number is $(176)_8$.

Since at the time of grouping the three digits in Example 1.14 starting from the LSB, we find that the third group cannot be completed, since only one 1 is left out in the third group, so we complete the group by adding two 0s in the MSB side. This is called left-padding of the number with 0. Now if the number has a fractional part then there will be two different classes of groups—one for the integer part starting from the left of the decimal point and proceeding toward the left and the second one starting from the right of the decimal point and proceeding toward the right. If, for the second class, any 1 is left out, we complete the group by adding two 0s on the right side. This is called right-padding.

Example 1.15. *Convert 1101.0111_2 into an equivalent octal number.*

Solution.
The binary number given is	1101.0111
Grouping 3 bits	001 101. 011 100
Octal equivalent:	1 5 3 4

Hence the octal number is $(15.34)_8$.

Now if the octal number is given and you're asked to convert it into its binary equivalent, then each octal digit is converted into a 3-bit-equivalent binary number and—combining all those digits we get the final binary equivalent.

Example 1.16. *Convert 235₈ into an equivalent binary number.*

Solution.

The octal number given is	2	3	5
3-bit binary equivalent	010	011	101

Hence the binary number is $(010011101)_2$.

Example 1.17. *Convert 47.321₈ into an equivalent binary number.*

Solution.

The octal number given is	4	7	3	2	1
3-bit binary equivalent	100	111	011	010	001

Hence the binary number is $(100111.011010001)_2$.

1.3.9 Conversion from a Binary to Hexadecimal Number and Vice Versa

We know that the maximum digit in a hexadecimal system is 15, which can be represented by 1111_2 in a binary system. Hence, starting from the LSB, we group four digits at a time and replace them with the hexadecimal equivalent of those groups and we get the final hexadecimal number.

Example 1.18. *Convert 11010110₂ into an equivalent hexadecimal number.*

Solution.

The binary number given is	11010110	
Starting with LSB and grouping 4 bits	1101	0110
Hexadecimal equivalent	D	6

Hence the hexadecimal equivalent number is $(D6)_{16}$.

Example 1.19. *Convert 110011110₂ into an equivalent hexadecimal number.*

Solution.

The binary number given is	110011110		
Starting with LSB and grouping 4 bits	0001	1001	1110
Hexadecimal equivalent	1	9	E

Hence the hexadecimal equivalent number is $(19E)_{16}$.

Since at the time of grouping of four digits starting from the LSB, in Example 1.19 we find that the third group cannot be completed, since only one 1 is left out, so we complete the group by adding three 0s to the MSB side. Now if the number has a fractional part, as in the case of octal numbers, then there will be two different classes of groups—one for the integer part starting from the left of the decimal point and proceeding toward the left and the second one starting from the right of the decimal point and proceeding toward the right. If, for the second class, any uncompleted group is left out, we complete the group by adding 0s on the right side.

Example 1.20. *Convert 111011.011₂ into an equivalent hexadecimal number.*

Solution.

The binary number given is	111011.011		
Grouping 4 bits	0011	1011.	0110
Hexadecimal equivalent	3	B	6

Hence the hexadecimal equivalent number is $(3B.6)_{16}$.

Now if the hexadecimal number is given and you're asked to convert it into its binary equivalent, then each hexadecimal digit is converted into a 4-bit-equivalent binary number and by combining all those digits we get the final binary equivalent.

Example 1.21. *Convert $29C_{16}$ into an equivalent binary number.*

Solution. The hexadecimal number given is   2   9   C

4-bit binary equivalent   0010   1001   1100

Hence the equivalent binary number is $(001010011100)_2$.

Example 1.22. *Convert $9E.AF2_{16}$ into an equivalent binary number.*

Solution. The hexadecimal number given is   9   E   A   F   2

4-bit binary equivalent   1001   1110   1010   1111   0010

Hence the equivalent binary number is $(10011110.101011110010)_2$.

1.3.10 Conversion from an Octal to Hexadecimal Number and Vice Versa

Conversion from octal to hexadecimal and vice versa is sometimes required. To convert an octal number into a hexadecimal number the following steps are to be followed:

(*i*) First convert the octal number to its binary equivalent (as already discussed above).

(*ii*) Then form groups of 4 bits, starting from the LSB.

(*iii*) Then write the equivalent hexadecimal number for each group of 4 bits.

Similarly, for converting a hexadecimal number into an octal number the following steps are to be followed:

(*i*) First convert the hexadecimal number to its binary equivalent.

(*ii*) Then form groups of 3 bits, starting from the LSB.

(*iii*) Then write the equivalent octal number for each group of 3 bits.

Example 1.23. *Convert the following hexadecimal numbers into equivalent octal numbers.*

     (*a*) A72E   (*b*) 4.BF85

Solution.

(*a*)   Given hexadecimal number is   A   7   2   E

Binary equivalent is   1010   0111   0010   1110

= 1010011100101110

Forming groups of 3 bits from the LSB   001   010   011   100   101   110

Octal equivalent   1   2   3   4   5   6

Hence the octal equivalent of $(A72E)_{16}$ is $(123456)_8$.

(*b*)   Given hexadecimal number is   4   B   F   8   5

Binary equivalent is   0100   1011   1111   1000   0101

= 0100.1011111110000101

Forming groups of 3 bits   100. 101   111   111   000   010   100

Octal equivalent   4   5   7   7   0   2   4

Hence the octal equivalent of $(4.BF85)_{16}$ is $(4.577024)_8$.

Example 1.24. *Convert* $(247)_8$ *into an equivalent hexadecimal number.*

Solution.

Given octal number is	2	4	7
Binary equivalent is	010	100	111

$$= 010100111$$

Forming groups of 4 bits from the LSB	1010	0111
Hexadecimal equivalent	A	7

Hence the hexadecimal equivalent of $(247)_8$ is $(A7)_{16}$.

Example 1.25. *Convert* $(36.532)_8$ *into an equivalent hexadecimal number.*

Solution.

Given octal number is	3	6	5	3	2
Binary equivalent is	011	110	101	011	010

$$=011110.101011010$$

Forming groups of 4 bits	0001	1110.	1010	1101
Hexadecimal equivalent	1	E.	A	D

Hence the hexadecimal equivalent of $(36.532)_8$ is $(1E.AD)_{16}$.

1.4 COMPLEMENTS

Complements are used in digital computers for simplifying the subtraction operation and for logical manipulations. There are two types of complements for each number system of base-r: the r's complement and the $(r-1)$'s complement. When we deal with a binary system the value of r is 2 and hence the complements are 2's and 1's complements. Similarly for a decimal system the value of r is 10 and we get 10's and 9's complements. With the same logic if the number system is octal we get 8's and 7's complement, while it is 16's and 15's complements for hexadecimal system.

1.4.1 The r's Complement

If a positive number N is given in base r with an integer part of n digits, the r's complement of N is given as $r^n - N$ for $N \neq 0$ and 0 for $N = 0$. The following examples will clarify the definition.

The 10's complement of $(23450)_{10}$ is $\quad 10^5 - 23450 = 76550.$

The number of digits in the number is $\quad n = 5.$

The 10's complement of $(0.3245)_{10}$ is $\quad 10^0 - 0.3245 = 0.6755.$

Since the number of digits in the integer part of the number is $n = 0$, we have $10^0 = 1$.

The 10's complement of $(23.324)_{10}$ is $\quad 10^2 - 23.324 = 76.676.$

The number of digits in the integer part of the number is $n = 2$.

Now if we consider a binary system, then $r = 2$.

The 2's complement of $(10110)_2$ is $\quad (2^5)_{10} - (10110)_2 = (100000 - 10110)_2 = 01010.$

The 2's complement of $(0.1011)_2$ is $\quad (2^0)_{10} - (0.1011)_2 = (1 - 0.1011)_2 = 0.0101.$

Now if we consider an octal system, then $r = 8$.

The 8's complement of $(2450)_8$ is $\quad (8^4)_{10} - (2450)_8.$

$$= (4096_{10} - 2450_8)$$
$$= (4096_{10} - 1320_{10})$$
$$= 2776_{10}.$$
$$= 5330_8.$$

Now if we consider a hexadecimal system, then $r = 16$.

The 16's complement of $(4A30)_{16}$ is $\qquad (16^4)_{10} - (4A30)_{16}$

$$= (65536_{10} - 4A30_{16})$$
$$= (65536_{10} - 18992_{10})$$
$$= 46544_{10}$$
$$= B5D0_{16}.$$

From the above examples, it is clear that to find the 10's complement of a decimal number all of the bits until the first significant 0 is left unchanged and the first nonzero least-significant digit is subtracted from 10 and the rest of the higher significant digits are subtracted from 9. With a similar reasoning, the 2's complement of a binary number can be obtained by leaving all of the least significant zeros and the first nonzero digit unchanged, and then replacing 1's with 0's and 0's with 1's. Similarly the 8's complement of an octal number can be obtained by keeping all the bits until the first significant 0 is unchanged, and the first nonzero least-significant digit is subtracted from 8 and the rest of the higher significant digits are subtracted from 7. Similarly, the 16's complement of a hexadecimal number can be obtained by keeping all the bits until the first significant 0 is unchanged, and the first nonzero least-significant digit is subtracted from 16 and the rest of the higher significant digits are subtracted from 15.

Since r's complement is a general term, r can take any value $e.g.$, $r = 11$. Then we will have 11's complement for r's complement case and 10's complement for $(r - 1)$'s complement case.

1.4.2 The (r–1)'s Complement

If a positive number N is given in base r with an integer part of n digits and a fraction part of m digits, then the $(r - 1)$'s complement of N is given as $(r^n - r^{-m} - N)$ for $N \neq 0$ and 0 for $N = 0$. The following examples will clarify the definition.

The 9's complement of $(23450)_{10}$ is $\qquad 10^5 - 10^0 - 23450 = 76549.$

Since there is no fraction part, $\qquad 10^{-m} = 10^0 = 1.$

The 9's complement of $(0.3245)_{10}$ is $\qquad 10^0 - 10^{-4} - 0.3245 = 0.6754.$

Since there is no integer part, $\qquad 10^n = 10^0 = 1.$

The 9's complement of $(23.324)_{10}$ is $\qquad 10^2 - 10^{-3} - 23.324 = 76.675.$

Now if we consider a binary system, then $r = 2$, $i.e.$, $(r - 1) = 1$.

The 1's complement of $(10110)_2$ is $\qquad (2^5-1)_{10} - (10110)_2 = 01001.$

The 1's complement of $(0.1011)_2$ is $\qquad (1-2^{-4})_{10} - (0.1011)_2 = 0.0100.$

Now if we consider an octal system, then $r = 8$, $i.e.$, $(r - 1) = 7$.

The 7's complement of $(2350)_8$ is $\qquad 8^4 - 8^0 - 2350_8$

$$= 4095_{10} - 1256_{10}$$

$$= 2839_{10}$$

$$= 5427_8.$$

The 15's complement of $(A3E4)_{16}$ is

$$16^4 - 16^0 - A3E4_{16}$$

$$= 65535_{10} - 41956_{10}$$

$$= 23579_{10}$$

$$= 5C1B_{16}.$$

From the above examples, it is clear that to find the 9's complement of a decimal number each of the digits can be separately subtracted from 9. The 1's complement of a binary number can be obtained by changing 1s into 0s and 0s into 1s. Similarly, to find the 7's complement of a decimal number each of the digits can be separately subtracted from 7. Again, to find the 15's complement of a decimal number each of the digits can be separately subtracted from 15.

Example 1.26. *Find out the 11's and 10's complement of the number* $(576)_{11}$.

Solution.

The number in base is 11. So to find 11's complement we have to follow the r's complement rule and in order to get 10's complement the $(r - 1)$'s complement rule is to be followed.

11's complement:

$$r^n - N = 11^3 - 576_{11}$$

$$= (1331)_{10} - (576)_{11}$$

Now, 576_{11} $= 5 \times 11^2 + 7 \times 11^1 + 6 \times 11^0$

$$= 605 + 77 + 6$$

$$= 688_{10}$$

Therefore, 11's complement is $1331_{10} - 688_{10} = 643_{10}$

Now, the decimal number has to be changed in the number system of base 11.

Division	Quotient	Generated remainder
$\dfrac{643}{11}$	58	5
$\dfrac{58}{11}$	5	3
$\dfrac{5}{11}$	0	5

Hence the 11's complement number is $(535)_{11}$.

10's complement:

$$r^n - r^{-m} - N = 11^3 - 11^0 - 576_{11}$$

$$= (1331)_{10} - (1)_{10} - (576)_{11}$$

Therefore, 10's complement is $1331_{10} - 1_{10} - 688_{10} = 642_{10}$

Now, the decimal number has to be changed in the number system of base 11.

Division	Quotient	Generated remainder
$\dfrac{642}{11}$	58	4
$\dfrac{58}{11}$	5	3
$\dfrac{5}{11}$	0	5

Hence the 10's complement number is $(534)_{11}$.

1.5 BINARY ARITHMETIC

We are very familiar with different arithmetic operations, *viz.* addition, subtraction, multiplication, and division in a decimal system. Now we want to find out how those same operations may be performed in a binary system, where only two digits, *viz.* 0 and 1 exist.

1.5.1 Binary Addition

The rules of binary addition are given in Table 1.1.

Table 1.1

Augend	Addend	Sum	Carry	Result
0	0	0	0	0
0	1	1	0	1
1	0	1	0	1
1	1	0	1	10

The procedure of adding two binary numbers is same as that of two decimal numbers. Addition is carried out from the LSB and it proceeds to higher significant bits, adding the carry resulting from the addition of two previous bits each time.

Example 1.27. *Add the binary numbers:*

    (a) 1010 and 1101   *(b) 0110 and 1111*

Solution.

(a)
```
        1   0   1   0
  (+)   1   1   0   1
  1 0   1   1   1
  ↑
  Carry
```

(b)
```
       (1) (1)  ←── Carry
        0   1   1   0
  (+)   1   1   1   1
  1 0   1   0   1
  ↑
  Carry
```

1.5.2 Binary Subtraction

The rules of binary subtraction are given in Table 1.2.

Table 1.2

Minuend	Subtrahend	Difference	Borrow
0	0	0	0
0	1	1	1
1	0	1	0
1	1	0	0

Binary subtraction is also carried out in a similar method to decimal subtraction. The subtraction is carried out from the LSB and proceeds to the higher significant bits. When borrow is 1, as in the second row, this is to be subtracted from the next higher binary bit as it is performed in decimal subtraction.

Actually, the subtraction between two numbers can be performed in three ways, *viz.*

(*i*) the direct method,

(*ii*) the r's complement method, and

(*iii*) the $(r - 1)$'s complement method.

Subtraction Using the Direct Method

The direct method of subtraction uses the concept of borrow. In this method, we borrow a 1 from a higher significant position when the minuend digit is smaller than the corresponding subtrahend digit.

Example 1.28. *Using the direct method to perform the subtraction*

$1001 - 1000.$

Solution:

$$
\begin{array}{r}
1\ 0\ 0\ 1 \\
(-)\quad \underline{1\ 0\ 0\ 0} \\
0\ 0\ 0\ 1
\end{array}
$$

Example 1.29. *Using the direct method to perform the subtraction*

$1000 - 1001.$

Solution.

$$
\begin{array}{r}
1\ 0\ 0\ 0 \\
(-)\quad \underline{1\ 0\ 0\ 1}
\end{array}
$$

End carry $\longrightarrow$ 1 1 1 1 1

End carry has to be ignored.

Answer: 1111 = (2's complement of 0001).

When the minuend is smaller than the subtrahend the result of subtraction is negative and in the direct method the result obtained is in 2's complement form. So to get back the actual result we have to perform the 2's complement again on the result thus obtained.

But to tackle the problem shown in Example 1.29 we have applied a trick. When a digit is smaller in the minuend than that in the subtrahend we add 2 (the base of the binary system) to the minuend digit mentally and we perform the subtraction (in this case 1 from 2) in decimal and write down the result in the corresponding column. Since we have added 2 to the column, we have to add 1 to the subtrahend digit in the next higher order column. This process is to be carried on for all of the columns whenever the minuend digit is smaller than the corresponding subtrahend digit.

The rest of the two binary subtraction methods, *i.e.*, the *r*'s complement and the (*r* − 1)'s complement methods will be discussed in due course.

		(+2)	(+2)	(+2)	(+2)	
		1	0	0	0	
		1	0	0	1	
	(+1)	(+1)	(+1)	(+1)		
End carry →	1	1	1	1	1	

End carry has to be ignored.

1.5.3 Binary Multiplication

Binary multiplication is similar to decimal multiplication but much simpler than that. In a binary system each partial product is either zero (multiplication by 0) or exactly the same as the multiplicand (multiplication by 1). The rules of binary multiplication are given in Table 1.3.

Table 1.3

Multiplicand	Multiplier	Result
0	0	0
0	1	0
1	0	0
1	1	1

Actually, in a digital circuit, the multiplication operation is done by repeated additions of all partial products to obtain the full product.

Example 1.30. *Multiply the following binary numbers:*

(a) 0111 and 1101 and (b) 1.011 and 10.01.

Solution.

(a)   0111 × 1101

			0	1	1	1	Multiplicand
		×	1	1	0	1	Multiplier
			0	1	1	1	
		0	0	0	0		Partial
	0	1	1	1			Products
0	1	1	1				
1	0	1	1	0	1	1	Final Product

(b) 1.011 × 10.01

		1.	0	1	1		Multiplicand
		× 1	0.	0	1		Multiplier
		1	0	1	1		
	0	0	0	0			Partial
0	0	0	0				Products
1	0	1	1				
1	1 .	0	0	0	1	1	Final Product

1.5.4 Binary Division

Binary division follows the same procedure as decimal division. The rules regarding binary division are listed in Table 1.4.

Table 1.4

Dividend	Divisor	Result
0	0	Not allowed
0	1	0
1	0	Not allowed
1	1	1

Example 1.31. *Divide the following binary numbers:*

(a) 11001 and 101 and (b) 11110 and 1001.

Solution.

(a) 11001 ÷ 101

```
                                1   0   1
              1     0     1 | 1   1   0   0   1
                             1   0   1
                             0   0   1   0   1
                                     1   0   1
                                     0   0   0
```

Answer: 101

(b) 11110 ÷ 1001

```
                                1   1.  0   1   0
          1   0   0   1 | 1   1   1   1   0
                          1   0   0   1
                        0   1   1   0   0
                          1   0   0   1
```

	1	0	0	0	0
		1	0	0	1
			1	1	0
		1	0	0	1
			1	0	1

Answer: 11.010

1.6 1'S AND 2'S COMPLEMENT ARITHMETIC

Digital circuits perform binary arithmetic operations. It is possible to use the circuits designed for binary addition to perform binary subtraction. Only we have to change the problem of subtraction into an equivalent addition problem. This can be done if we make use of 1's and 2's complement form of the binary numbers as we have already discussed.

1.6.1 Subtraction Using 1's Complement

Binary subtraction can be performed by adding the 1's complement of the subtrahend to the minuend. If a carry is generated, remove the carry, add it to the result. This carry is called the end-around carry. Now if the subtrahend is larger than the minuend, then no carry is generated. The answer obtained is 1's complement of the true result and opposite in sign.

Example 1.32. *Subtract* $(1001)_2$ *from* $(1101)_2$ *using the 1's complement method. Also subtract using the direct method and compare.*

Solution.

Direct Subtraction		1's complement method	
1 1 0 1		1 1 0 1	(+)
$-$ 1 0 0 1	1's complement $\rightarrow$	0 1 1 0	
0 1 0 0	Carry $\rightarrow$	1 0 0 1 1	
	Add Carry $\rightarrow$	1	
		0 1 0 0	

Example 1.33. *Subtract* $(1100)_2$ *from* $(1001)_2$ *using the 1's complement method. Also subtract using the direct method and compare.*

Solution.

	Direct Subtraction		1's complement method	
	1 0 0 1		1 0 0 1	(+)
	$-$ 1 1 0 0	1's complement $\rightarrow$	0 0 1 1	
Carry $\rightarrow$	1 1 1 0 1		1 1 0 0	
2's complement	0 0 1 1	1's complement $\rightarrow$	0 0 1 1	
True result	0 0 1 1	True result	$-$ 0 0 1 1	

In the direct method, whenever a larger number is subtracted from a smaller number, the result obtained is in 2's complement form and opposite in sign. To get the true result we have to discard the carry and make the 2's complement of the result obtained and put a negative sign before the result.

In the 1's complement subtraction, no carry is obtained and the result obtained is in 1's complement form. To get the true result we have to make the 1's complement of the result obtained and put a negative sign before the result.

1.6.2 Subtraction Using 2's Complement

Binary subtraction can be performed by adding the 2's complement of the subtrahend to the minuend. If a carry is generated, discard the carry. Now if the subtrahend is larger than the minuend, then no carry is generated. The answer obtained is in 2's complement and is negative. To get a true answer take the 2's complement of the number and change the sign. The advantage of the 2's complement method is that the end-around carry operation present in the 1's complement method is not present here.

Example 1.34. *Subtract* $(0111)_2$ *from* $(1101)_2$ *using the 2's complement method. Also subtract using the direct method and compare.*

Solution.

Direct Subtraction		1's complement method
1 1 0 1		1 1 0 1 (+)
– 0 1 1 1	2's complement ⟶	1 0 0 1
0 1 1 0	Carry ⟶	1 0 1 1 0
	Discard Carry	0 1 1 0 (Result)

Example 1.35. *Subtract* $(1010)_2$ *from* $(1001)_2$ *using the 1's complement method. Also subtract using the direct method and compare.*

Solution.

	Direct Subtraction		1's complement method
	1 0 0 1		1 0 0 1 (+)
	– 1 0 1 0	2's complement ⟶	0 1 1 0
Carry ⟶	1 1 1 1 1		1 1 1 1
2's complement	0 0 0 1	2's complement ⟶	0 0 0 1
True result	–0001	True result	–0001

In the direct method, whenever a larger number is subtracted from a smaller number, the result obtained is in 2's complement form and opposite in sign. To get the true result we have to discard the carry and make the 2's complement of the result obtained and put a negative sign before the result.

In the 2's complement subtraction, no carry is obtained and the result obtained is in 2's complement form. To get the true result we have to make the 2's complement of the result obtained and put a negative sign before the result.

1.6.3 Comparison between 1's and 2's Complements

A comparison between 1's and 2's complements reveals the advantages and disadvantages of each.

(*i*) The 1's complement has the advantage of being easier to implement by digital components (*viz.* inverter) since the only thing to be done is to change the 1s to 0s and vice versa. To implement 2's complement we can follow two ways: (1) by finding out the 1's complement of the number and then adding 1 to the LSB of the 1's complement, and (2) by leaving all leading 0s in the LSB positions and the first 1 unchanged, and only then changing all 1's to 0s and vice versa.

(*ii*) During subtraction of two numbers by a complement method, the 2's complement is advantageous since only one arithmetic addition is required. The 1's complement requires two arithmetic additions when an end-around carry occurs.

(*iii*) The 1's complement has an additional disadvantage of having two arithmetic zeros: one with all 0s and one with all 1s. The 2's complement has only one arithmetic zero. The fact is illustrated below:

We consider the subtraction of two equal binary numbers 1010 – 1010.

Using 1's complement:

$$1010$$
$$+ \ \underline{0101} \ \text{(1's complement of 1010)}$$
$$+ \ 1111 \ \text{(negative zero)}$$

We complement again to obtain (– 0000) (positive zero).

Using 2's complement:

$$1010$$
$$+ \ \underline{0110} \ \text{(2's complement of 1010)}$$
$$+ \ 0000$$

In this 2's complement method no question of negative or positive zero arises.

1.7 SIGNED BINARY NUMBERS

So far whatever discussions were made, there was no consideration of sign of the numbers. But in real life one may have to face a situation where both positive and negative numbers may arise. So we have to know how the positive and negative binary numbers may be represented. Basically there are three types of representations of signed binary numbers— sign-magnitude representation, 1's complement representation, and 2's complement representations, which are discussed below.

1.7.1 Sign-magnitude Representation

In decimal system, generally a plus (+) sign denotes a positive number whereas a minus (–) sign denotes a negative number. But, the plus sign is usually dropped, and no sign means the number is positive. This type of representation of numbers is known as *signed numbers*. But in digital circuits, there is no provision to put a plus or minus sign, since everything in digital circuits have to be represented in terms of 0 and 1. Normally an additional bit is used as the *sign bit*. This sign bit is usually placed as the MSB. Generally a 0 is reserved for a positive number and a 1 is reserved for a negative number. For example, an 8-bit signed binary number 01101001 represents a positive number whose magnitude is $(1101001)_2 = (105)_{10}$. The MSB is 0, which indicates that the number is positive. On the other hand, in the signed binary form, 11101001 represents a negative number whose magnitude is $(1101001)_2 = (105)_{10}$. The 1 in the MSB position indicates that the number is negative and the other seven bits give its magnitude. This kind of representation of binary numbers is called *sign-magnitude representation*.

Example 1.36. *Find the decimal equivalent of the following binary numbers assuming the binary numbers have been represented in sign-magnitude form.*

(a) 0101100   (b) 101000   (c) 1111   (d) 011011

Solution.

(a)   Sign bit is 0, which indicates the number is positive.

Magnitude $101100 = (44)_{10}$

Therefore $(0101100)_2 = (+44)_{10}.$

(b)   Sign bit is 1, which indicates the number is negative.

Magnitude $01000 = (8)_{10}$

Therefore $(101000)_2 = (-8)_{10}.$

(c)   Sign bit is 1, which indicates the number is negative.

Magnitude $111 = (7)_{10}$

Therefore $(1111)_2 = (-7)_{10}.$

(d)   Sign bit is 0, which indicates the number is positive.

Magnitude $11011 = (27)_{10}$

Therefore $(011011)_2 = (+27)_{10}.$

1.7.2 1's Complement Representation

In 1's complement representation, both numbers are a complement of each other. If one of the numbers is positive, then the other will be negative with the same magnitude and vice versa. For example, $(0111)_2$ represents $(+ 7)_{10}$, whereas $(1000)_2$ represents $(- 7)_{10}$ in 1's complement representation. Also, in this type of representation, the MSB is 0 for positive numbers and 1 for negative numbers.

Example 1.37. *Represent the following numbers in 1's complement form.*

(a) +5 and –5 (b) +9 and –9 (c) +15 and –15

Solution.

(a) $(+5)_{10} = (0101)_2$

and $(-5)_{10} = (1010)_2$

(b) $(+9)_{10} = (01001)_2$

and $(-9)_{10} = (10110)_2$

(c) $(+15)_{10} = (01111)_2$

and $(-15)_{10} = (10000)_2$

From the above examples it can be observed that for an *n*-bit number, the maximum positive number which can be represented in 1's complement form is $(2^{n-1}-1)$ and the maximum negative number is $-(2^{n-1} - 1)$.

1.7.3 2's Complement Representation

If 1 is added to 1's complement of a binary number, the resulting number is 2's complement of that binary number. For example, $(0110)_2$ represents $(+6)_{10}$, whereas $(1010)_2$ represents $(-6)_{10}$ in 2's complement representation. Also, in this type of representation, the MSB is 0 for positive numbers and 1 for negative numbers. For an n-bit number, the maximum positive number which can be represented in 2's complement form is $(2^{n-1}-1)$ and the maximum negative number is -2^{n-1}.

Example 1.38. *Represent the following numbers in 2's complement form.*

        (a) +11 and –11   *(b) +9 and –9*   *(c) +18 and –18*

Solution.

    (a) $(+11)_{10} = (01011)_2$

        and $(-11)_{10} = (10101)_2$

    (b) $(+9)_{10} = (01001)_2$

        and $(-9)_{10} = (10111)_2$

    (c) $(+18)_{10} = (010010)_2$

        and $(-18)_{10} = (101110)_2$

Example 1.39. *Represent (–19) in*

    (a) Sign-magnitude,

    (b) one's complement, and

    (c) two's complement representation.

Solution.

The minimum number of bits required to represent $(+19)_{10}$ in signed number format is six.

Therefore, $(+19)_{10} = (010011)_2$

Therefore, $(-19)_{10}$ is represented by

    (a) 110011 in sign-magnitude representation.

    (b) 101100 in 1's complement representation.

    (c) 101101 in 2's complement representation.

1.8 7's AND 8's COMPLEMENT ARITHMETIC

The 7's complement of an octal number can be found by subtracting each digit in the number from 7. The 8's complement can be obtained by subtracting the LSB from 8 and the rest of each digit in the number from 7. The 7's and 8's complement of the octal digits 0 to 7 is shown in Table 1.5.

The method of subtraction using 7's complement method is the same as 1's complement method in binary system. Here also the carry obtained is added to the result to get the true result. And as in the previous cases, if the minuend is larger than the subtrahend, no carry is obtained and the result is obtained in 7's complement form. To get the true result we have to again get the 7's complement of the result obtained and put a negative sign before it.

Similarly, the method of subtraction using 8's complement method is the same as 2's complement method in a binary system. Here also the carry obtained is discarded to get the true result. And as in the previous cases, if the minuend is larger than the subtrahend, no carry is obtained and the result is obtained in 8's complement form. To get the true result we have to again get the 8's complement of the result obtained and put a negative sign before it.

Table 1.5

Octal digit	7's complement	8's complement
0	7	8
1	6	7
2	5	6
3	4	5
4	3	4
5	2	3
6	1	2
7	0	1

1.8.1 Subtraction Using 7's Complement

Example 1.40. *Subtract $(372)_8$ from $(453)_8$ using the 7's complement method. Also subtract using the direct method and compare.*

Solution.

Direct Subtraction

```
    4 5 3
  – 3 7 2
    6 1
```

7's complement method

```
                       4 5 3   (+)
7's complement  →       4 0 5
                      1 0 6 0
Add Carry       →           1
                    (6 1)₈ (Result)
```

$$\text{7's complement method}$$
$$4\,5\,3 \quad (+)$$
$$\text{7's complement} \rightarrow 4\,0\,5$$
$$1\,0\,6\,0$$
$$\text{Add Carry} \rightarrow 1$$
$$(6\ 1)_8 \text{ (Result)}$$

Example 1.41. *Subtract $(453)_8$ from $(372)_8$ using the 7's complement method. Also subtract using the direct method and compare.*

Solution.

Direct Subtraction

```
                        3 7 2
                        4 5 3
                      1 7 1 7
Discard Carry           7 1 7
8's complement          6 1
True result           (–61)₈
```

7's complement method

```
                          3 7 2   (+)
7's complement  →         3 2 4
                          7 1 6
7's complement  →           6 1
True result            (–61)₈
```

In the direct method, whenever a larger number is subtracted from a smaller number, the result obtained is in 8's complement form and opposite in sign. To get the true result we have to discard the carry and make the 8's complement of the result obtained and put a negative sign before the result.

1.8.2 Subtraction Using 8's Complement

Example 1.42. *Subtract $(256)_8$ from $(461)_8$ using the 8's complement method. Also subtract using the direct method and compare.*

Solution.

Direct Subtraction		8's complement method
4 6 1		4 6 1 (+)
– 2 5 6	8's complement →	5 2 2
2 0 3	Carry →	1 2 0 3
	Discard Carry	(2 0 3)$_8$ (Result)

Example 1.43. *Subtract (461)$_8$ from (256)$_8$ using the 8's complement method. Also subtract using the direct method and compare.*

Solution.

Direct Subtraction		8's complement method
2 5 6		2 5 6 (+)
– 4 6 1	8's complement →	3 1 7
1 5 7 5		5 7 5
Discard Carry 5 7 5	8's complement →	2 0 3
8's complement 2 0 3		
True result (–203)$_8$	True result	(–203)$_8$

In the direct method, whenever a larger number is subtracted from a smaller number, the result obtained is in 8's complement form and opposite in sign. To get the true result we have to discard the carry and make the 8's complement of the result obtained and put a negative sign before the result.

1.9 9's AND 10's COMPLEMENT ARITHMETIC

The 9's complement of a decimal number can be found by subtracting each digit in the number from 9. The 10's complement can be obtained by subtracting the LSB from 10 and the rest of each digit in the number from 9. The 9's and 10's complement of the decimal digits 0 to 9 is shown in Table 1.6.

Table 1.6

Decimal digit	9's complement	10's complement
0	9	10
1	8	9
2	7	8
3	6	7
4	5	6
5	4	5
6	3	4
7	2	3
8	1	2
9	0	1

The method of subtraction using 9's complement method is the same as 1's complement method in a binary system. Here also the carry obtained is added to the result to get the true result. And as in the previous cases, if the minuend is larger than the subtrahend, no carry is obtained and the result is obtained in 9's complement form. To get the true result we have to again get the 9's complement of the result obtained and put a negative sign before it.

Similarly, the method of subtraction using 10's complement method is the same as 2's complement method in a binary system. Here also the carry obtained is discarded to get the true result. And as in the previous cases, if the minuend is larger than the subtrahend, no carry is obtained and the result is obtained in 10's complement form. To get the true result we have to again get the 10's complement of the result obtained and put a negative sign before it.

1.9.1 Subtraction Using 9's Complement

Example 1.44. *Subtract* $(358)_{10}$ *from* $(592)_{10}$ *using the 9's complement method. Also subtract using the direct method and compare.*

Solution.

Direct Subtraction			9's complement method
5 9 2			5 9 2 (+)
− 3 5 8	9's complement	→	6 4 1
2 3 4			1 2 3 3
	Add Carry	→	1
			(2 3 4)$_{10}$ (Result)

Example 1.45. *Subtract* $(592)_{10}$ *from* $(358)_{10}$ *using the 9's complement method. Also subtract using the direct method and compare.*

Solution.

Direct Subtraction			9's complement method
3 5 8			3 5 8 (+)
− 5 9 2	9's complement	→	4 0 7
− 1 7 6 6			7 6 5
Discard carry 7 7 6	9's complement	→	2 3 4
10's complement 2 3 4	True result		(−234)$_{10}$
True result (−234)$_{10}$			

1.9.2 Subtraction Using 10's Complement

Example 1.46. *Subtract* $(438)_{10}$ *from* $(798)_{10}$ *using the 10's complement method. Also subtract using the direct method and compare.*

Solution.

Direct Subtraction			10's complement method
7 9 8			7 9 8 (+)
− 4 3 8	10's complement	→	5 6 2
3 6 0	Carry	→	1 3 6 0
	Discard Carry		(3 6 0)$_{10}$ (Result)

Example 1.47. *Subtract* $(798)_{10}$ *from* $(438)_{10}$ *using the 10's complement method. Also subtract using the direct method and compare.*

Solution.

Direct Subtraction		10's complement method	

	Direct Subtraction		10's complement method
	4 3 8		4 3 8 (+)
	− 7 9 8	10's complement ⟶	2 0 2
	1 6 4 0		6 4 0
Discard carry	6 4 0	10's complement ⟶	3 6 0
10's complement	3 6 0	True result	$(-360)_{10}$
True result	$(-360)_{10}$		

1.10 15's AND 16's COMPLEMENT ARITHMETIC

The 15's complement of a hexadecimal number can be found by subtracting each digit in the number from 15. The 16's complement can be obtained by subtracting the LSB from 16 and the rest of each digit in the number from 15. The 15's and 16's complement of the hexadecimal digits 0 to F is shown in Table 1.7.

Table 1.7

Hexadecimal digit	15's complement	16's complement
0	15	16
1	14	15
2	13	14
3	12	13
4	11	12
5	10	11
6	9	10
7	8	9
8	7	8
9	6	7
A	5	6
B	4	5
C	3	4
D	2	3
E	1	2
F	0	1

The method of subtraction using 15's complement method is the same as 9's complement method in a decimal system. Here also the carry obtained is added to the result to get the true result. And as in the previous cases, if the minuend is larger than the subtrahend, no carry is obtained and the result is obtained in 15's complement form. To get the true result we have to again get the 15's complement of the result obtained and put a negative sign before it.

Similarly, the method of subtraction using 16's complement method is the same as 10's complement method in a decimal system. Here also the carry obtained is discarded to get the true result. And as in the previous cases, if the minuend is larger than the subtrahend, no carry is obtained and the result is obtained in 16's complement form. To get the true result we have to again get the 16's complement of the result obtained and put a negative sign before it.

1.10.1 Subtraction Using 15's Complement

Example 1.48. *Subtract $(2B1)_{16}$ from $(3A2)_{16}$ using the 15's complement method. Also subtract using the direct method and compare.*

Solution.

Direct Subtraction		15's complement method	
3 A 2		3 A 2	(+)
− 2 B 1	15's complement ⟶	D 4 E	
F 1		1 0 F 0	
	Add Carry ⟶	1	
		$(F\ 1)_{16}$ (Result)	

Example 1.49. *Subtract $(3A2)_{16}$ from $(2B1)_{16}$ using the 15's complement method. Also subtract using the direct method and compare.*

Solution.

Direct Subtraction		15's complement method	
2 B 1		2 B 1	(+)
− 3 A 2	15's complement ⟶	C 5 D	
1 F 0 F		F 0 E	
Discard Carry F 0 F	15's complement ⟶	F 1	
16's complement F 1			
True result $(-F1)_{16}$		True result $(-F1)_{16}$	

In the direct method, whenever a larger number is subtracted from a smaller number, the result obtained is in 16's complement form and opposite in sign. To get the true result we have to discard the carry and make the 16's complement of the result obtained and put a negative sign before the result.

1.10.2 Subtraction Using 16's Complement

Example 1.50. *Subtract $(1FA)_{16}$ from $(2DC)_{16}$ using the 16's complement method. Also subtract using the direct method and compare.*

Solution.

Direct Subtraction		16's complement method	
2 D C		2 D C	(+)
− 1 F A	16's complement ⟶	E 0 6	
E 2	Carry	1 0 E 2	
	Discard Carry	$(E\ 2)_{16}$ (Result)	

Example 1.51. *Subtract $(2DC)_{16}$ from $(1FA)_{16}$ using the 16's complement method. Also subtract using the direct method and compare.*

Solution.

Direct Subtraction				16's complement method	
	1 F A			1 F A	(+)
	–2 D C	16's complement	→	D 2 4	
	1 0 1 E			F 1 E	
Discard carry	1 E	16's complement	→	E 2	
16's complement	E 2				
True result	$(-E2)_{16}$		True result	$(-E2)_{16}$	

1.11 BCD ADDITION

The full form of BCD is Binary Coded Decimal. We will discuss this in detail in the next chapter. The only thing we want to mention here is that, in this code, each decimal digit from 1 to 9 is coded in 4-bit binary numbers. But with 4-bit binary sixteen different groups can be obtained, whereas we require only ten groups to write BCD code. The other six groups are called forbidden codes in BCD and they are invalid for BCD. BCD is a numerical code. Many applications require arithmetic operation. Addition is the most important of these because the other three operations, *viz.* subtraction, multiplication, and division, can be performed using addition.

There are certain rules to be followed in BCD addition as given below.

(*i*) First add the two numbers using normal rules for binary addition.

(*ii*) If the 4-bit sum is equal to or less than 9, it becomes a valid BCD number.

(*iii*) If the 4-bit sum is greater than 9, or if a carry-out of the group is generated, it is an invalid result. In such a case, add $(0110)_2$ or $(6)_{10}$ to the 4-bit sum in order to skip the six invalid states and return the code to BCD. If a carry results when 6 is added, add the carry to the next 4-bit group.

Example 1.52. *Add the following BCD numbers:*

(a) 0111 and 1001 and (b) 10010010 and 01011000.

Solution.

(*a*)

```
        0  1  1  1
     +  1  0  0  1
        1  0  0  0   →    Invalid BCD number
     +  0  1  1  0   →    Add 6
  0  0  0  1  0  1  1  0  →    Valid BCD number
        1        6
```

(b)
$$\begin{array}{c@{\qquad}c}
\begin{array}{r}
1\ 0\ 0\ 1 \\
+\ 0\ 1\ 0\ 1 \\
\hline
1\ 1\ 1\ 0 \\
\end{array} &
\begin{array}{r}
0\ \ 0\ \ 1\ \ 0 \\
1\ \ 0\ \ 0\ \ 0 \\
\hline
1\ \ 0\ \ 1\ \ 0 \\
\end{array}
\end{array}$$

```
            1 0 0 1      0  0  1  0
          + 0 1 0 1      1  0  0  0
          ---------      ----------
            1 1 1 0      1  0  1  0  ──→  Both groups are invalid
          + 0 1 1 0      0  1  1  0  ──→  Add 6
  0 0 0 1  0 1 0 1       0  0  0  0  ──→  Valid BCD number
  \___ ___/ \___ ___/    \____ ____/
      1         5              0
```

1.12 BCD SUBTRACTION

There are two methods that can be followed for BCD subtraction.

METHOD 1. In order to subtract any number from another number we have to add the 9's complement of the subtrahend to the minuend. We can use the 10's complement also to perform the subtraction operation.

Example 1.53. *Carry out BCD subtraction for (893) – (478) using 9's complement method.*

Solution.

9's complement of 478 is
$$\begin{array}{r} 999 \\ -\ 478 \\ \hline 521 \end{array}$$

Direct method
$$\begin{array}{r} 893 \\ -\ 478 \\ \hline 415 \end{array}$$

Now in BCD form we may write

```
          1000  1001  0011
        + 0101  0010  0001
        -------------------
          1101  1011  0100   Left and middle groups are invalid
        + 0110  0110          Add 6
        -------------------
     1    0100  0001  0100
          ──────────────────→ 1   End around carry
        -------------------
          0100  0001  0101
```

Hence, the final result is $(0100\ 0001\ 0101)_2$ or $(415)_{10}$.

Example 1.54. *Carry out BCD subtraction for (768) – (274) using 10's complement method.*

Solution.

10's complement of 274 is
$$\begin{array}{r} 9910 \\ -\ 274 \\ \hline 726 \end{array}$$

Direct method
$$\begin{array}{r} 768 \\ -\ 274 \\ \hline 494 \end{array}$$

Now in BCD form we may write   0111   0110   1000

            + 0111   0010   0110

               1110   1000   1110 Left and right groups are invalid

Therefore   1110   1000   1110

            + 0110   0110 Add 6

Ignore Carry ⟶ 1 0100   1001   0100

Hence, the final result is $(0100\ 1001\ 0100)_2$ or $(494)_{10}$.

METHOD 2   . Table 1.8 shows an algorithm for BCD subtraction.

Table 1.8

Decade result	Sign of total result	
	(+) End around carry = 1	(–) End around carry = 0
$C_n = 1$ $C_n = 0$	Transfer true results of adder 1 0000 is added in adder 2 1010 is added in adder 2	Transfer 1's complement of result of adder 1 1010 is added in adder 2 0000 is added in adder 2

Total result positive

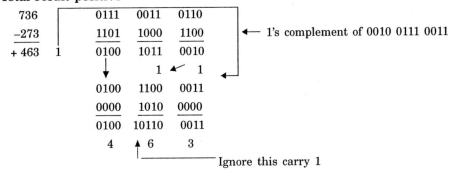

```
  736  │ 0111   0011   0110
 –273  │ 1101   1000   1100   ◄─ 1's complement of 0010 0111 0011
+ 463 1│ 0100   1011   0010
       │   │       1 ◄─ 1   ◄─┘
       │   ▼
         0100   1100   0011
         0000   1010   0000
         0100  10110   0011
           4  ▲ 6       3
              │_____ Ignore this carry 1
```

Total result negative

```
  427    0100   0010   0111
 –572    1010   1000   1101   ◄─ 1's complement of 0101 0111 0010
  145    1110   1010   0100
           │      1 ◄─┘
           ▼    ────
                1011     ▼
         0001   0100   1011   Transfer 1's complement of adder 1 output
         0000   0000   1010
         0001   0100  10101
           1      4  ▲ 5
                     │_____ Ignore this carry 1
```

Example 1.55. *Determine the base of the following arithmetic operation:*

$$1234 + 5432 = 6666$$

Solution. Let us assume that the base of the system is x.

Hence we may write,

$$(1 \times x^3 + 2 \times x^2 + 3 \times x^1 + 4 \times x^0) + (5 \times x^3 + 4 \times x^2 + 3 \times x^1 + 2 \times x^0)$$
$$= (6 \times x^3 + 6 \times x^2 + 6 \times x^1 + 6 \times x^0)$$

or, $x = 0$.

Hence, the value of the base can be any number greater than or equal to 7. Since the maximum digit in the problem is 6, the base cannot be less than 7.

Example 1.56. *Determine the base of the following arithmetic operation:*

$$\frac{302}{20} = 12.1$$

Solution. Let us assume that the base of the system is x.

Hence we may write,

$$\frac{3 \times x^2 + 0 \times x^1 + 2 \times x^0}{2 \times x^1 + 0 \times x^0} = 1 \times x^1 + 2 \times x^0 + 1 \times x^1$$

or, $$\frac{3x^2 + 2}{2x} = x + 2 + \frac{1}{x}$$

or, $x^2 - 4x = 0$

or, $x(x - 4) = 0$

$\therefore$ $x = 0$, or, $x = 4$

Now, the value of the base of a number system cannot be 0. Hence the value of the base is 4.

REVIEW QUESTIONS

1.1 Convert the decimal number 247.8 to base 3, base 4, base 5, base 11, and base 16.

1.2 Convert the following decimal numbers to binary: 12.345, 103, 45.778, and 9981.

1.3 Convert the following binary numbers to decimal: 11110001, 00101101, 1010001, and 1001110.

1.4 Perform the subtractions with the following binary numbers using (1) 1's complement and (2) 2's complement. Check the answer by straight binary subtractions.

(a) 10011 – 10001,   (b) 10110 – 11000, and   (c) 100111011 – 10001.

1.5 Perform the subtractions with the following decimal numbers using (1) 9's complement and (2) 10's complement. Check the answer by straight subtractions.

(a) 1045 – 567,   (b) 4587 – 5668, and   (c) 763 – 10001.

1.6 Perform the BCD addition of the following numbers:

(a) 234 + 146,   (b) 67 + 39, and   (c) 9234 + 4542.

1.7 Each of the following arithmetic operations is correct in at least one number system. Determine the bases in each operation:

(a) $\dfrac{41}{3} = 13$ (b) $\sqrt{41} = 5$ and   (c) 23 + 44 + 14 + 32 = 223

1.8 Add and multiply the following numbers in the given base without converting to decimal.

(a) $(1231)_4$ and $(32)_4$,   (b) $(135.3)_6$ and $(42.3)_6$ and,   (c) $(376)_8$ and $(157)_8$.

1.9 Find the 10's complement of $(349)_{11}$.

1.10 Explain how division and multiplication can be performed in digital systems.

□ □ □

CODES AND THEIR CONVERSIONS

Chapter **2**

2.1 INTRODUCTION

As we have discussed, digital circuits use binary signals but are required to handle data which may be alphabetic, numeric, or special characters. Hence the signals that are available in some other form other than binary have to be converted into suitable binary form before they can be processed further by digital circuits. This means that in whatever format the information may be available it must be converted into binary format. To achieve this, a process of coding is required where each letter, special character, or numeral is coded in a unique combination of 0s and 1s using a coding scheme known as *code*.

In digital systems a variety of codes are used to serve different purposes, such as data entry, arithmetic operation, error detection and correction, etc. Selection of a particular code depends on the requirement. Even in a single digital system a number of different codes may be used for different operations and it may even be necessary to convert data from one type of code to another. For conversion of data, code converter circuits are required, which will be discussed in due time.

Codes can be broadly classified into five groups, *viz.* (*i*) Weighted Binary Codes, (*ii*) Nonweighted Codes, (*iii*) Error-detection Codes, (*iv*) Error-correcting Codes, and (*v*) Alphanumeric Codes.

2.2 CODES

Computers and other digital circuits process data in binary format. Various binary codes are used to represent data which may be numeric, alphabetic or special characters. Codes are also used for error detection and error correction in digital systems. Although, in digital systems in every code used, the information is represented in binary form, but the interpretation of the data is only possible if the code in which the data is being represented is known. For example, the binary number 1000010 represents 66 (decimal) in straight binary, 42 (decimal) in BCD, and letter B in ASCII code. Hence, while interpreting the data, one must be very careful regarding the code used. Some of the commonly used codes are discussed below.

31

2.2.1 Weighted Binary Codes

If each position of a number represents a specific weight then the coding scheme is called weighted binary code. In such coding the bits are multiplied by their corresponding individual weight, and then the sum of these weighted bits gives the equivalent decimal digit.

BCD Code or 8421 Code

The full form of BCD is 'Binary-Coded Decimal.' Since this is a coding scheme relating decimal and binary numbers, four bits are required to code each decimal number. For example, $(35)_{10}$ is represented as 0011 0101 using BCD code, rather than $(100011)_2$. From the example it is clear that it requires more number of bits to code a decimal number using BCD code than using the straight binary code. However, inspite of this disadvantage it is convenient to use BCD code for input and output operations in digital systems.

The code is also known as 8-4-2-1 code. This is because 8, 4, 2, and 1 are the weights of the four bits of the BCD code. The weight of the LSB is 2^0 or 1, that of the next higher order bit is 2^1 or 2, that of the next higher order bit is 2^2 or 4, and that of the MSB is 2^3 or 8. Therefore, this is a *weighted* code and arithmetic operations can be performed using this code, which will be discussed later on. The bit assignment 0101, for example, can be interpreted by the weights to represent the decimal digit 5 because $0 \times 8 + 1 \times 4 + 0 \times 2 + 1 \times 1 = 5$. Since four binary bits are used the maximum decimal equivalent that may be coded is 15_{10} (*i.e.*, 1111_2). But the maximum decimal digit available is 9_{10}. Hence the binary codes 1010, 1011, 1100, 1101, 1110, 1111, representing 10, 11, 12, 13, 14, and 15 in decimal are never being used in BCD code. So these six codes are called forbidden codes and the group of these codes is called the forbidden group in BCD code. BCD code for decimal digits 0 to 9 is shown in Table 2.1.

Example 2.1. *Give the BCD equivalent for the decimal number 589.*

Solution.   The decimal number is 589

BCD code is   0101   1000   1001

Hence, $(589)_{10} = (010110001001)_{BCD}$

Example 2.2. *Give the BCD equivalent for the decimal number 69.27.*

Solution.   The decimal number   6   92   7

BCD code is   0110   1001   0010   0111

Hence, $(69.27)_{10} = (01101001.00100111)_{BCD}$

84-2-1 Code

It is also possible to assign negative weights to decimal code, as shown by the 84-2-1 code. In this case the bit combination 0101 is interpreted as the decimal digit 3, as obtained from $0 \times 8 + 1 \times 4 + 0 \times (-2) + 1 \times (-1) = 3$. This is a self-complementary code, that is, the 9's complement of the decimal number is obtained just by changing the 1s to 0s and 0s to 1s, or in effect by getting the 1's complement of the corresponding number. For example, if we change the 1s to 0s and 0s to 1s in the previous example we have 1010, which is interpreted as decimal 6, as obtained from $1 \times 8 + 0 \times 4 + 1 \times (-2) + 0 \times (-1) = 6$. And 6 is the 9's complement of 3. This property is useful when arithmetic operations are done internally with decimal numbers (in a binary code) and subtraction is calculated by means of 9's complement.

2421 Code

Another weighted code is 2421 code. The weights assigned to the four digits are 2, 4, 2, and 1. The 2421 code is the same as that in BCD from 0 to 4; however, it varies from 5 to 9. For example, in this case the bit combination 0100 represents decimal 4; whereas the bit combination 1101 is interpreted as the decimal 7, as obtained from $2 \times 1 + 1 \times 4 + 0 \times 2 + 1 \times 1 = 7$. This is also a self-complementary code, that is, the 9's complement of the decimal number is obtained by changing the 1s to 0s and 0s to 1s. The 2421 codes for decimal numbers 0 through 9 are shown in Table 2.1.

2.2.2 Nonweighted Codes

These codes are not positionally weighted. It basically means that each position of the binary number is not assigned a fixed value. Excess-3 codes and Gray codes are such non-weighted codes.

Excess-3 Code

A decimal code that has been used in some old computers is Excess-3 code. This is a nonweighted code. This code assignment is obtained from the corresponding value of 4-bit binary code after adding 3 to the given decimal digit. Here the maximum value may be 1100_2. Since the maximum decimal digit is 9 we have to add 3 to 9 and then get the BCD equivalent. Like 84-2-1 and 2421 codes Excess-3 is also a self-complementary code, that is, the 9's complement of the decimal number is obtained by changing the 1s to 0s and 0s to 1s. This self-complementary property of the code helps considerably in performing subtraction operation in digital systems.

Example 2.3. *Convert* $(367)_{10}$ *into its Excess-3 code.*

Solution.

The decimal number is	3	6	7
Add 3 to each bit	+3	+3	+3
Sum	6	9	10

Converting the above sum into 4-bit binary equivalent, we have a

4-bit binary equivalent of   0110   1001   1010

Hence, the Excess-3 code for $(367)_{10}$ = 0110 1001 1010

Example 2.4. *Convert* $(58.43)_{10}$ *into its Excess-3 code.*

Soluton.

The decimal number is	5	8	4	3
Add 3 to each bit	+3	+3	+3	+3
Sum	8	11	7	6

Converting the above sum into 4-bit binary equivalent, we have a

4-bit binary equivalent of   1000   1011   0111   0110

Hence, the Excess-3 code for $(367)_{10}$ = 10001011.01110110

Table 2.1 Binary codes for decimal digits

Decimal digit	(BCD) 8421	84-2-1	2421	Excess-3
0	0000	0000	0000	0011
1	0001	0111	0001	0100
2	0010	0110	0010	0101
3	0011	0101	0011	0110
4	0100	0100	0100	0111
5	0101	1011	1011	1000
6	0110	1010	1100	1001
7	0111	1001	1101	1010
8	1000	1000	1110	1011
9	1001	1111	1111	1100

Gray Code

Gray code belongs to a class of code known as minimum change code, in which a number changes by only one bit as it proceeds from one number to the next. Hence this code is not useful for arithmetic operations. This code finds extensive use for shaft encoders, in some types of analog-to-digital converters, etc. Gray code is reflected code and is shown in Table 2.3. The Gray code may contain any number of bits. Here we take the example of 4-bit Gray code. The code shown in Table 2.3 is only one of many such possible codes. To obtain a different reflected code, one can start with any bit combination and proceed to obtain the next bit combination by changing only one bit from 0 to 1 or 1 to 0 in any desired random fashion, as long as two numbers do not have identical code assignments. The Gray code is not a weighted code.

Table 2.2 Four-bit reflected code

Reflected Code	Decimal Equivalent
m4 0000	0
m3 0001	1
0011	2
m2 0010	3
0110	4
0111	5
0101	6
m1 0100	7
1100	8
1101	9
1111	10
m5 1110	11
1010	12
m6 1011	13
m7 1001	14
1000	15

Now we try to analyze the name "Reflected Code." If we look at the Table 2.3 we can consider seven virtual mirrors m1, m2, m3, m4, m5, m6, and m7 placed. Now, for mirror m1, if we consider the MSB as the refractive index of the input and output medium then leaving out the MSB we can see that all of the eight combinations of three bits each have their corresponding reflected counterparts. Similarly, for mirrors m2 and m5, if we now leave the actual MSB we can consider the combination of three bits where now we consider the third bit as the new MSB. And similar arguments follow for mirror m1. Similarly, we may analyze the cases for mirrors m3, m4, m6, and m7.

Table 2.3 Binary and Gray codes

Decimal numbers	Binary code	Gray code
0	0000	0000
1	0001	0001
2	0010	0011
3	0011	0010
4	0100	0110
5	0101	0111
6	0110	0101
7	0111	0100
8	1000	1100
9	1001	1101
10	1010	1111
11	1011	1110
12	1100	1010
13	1101	1011
14	1110	1001
15	1111	1000

Conversion of a Binary Number into Gray Code

Any binary number can be converted into equivalent Gray code by the following steps:

(*i*) the MSB of the Gray code is the same as the MSB of the binary number;

(*ii*) the second bit next to the MSB of the Gray code equals the Ex-OR of the MSB and second bit of the binary number; it will be 0 if there are same binary bits or it will be 1 for different binary bits;

(*iii*) the third bit for Gray code equals the exclusive-OR of the second and third bits of the binary number, and similarly all the next lower order bits follow the same mechanism.

Example 2.5. *Convert* $(101011)_2$ *into Gray code.*

Solution.

Step 1.   The MSB of the Gray code is the same as the MSB of the binary number.

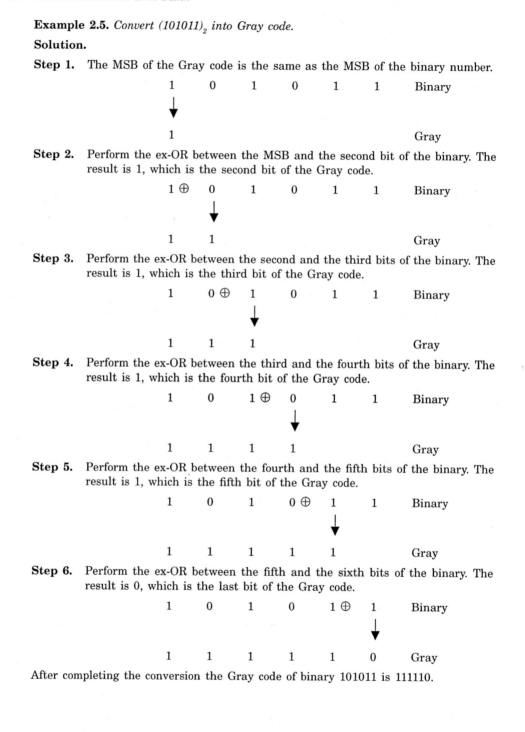

Step 2.   Perform the ex-OR between the MSB and the second bit of the binary. The result is 1, which is the second bit of the Gray code.

Step 3.   Perform the ex-OR between the second and the third bits of the binary. The result is 1, which is the third bit of the Gray code.

Step 4.   Perform the ex-OR between the third and the fourth bits of the binary. The result is 1, which is the fourth bit of the Gray code.

Step 5.   Perform the ex-OR between the fourth and the fifth bits of the binary. The result is 1, which is the fifth bit of the Gray code.

Step 6.   Perform the ex-OR between the fifth and the sixth bits of the binary. The result is 0, which is the last bit of the Gray code.

After completing the conversion the Gray code of binary 101011 is 111110.

Example 2.6. *Convert* $(564)_{10}$ *into Gray code.*

Solution.

Step 1.   Convert the decimal 564 into equivalent binary.

Decimal number 564

Binary number 1000110100

Step 2.   Convert the binary number into equivalent Gray code.

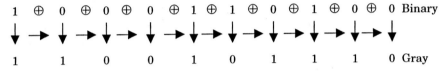

Conversion of Gray Code into a Binary Number

Any Gray code can be converted into an equivalent binary number by the following steps:

(*i*) the MSB of the binary number is the same as the MSB of the Gray code;

(*ii*) the second bit next to the MSB of the binary number equals the Ex-OR of the MSB of the binary number and second bit of the Gray code; it will be 0 if there are same binary bits or it will be 1 for different binary bits;

(*iii*) the third bit for the binary number equals the exclusive-OR of the second bit of the binary number and third bit of the Gray code, and similarly all the next lower order bits follow the same mechanism.

Example 2.7. *Convert the Gray code 101101 into a binary number.*

Solution.

Step 1. The MSB of the binary number is the same as the MSB of the Gray code.

Step 2. Perform the ex-OR between the MSB of the binary number and the second bit of the Gray code. The result is 1, which is the second bit of the binary number.

Step 3. Perform the ex-OR between the second bit of the binary number and the third bit of the Gray code. The result is 0, which is the third bit of the binary number.

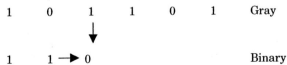

Step 4. Perform the ex-OR between the third bit of the binary number and the fourth bit of the Gray code. The result is 1, which is the fourth bit of the binary number.

Step 5. Perform the ex-OR between the fourth bit of the binary number and the fifth bit of the Gray code. The result is 1, which is the fifth bit of the binary number.

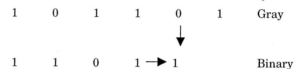

Step 6. Perform the ex-OR between the fifth bit of the binary number and the sixth bit of the Gray code. The result is 0, which is the last bit of the binary number.

| 1 | 0 | 1 | 1 | 0 | 1 | Gray |

| 1 | 1 | 0 | 1 | 1 → | 0 | Binary |

After completing the conversion, the binary number of the Gray code 101101 is 110110.

2.2.3 Error-detection Codes

Parity Bit Coding Technique

Binary information may be transmitted through some form of communication medium such as wires or radio waves or fiber optic cables, etc. Any external noise introduced into a physical communication medium changes bit values from 0 to 1 or vice versa. An error-detection code can be used to detect errors during transmission. The detected error cannot be corrected, but its presence is indicated.

A parity bit is an extra bit included with a message to make the total number of 1s either odd or even. A message of four bits and a parity bit, P, are shown in Table 2.4. In (a), P is chosen so that the sum of all 1s is odd (including the parity bit). In (b), P is chosen so that the sum of all 1s is even (including the parity bit). In the sending end, the message (in this case the first four bits) is applied to a "parity generation" circuit where the required P bit is generated. The message, including the parity bit, is transferred to its destination. In the receiving end, all the incoming bits (in this case five) are applied to a "parity check" circuit to check the proper parity adopted. An error is detected if the checked parity does not correspond to the adopted one. The parity method detects the presence of one, three, five, or any odd combination of errors. An even combination of errors is undetectable since an even number of errors will not change the parity of the bits. The parity bit may be included with the message bits either on the MSB or on the LSB side. Hence, in such cases, some other coding scheme is to be adopted. Such a coding technique is Check Sums, which will be discussed next.

Table 2.4 Parity bit

(a) Message	P (odd)	(b) Message	P (even)
0000	1	0000	0
0001	0	0001	1
0010	0	0010	1
0011	1	0011	0
0100	0	0100	1
0101	1	0101	0
0110	1	0110	0
0111	0	0111	1
1000	0	1000	1
1001	1	1001	0
1010	1	1010	0
1011	0	1011	1
1100	1	1100	0
1101	0	1101	1
1110	0	1110	1
1111	1	1111	0

Check Sums

As we have discussed aboves the parity bit technique fails for double errors, hence we use the Check Sums method in such case. Initially any word A 10010011 is transmitted; next another word B 01110110 is transmitted. The binary digits in the two words are added and the sum obtained is retained in the transmitter. Then any other word C is transmitted and added to the previous sum retained in the transmitter and the new sum is now retained. In a similar manner, each word is added to the previous sum already retained; after transmitting all the words, the final sum, which is called the Check Sum, is also transmitted. The same operation is done at the receiving end and the final sum, which has been obtained here, is being checked against the transmitted Check Sum. There is no error if the two sums are equal.

2.2.4 Error-correcting Codes

We have already discussed two coding techniques that may be used in transmission to detect errors. But, unfortunately, those discussed above are not capable of correcting the errors. For correction of errors we will now discuss a code called the Hamming code.

Hamming Code

This coding had been developed by R. W. Hamming where one or more parity bits are added to a data character methodically in order to detect and correct errors. The number of bits changed from one code word to another is known as *Hamming distance*.

Let us consider A_i and A_j to be any two code words in any particular block code. Now the Hamming distance d_{ij} between the two vectors A_i and A_j is defined by the number of components in which they differ. Assuming that d_{ij} is determined for each pair of code words, the minimum value of d_{ij} is called the minimum Hamming distance, d_{min}.

For example,

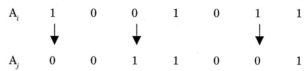

Here, these code words differ in the MSB and in the third and sixth bit positions from the left. Hence, d_{ij} is 3.

From Hamming's analysis of code distances, some important properties have been derived:

(*i*) For detection of a single error d_{min} should be at least two.

(*ii*) For single error correction, d_{min} should be at least three, since the number of errors, $E \le [(d_{min} - 1)/2]$.

(*iii*) Greater values of d_{min} will provide detection and/or correction of more number of errors.

The 7-bit Hamming (7, 4) code word $h_1\,h_2\,h_3\,h_4\,h_5\,h_6\,h_7$ associated with a 4-bit binary number $b_3\,b_2\,b_1\,b_0$ is:

$$h_1 = b_3 \oplus b_2 \oplus b_0$$
$$h_2 = b_3 \oplus b_1 \oplus b_0$$
$$h_4 = b_2 \oplus b_1 \oplus b_0$$
$$h_3 = b_3$$
$$h_5 = b_2$$
$$h_6 = b_1$$
$$h_7 = b_0$$

Bits h_1, h_2, and h_4 produce even parity bits for the bit fields $b_3\,b_2\,b_0$, $b_3\,b_1\,b_0$, and $b_2\,b_1\,b_0$ respectively. Generally the parity bits (h_1, h_2, h_4, h_8, h_{16}...) are located in the positions corresponding to ascending powers of two (*i.e.*, 2^0, 2^1, 2^2, 2^3, 2^4... = 1, 2, 4, 8,16...).

The h_1 parity bit has a 1 in the LSB position of its binary representation. Therefore it can check all the bit positions, including those that have 1s in the LSB position in the binary representation (*i.e.*, h_1, h_3, h_5, and h_7). The binary representation of h_2 has a 1 in the middle bit position. Therefore it can check all the bit positions, including those that have 1s in the middle bit position in the binary representation (*i.e.*, h_2, h_3, h_6, and h_7). The h_4 parity bit has a 1 in the MSB position of its binary representation. Therefore it can check all the bit positions, including those that have 1s in the MSB position in the binary representation (*i.e.*, h_4, h_5, h_6, and h_7).

To decode a Hamming code, checking needs to be done for odd parity over the bit fields in which even parity was previously established. For example, a single bit error is indicated by a nonzero parity word $a_4\,a_2\,a_1$, where

$$a_1 = h_1 \oplus h_3 \oplus h_5 \oplus h_7$$

$$a_2 = h_2 \oplus h_3 \oplus h_6 \oplus h_7$$
$$a_4 = h_4 \oplus h_5 \oplus h_6 \oplus h_7$$

If $a_4\, a_2\, a_1 = 000$, we conclude there is no error in the Hamming code. On the other hand, if it has a nonzero value, it indicates the bit position in error. For example, if $a_4\, a_2\, a_1 = 110$, then bit 6 is in error. To correct this error, bit 6 has to be complemented.

Example 2.8. *Encode data bits 0110 into a 7-bit even parity Hamming code.*

Solution. Given

$$b_3\, b_2\, b_1\, b_0 = 0\ 1\ 1\ 0$$

Therefore,

$$h_1 = b_3 \oplus b_2 \oplus b_0 = 0 \oplus 1 \oplus 0 = 1$$
$$h_2 = b_3 \oplus b_1 \oplus b_0 = 0 \oplus 1 \oplus 0 = 1$$
$$h_4 = b_2 \oplus b_1 \oplus b_0 = 1 \oplus 1 \oplus 0 = 0$$
$$h_3 = b_3 = 0$$
$$h_5 = b_2 = 1$$
$$h_6 = b_1 = 1$$
$$h_7 = b_0 = 0$$

h_1	h_2	h_3	h_4	h_5	h_6	h_7
1	1	0	0	1	1	0

Example 2.9. *A 7-bit Hamming code is received as 0110110. What is its correct code?*

Solution.

h_1	h_2	h_3	h_4	h_5	h_6	h_7
0	1	1	0	1	1	0

Now, to find the error,

$$a_1 = h_1 \oplus h_3 \oplus h_5 \oplus h_7 = 0 \oplus 1 \oplus 1 \oplus 0 = 0$$
$$a_2 = h_2 \oplus h_3 \oplus h_6 \oplus h_7 = 1 \oplus 1 \oplus 1 \oplus 0 = 1$$
$$a_4 = h_4 \oplus h_5 \oplus h_6 \oplus h_7 = 0 \oplus 1 \oplus 1 \oplus 0 = 0$$

Thus, $a_4\, a_2\, a_1 = 010$. Therefore, bit 2 is in error and the corrected code can be obtained by complementing the second bit in the received as 00 10110.

2.2.5 Alphanumeric Codes

Many applications of the computer require not only handling of numbers, but also of letters. To represent letters it is necessary to have a binary code for the alphabet. In addition, the same binary code must represent the decimal numbers and some other special characters. An alphanumeric code is a binary code of a group of elements consisting of ten decimal digits, the 26 letters of the alphabet (both in uppercase and lowercase), and a certain number of special symbols such as #, /, &, %, etc. The total number of elements in an alphanumeric code is greater than 36. Therefore it must be coded with a minimum number of 6 bits ($2^6 = 64$, but $2^5 = 32$ is insufficient). One possible 6-bit alphanumeric code is given in Table 2.5. It is used in many computers to represent alphanumeric characters and symbols internally and therefore can be called "internal code." Frequently there is a need to represent more than 64 characters, including the lowercase letters and special control characters. For this reason the following two codes are normally used.

ASCII

The full form of ASCII (pronounced "as-kee") is "American Standard Code for Information Interchange," used in most microcomputers. It is actually a 7-bit code, where a character is represented with seven bits. The character is stored as one byte with one bit remaining unused. But often the extra bit is used to extend the ASCII to represent an additional 128 characters. Some of the codes are shown in Table 2.5.

EBCDIC

The full form of EBCDIC is "Extended Binary Coded Decimal Interchange Code." It is also an alphanumeric code generally used in IBM equipment and in large computers for communicating alphanumeric data. For the different alphanumeric characters the code grouping in this code is different from the ASCII code. It is actually an 8-bit code and a ninth bit is added as the parity bit.

Hollerith Code

Generally this code is used in punched cards. A punched card consists of 12 rows and 80 columns. An alphanumeric character is represented by each column of 12 bits each by punching holes in the appropriate rows. The presence of a hole represents a 1 and its absence indicates 0. The 12 rows are marked starting from the top, as 12, 11, 0, 1, 2, 3, 4, 5, 6, 7, 8, and 9. The first three rows are called the zone punch and the last nine are called the numeric punch. The code used here is called the Hollerith code. The letters are represented as two holes in a column, one in zone punch and the other in numeric punch; decimal digits are represented as a single hole in a numeric punch. Special characters are represented as one, two, or three holes in a column; while the zone is always used, the other two holes, if used, are in a numeric punch with the eighth punch being commonly used. The Hollerith code is BCD and hence the transition from EBCDIC is simple. The Hollerith code is used in the card readers and punches of large computers, while EBCDIC may be used within the computer.

Table 2.5 Partial list of alphanumeric codes

Character	6-bit Internal code	7-bit ASCII code	8-bit EBCDIC code	12-bit Hollerith code
A	010001	1000001	11000001	12,1
B	010010	1000010	11000010	12,2
C	010011	1000011	11000011	12,3
D	010100	1000100	11000100	12,4
E	010101	1000101	11000101	12,5
F	010110	1000110	11000110	12,6
G	010111	1000111	11000111	12,7
H	011000	1001000	11001000	12,8
I	011001	1001001	11001001	12,9
J	100001	1001010	11010001	11,1
K	100010	1001011	11010010	11,2
L	100011	1001100	11010011	11,3
M	100100	1001101	11010100	11,4

N	100101	1001110	11010101	11,5
O	100110	1001111	11010110	11,6
P	100111	1010000	11010111	11,7
Q	101000	1010001	11011000	11,8
R	101001	1010010	11011001	11,9
S	110010	1010011	11100010	0,2
T	110011	1010100	11100011	0,3
U	110100	1010101	11100100	0,4
V	110101	1010110	11100101	0,5
W	110110	1010111	11100110	0,6
X	110111	1011000	11100111	0,7
Y	111000	1011001	11101000	0,8
Z	111001	1011010	11101001	0,9
0	000000	0110000	11110000	0
1	000001	0110001	11110001	1
2	000010	0110010	11110010	2
3	000011	0110011	11110011	3
4	000100	0110100	11110100	4
5	000101	0110101	11110101	5
6	000110	0110110	11110110	6
7	000111	0110111	11110111	7
8	001000	0111000	11111000	8
9	001001	0111001	11111001	9
Blank	110000	0100000	01000000	No punch
.	011011	0101110	01001011	12,3,8
(	111100	0101000	01001101	12,5,8
+	010000	0101011	01001110	12,6,8
*	101100	0101010	01011100	11,4,8
$	101011	0100100	01011011	11,3,8
)	011100	0101001	01011101	11,5,8
/	110001	0101111	01100001	0,1
,	111011	0111100	01101011	0,3,8
=	001011	0111101	01111110	6,8
−	100000	0101101	01100000	11

2.3 SOLVED PROBLEMS

Example 2.10. *Encode the following decimal numbers in BCD code:*

(a) 45 (b) 273.98 (c) 62.905

Solution. (a) Decimal number is 4 5

BCD code is 0100 0101

Hence the BCD coded form of 45_{10} is 0100 0101

(b) Decimal number is 2 7 3 9 8

BCD code is 0010 0111 0011 1001 1000

Hence the BCD coded form of 273.98_{10} is 0010 0111 0011.1001 1000

(c) Decimal number is 6 2 9 0 5

BCD code is 0110 0010 1001 0000 0101

Hence the BCD coded form of 62.905_{10} is 0110 0010.1001 0000 0101

Example 2.11. *Write down the decimal numbers represented by the following BCD codes:*

(a) 100101001 (b) 100010010011 (c) 01110001001.10010010

Solution. (a) BCD code is 1 0010 1001

By padding up the first number with 3 zeros 0001 0010 1001

Decimal number is 1 2 9

Hence the decimal number is 129.

(b) BCD code is 1000 1001 0011

Decimal number is 8 9 3

Hence the decimal number is 893.

(c) BCD code is 011 1000 1001 1001 0010

By padding up the first number with 1 zero 0011 1000 1001 1001 0010

Decimal number is 3 8 9 9 2

Hence the decimal number is 389.92.

Example 2.12. *Encode the following decimal numbers to Excess-3 code:*

(a) 38 (b) 471.78 (c) 23.105

Solution. (a) Decimal number is 3 8

BCD code is 0011 1000

Now adding 3 +0011 +0011

Excess-3 code is 0110 1011

Hence the Excess-3 coded form of 38_{10} is 0110 1011

(b) Decimal number is 4 7 1 7 8

BCD code is 0100 0111 0001 0111 1000

Now adding 3 +0011 +0011 +0011 +0011 +0011

Excess-3 code is 0111 1010 0100 1010 1011

Hence the Excess-3 coded form of 471.78_{10} is 0111 1010 0100.1010 1011

(c) Decimal number is   2   3   1   0   5

Decimal number is	2	3	1	0	5
BCD code is	0010	0011	0001	0000	0101
Now adding 3	+0011	+0011	+0011	+0011	+0011
Excess-3 code is	0101	0110	0100	0011	1000

Hence the Excess-3 coded form of 23.105_{10} is 0101 0110.0100 0011 1000

Example 2.13. *Express the following Excess-3 codes as decimal numbers:*

    (a) 0101 1011 1100 0111   *(b) 0011 1000 1010 0100*   *(c) 0101 1001 0011*

Solution. (a)

Excess-3 code is	0101	1011	1100	0111
Subtracting 3 from each digit	−0011	−0011	−0011	−0011
BCD number is	0010	1000	1001	0100
Decimal number is	2	8	9	4

Hence the decimal number is 2894.

  (b)

Excess-3 code is	0011	1000	1010	0100
Subtracting 3 from each digit	−0011	−0011	−0011	−0011
BCD number is	0000	0101	0111	0001
Decimal number is	0	5	7	1

Hence the decimal number is 571.

  (c)

Excess-3 code is	0101	1001	0011
Subtracting 3 from each digit	−0011	−0011	−0011
BCD number is	0010	0110	0000
Decimal number is	2	6	0

Hence the decimal number is 260.

Example 2.14. *Encode the following decimal numbers to Gray codes:*

    (a) 61   *(b) 83*   *(c) 324*   *(d) 456*

Solution. (a)

Decimal number is	61
Binary code is	111101
Gray code is	100011

  (b)

Decimal number is	83
Binary code is	1010011
Gray code is	1111010

  (c)

Decimal number is	324
Binary code is	101000100
Gray code is	111100110

  (d)

Decimal number is	456
Binary code is	111001000
Gray code is	100101100

Example 2.15. *Express the following binary numbers as Gray codes:*

        (a) 10110   *(b) 0110111*   *(c) 101010011*

        (d) 101011100   *(e) 110110001*   *(f) 10001110110*

Solution.   (*a*)   Binary number is   10110

                    Gray code is   11101

      (*b*)   Binary number is   0110111

                    Gray code is   0101100

      (*c*)   Binary number is   101010011

                    Gray code is   111111010

      (*d*)   Binary number is   101011100

                    Gray code is   111110010

      (*e*)   Binary number is   110110001

                    Gray code is   101101001

      (*f*)   Binary number is   10001110110

                    Gray code is   11001001101

Example 2.16. *Express the following Gray codes as binary numbers:*

            (a) 10111   *(b) 0110101*   *(c) 10100011*

           (d) 100111100   *(e) 101010001*   *(f) 10110010101*

Solution.   (*a*)   Gray code is   10111

                    Binary number is   11010

      (*b*)   Gray code is   0110101

                    Binary number is   0100110

      (*c*)   Gray code is   10100011

                    Binary number is   11000010

      (*d*)   Gray code is   100111100

                    Binary number is   111010111

      (*e*)   Gray code is   101010001

                    Binary number is   110011110

      (*f*)   Gray code is   10110010101

                    Binary number is   11011100110

Example 2.17. *Encode the following binary numbers as 7-bit even Hamming codes:*

        (a) 1000   *(b) 0101*   *(c) 1011*

Solution. (*a*) Binary number is $b_3\, b_2\, b_1\, b_0 = 1000$

Now $h_1 = b_3 \oplus b_2 \oplus b_0 = 1 \oplus 0 \oplus 0 = 1$

                       $h_2 = b_3 \oplus b_1 \oplus b_0 = 1 \oplus 0 \oplus 0 = 1$

$$h_4 = b_2 \oplus b_1 \oplus b_0 = 0 \oplus 0 \oplus 0 = 0$$
$$h_3 = b_3 = 1$$
$$h_5 = b_2 = 0$$
$$h_6 = b_1 = 0$$
$$h_7 = b_0 = 0$$

h_1	h_2	h_3	h_4	h_5	h_6	h_7
1	1	1	0	0	0	0

(b) Binary number is $b_3\,b_2\,b_1\,b_0 = 0101$

Now $h_1 = b_3 \oplus b_2 \oplus b_0 = 0 \oplus 1 \oplus 1 = 0$

$$h_2 = b_3 \oplus b_1 \oplus b_0 = 0 \oplus 0 \oplus 1 = 1$$
$$h_4 = b_2 \oplus b_1 \oplus b_0 = 1 \oplus 0 \oplus 1 = 0$$
$$h_3 = b_3 = 0$$
$$h_5 = b_2 = 1$$
$$h_6 = b_1 = 0$$
$$h_7 = b_0 = 1$$

h_1	h_2	h_3	h_4	h_5	h_6	h_7
0	1	0	0	1	0	1

(c) Binary number is $b_3\,b_2\,b_1\,b_0 = 1011$

Now $h_1 = b_3 \oplus b_2 \oplus b_0 = 1 \oplus 0 \oplus 1 = 0$

$$h_2 = b_3 \oplus b_1 \oplus b_0 = 1 \oplus 1 \oplus 1 = 1$$
$$h_4 = b_2 \oplus b_1 \oplus b_0 = 0 \oplus 1 \oplus 1 = 0$$
$$h_3 = b_3 = 1$$
$$h_5 = b_2 = 0$$
$$h_6 = b_1 = 1$$
$$h_7 = b_0 = 1$$

h_1	h_2	h_3	h_4	h_5	h_6	h_7
0	1	1	0	0	1	1

Example 2.18. *Use the (a) 6-bit internal code, (b) 7-bit ASCII code, and (c) 8-bit EBCDIC code to represent the statement:*

$$P = 4*Q$$

Solution.

(a) P is encoded in 6-bit internal code as 100111

= is encoded in 6-bit internal code as 001011

4 is encoded in 6-bit internal code as 000100

* is encoded in 6-bit internal code as 101100

Q is encoded in 6-bit internal code as 101000

Hence the encoded form of P = 4*Q is 100111 001011 000100 101100 101000

(b) P is encoded in 7-bit ASCII code as   1010000

= is encoded in 7-bit ASCII code as   0111101

4 is encoded in 7-bit ASCII code as   0110100

* is encoded in 7-bit ASCII code as   0101010

Q is encoded in 7-bit ASCII code as   1010001

Hence the encoded form of P = 4*Q is   1010000 0111101 0110100 0101010 1010001

(c) P is encoded in 8-bit EBCDIC code as 11010111

= is encoded in 8-bit EBCDIC code as 01111110

4 is encoded in 8-bit EBCDIC code as 11110100

* is encoded in 8-bit EBCDIC code as 01011100

Q is encoded in 8-bit EBCDIC code as 11011000

Hence the encoded form of P = 4*Q is   11010111 01111110 11110100 01011100 11011000

Example 2.19. *Express the following decimal numbers as 2421 codes:*

(a) 168   (b) 254   (c) 6735   (d) 1973   (e) 9021

Solution. (a) Decimal number given is   1   6   8

Equivalent 2421 code is   0001   1100   1110

(b) Decimal number given is   2   5   4

Equivalent 2421 code is   0010   1011   0100

(c) Decimal number given is   6   7   3   5

Equivalent 2421 code is   1100   1101   0011   1011

(d) Decimal number given is   1   9   7   3

Equivalent 2421 code is   0001   1111   1101   0011

(e) Decimal number given is   9   0   2   1

Equivalent 2421 code is   1111   0000   0010   0001

Example 2.20. *Express the following 2421 codes as decimal numbers:*

(a) 1110 1011 1101   (b) 0010 1100 0001 (c) 1011 0100 1111 (d) 1101 1111 1011

Solution. (a) 2421 code given is   1110   1011   1101

Equivalent decimal number is   8   5   7

(b) 2421 code given is   0010   1100   0001

Equivalent decimal number is   2   6   1

(c) 2421 code given is   1011   0100   1111

Equivalent decimal number is   5   4   9

(d) 2421 code given is   1101   1111   1011

Equivalent decimal number is   7   9   5

REVIEW QUESTIONS

2.1 Express the following decimal numbers in Excess-3 code form:

(a) 245, (b) 739, (c) 4567, and (d) 532.

2.2 Express the following Excess-3 codes as decimals:

(a) 100000110110, (b) 0111110010010110, and (c) 110010100011.

2.3 Convert the following binary numbers to Gray codes:

(a) 10110, (b) 1110111, (c) 101010001, and (d) 1001110001110.

2.4 Express the following decimals in Gray code form:

(a) 5, (b) 27, (c) 567, and (d) 89345.

2.5 Write your first name and last name in an 8-bit code made up of the seven ASCII bits and an odd parity bit in the most significant position. Include blanks between names.

2.6 Express the following decimals in (1) 2,4,2,1 code and (2) 8, 4, −2, −1 code form:

(a) 35, (b) 7, (c) 566, and (d) 8945.

2.7 What is the difference between ASCII and EBCDIC codes? Why are EBCDIC codes used?

2.8 Why is Gray code called the reflected code? Explain.

2.9 What is Hamming code and how is it used?

2.10 Explain with an example how BCD addition is carried out?

❏ ❏ ❏

3 BOOLEAN ALGEBRA AND LOGIC GATES

3.1 INTRODUCTION

Binary logic deals with variables that have two discrete values—1 for TRUE and 0 for FALSE. A simple switching circuit containing active elements such as a diode and transistor can demonstrate the binary logic, which can either be ON (switch closed) or OFF (switch open). Electrical signals such as voltage and current exist in the digital system in either one of the two recognized values, except during transition.

The switching functions can be expressed with Boolean equations. Complex Boolean equations can be simplified by a new kind of algebra, which is popularly called Switching Algebra or Boolean Algebra, invented by the mathematician George Boole in 1854. Boolean Algebra deals with the rules by which logical operations are carried out.

3.2 BASIC DEFINITIONS

Boolean algebra, like any other deductive mathematical system, may be defined with a set of elements, a set of operators, and a number of assumptions and postulates. A set of elements means any collection of objects having common properties. If S denotes a set, and X and Y are certain objects, then $X \in S$ denotes X is an object of set S, whereas $Y \notin$ denotes Y is not the object of set S. A binary operator defined on a set S of elements is a rule that assigns to each pair of elements from S a unique element from S. As an example, consider this relation $X*Y = Z$. This implies that * is a binary operator if it specifies a rule for finding Z from the objects (X, Y) and also if all X, Y, and Z are of the same set S. On the other hand, * can not be binary operator if X and Y are of set S and Z is not from the same set S.

The postulates of a mathematical system are based on the basic assumptions, which make possible to deduce the rules, theorems, and properties of the system. Various algebraic structures are formulated on the basis of the most common postulates, which are described as follows.

1. **Closer:** A set is closed with respect to a binary operator if, for every pair of elements of S, the binary operator specifies a rule for obtaining a unique element of S. For example, the set of natural numbers N = {1, 2, 3, 4, ...} is said to be closed with respect

51

to the binary operator plus (+) by the rules of arithmetic addition, since for any $X, Y \in N$ we obtain a unique element $Z \in N$ by the operation $X + Y = Z$. However, note that the set of natural numbers is not closed with respect to the binary operator minus (–) by the rules of arithmetic subtraction because for $1 - 2 = -1$, where -1 is not of the set of naturals numbers.

2. **Associative Law:** A binary operator * on a set S is said to be associated whenever

$$(A*B)*C = A*(B*C) \qquad \text{for all } A, B, C \in S.$$

3. **Commutative Law:** A binary operator * on a set S is said to be commutative whenever

$$A*B = B*A \qquad \text{for all } A, B \in S.$$

4. **Identity Element:** A set S is to have an identity element with respect to a binary operation * on S, if there exists an element $E \in S$ with the property

$$E*A = A*X = A.$$

Example: The element 0 is an identity element with respect to the binary operator + on the set of integers $I = \{.... -4, -3, -2, -1, 0, 1, 2, 3, 4, ...\}$ as

$$A + 0 = 0 + A = A.$$

Similarly, the element 1 is the identity element with respect to the binary operator × as

$$A \times 1 = 1 \times A = A.$$

5. **Inverse:** If a set S has the identity element E with respect to a binary operator *, there exists an element $B \in S$, which is called the inverse, for every $A \in S$, such that $A*B = E$.

Example: In the set of integers I with $E = 0$, the inverse of an element A is $(-A)$ since $A + (-A) = 0$.

6. **Distributive Law:** If * and (.) are two binary operators on a set S, * is said to be distributive over (.), whenever

$$A*(B.C) = (A*B).(A*C).$$

If summarized, for the field of real numbers, the operators and postulates have the following meanings:

The binary operator + defines addition.

The additive identity is 0.

The additive inverse defines subtraction.

The binary operator (.) defines multiplication.

The multiplication identity is 1.

The multiplication inverse of A is 1/A, defines division *i.e.,* A. $1/A = 1$.

The only distributive law applicable is that of (.) over +

$$A . (B + C) = (A . B) + (A . C)$$

3.3 DEFINITION OF BOOLEAN ALGEBRA

In 1854 George Boole introduced a systematic approach of logic and developed an algebraic system to treat the logic functions, which is now called Boolean algebra. In 1938 C.E. Shannon developed a two-valued Boolean algebra called Switching algebra, and demonstrated that the properties of two-valued or bistable electrical switching circuits can be represented by this algebra. The postulates formulated by E.V. Huntington in 1904 are employed for the formal definition of Boolean algebra. However, Huntington postulates are not unique for defining Boolean algebra and other postulates are also used. The following Huntington postulates are satisfied for the definition of Boolean algebra on a set of elements S together with two binary operators (+) and (.).

1. (a) Closer with respect to the operator (+).

   (b) Closer with respect to the operator (.).

2. (a) An identity element with respect to + is designated by 0 i.e.,

   $$A + 0 = 0 + A = A.$$

   (b) An identity element with respect to . is designated by 1 i.e.,

   $$A.1 = 1. A = A.$$

3. (a) Commutative with respect to (+), i.e., $A + B = B + A$.

   (b) Commutative with respect to (.), i.e., $A.B = B.A$.

4. (a) (.) is distributive over (+), i.e., $A . (B+C) = (A . B) + (A . C)$.

   (b) (+) is distributive over (.), i.e., $A + (B .C) = (A + B) . (A + C)$.

5. For every element $A \in S$, there exists an element $A' \in S$ (called the complement of A) such that $A + A' = 1$ and $A . A' = 0$.

6. There exists at least two elements $A,B \in S$, such that A is not equal to B.

Comparing Boolean algebra with arithmetic and ordinary algebra (the field of real numbers), the following differences are observed:

1. Huntington postulates do not include the associate law. However, Boolean algebra follows the law and can be derived from the other postulates for both operations.

2. The distributive law of (+) over (.) i.e., $A+ (B.C) = (A+B) . (A+C)$ is valid for Boolean algebra, but not for ordinary algebra.

3. Boolean algebra does not have additive or multiplicative inverses, so there are no subtraction or division operations.

4. Postulate 5 defines an operator called Complement, which is not available in ordinary algebra.

5. Ordinary algebra deals with real numbers, which consist of an infinite set of elements. Boolean algebra deals with the as yet undefined set of elements S, but in the two-valued Boolean algebra, the set S consists of only two elements—0 and 1.

Boolean algebra is very much similar to ordinary algebra in some respects. The symbols (+) and (.) are chosen intentionally to facilitate Boolean algebraic manipulations by persons already familiar to ordinary algebra. Although one can use some knowledge from ordinary algebra to deal with Boolean algebra, beginners must be careful not to substitute the rules of ordinary algebra where they are not applicable.

It is important to distinguish between the elements of the set of an algebraic structure and the variables of an algebraic system. For example, the elements of the field of real numbers are numbers, the variables such as X, Y, Z, etc., are the symbols that stand for real numbers, which are used in ordinary algebra. On the other hand, in the case of Boolean algebra, the elements of a set S are defined, and the variables A, B, C, etc., are merely symbols that represent the elements. At this point, it is important to realize that in order to have Boolean algebra, the following must be shown.

1. The elements of the set S.

2. The rules of operation for the two binary operators.

3. The set of elements S, together with the two operators satisfies six Huntington postulates.

One may formulate many Boolean algebras, depending on the choice of elements of set S and the rules of operation. In the subsequent chapters, we will only deal with a two-valued Boolean algebra *i.e.*, one with two elements. Two-valued Boolean algebra has the applications in set theory and propositional logic. But here, our interest is with the application of Boolean algebra to gate-type logic circuits.

3.4 TWO-VALUED BOOLEAN ALGEBRA

Two-valued Boolean algebra is defined on a set of only two elements, S = {0,1}, with rules for two binary operators (+) and (.) and inversion or complement as shown in the following operator tables at Figures 3.1, 3.2, and 3.3 respectively.

A	B	A + B
0	0	0
0	1	1
1	0	1
1	1	1

A	B	A.B
0	0	0
0	1	0
1	0	0
1	1	1

A	A'
0	1
1	0

Figure 3.1   **Figure 3.2**   **Figure 3.3**

The rule for the complement operator is for verification of postulate 5.

These rules are exactly the same for as the logical OR, AND, and NOT operations, respectively. It can be shown that the Huntington postulates are applicable for the set S = {0,1} and the two binary operators defined above.

1. Closure is obviously valid, as form the table it is observed that the result of each operation is either 0 or 1 and 0,1 ∈ S.

2. From the tables, we can see that

    (*i*) $0 + 0 = 0$ $0 + 1 = 1 + 0 = 1$

    (*ii*) $1 . 1 = 1$ $0 . 1 = 1 . 0 = 0$

    which verifies the two identity elements 0 for (+) and 1 for (.) as defined by postulate 2.

3. The commutative laws are confirmed by the symmetry of binary operator tables.

4. The distributive laws of (.) over (+) *i.e.*, A . (B+C) = (A . B) + (A . C), and (+) over (.) *i.e.*, A + (B . C) = (A+B) . (A+C) can be shown to be applicable with the help of the truth tables considering all the possible values of A, B, and C as under.

From the complement table it can be observed that

(a) Operator (.) over (+)

A	B	C	B + C	A. (B + C)	A. B	A. C	(A. B) + (A.C)
0	0	0	0	0	0	0	0
0	0	1	1	0	0	0	0
0	1	0	1	0	0	0	0
0	1	1	1	0	0	0	0
1	0	0	0	0	0	0	0
1	0	1	1	1	0	1	1
1	1	0	1	1	1	0	1
1	1	1	1	1	1	1	1

Figure 3.4

(b) Operator (+) over (.)

A	B	C	B . C	A+(B . C)	A+B	A+C	(A+B).(A+C)
0	0	0	0	0	0	0	0
0	0	1	0	0	0	1	0
0	1	0	0	0	1	0	0
0	1	1	1	1	1	1	1
1	0	0	0	1	1	1	1
1	0	1	0	1	1	1	1
1	1	0	0	1	1	1	1
1	1	1	1	1	1	1	1

Figure 3.5

(c) $A + A' = 1$, since $0 + 0' = 1$ and $1 + 1' = 1$.

(d) $A . A' = 0$, since $0 . 0' = 0$ and $1 . 1' = 0$.

These confirm postulate 5.

5. Postulate 6 also satisfies two-valued Boolean algebra that has two distinct elements 0 and 1 where 0 is not equal to 1.

3.5 BASIC PROPERTIES AND THEOREMS OF BOOLEAN ALGEBRA

3.5.1 Principle of Duality

From Huntington postulates, it is evident that they are grouped in pairs as (a) and (b) and every algebraic expression deductible from the postulates of Boolean algebra remains valid if the operators and identity elements are interchanged. This means one expression can be obtained from the other in each pair by interchanging every element *i.e.*, every 0 with 1, every 1 with 0, as well as interchanging the operators *i.e.*, every (+) with (.) and every (.) with (+). This important property of Boolean algebra is called principle of duality.

3.5.2 DeMorgan's Theorem

Two theorems that were proposed by DeMorgan play important parts in Boolean algebra.

The first theorem states that the complement of a product is equal to the sum of the complements. That is, if the variables are A and B, then

$$(A.B)' = A' + B'$$

The second theorem states that the complement of a sum is equal to the product of the complements. In equation form, this can be expressed as

$$(A + B)' = A' . B'$$

The complements of Boolean logic function or a logic expression may be simplified or expanded by the following steps of DeMorgan's theorem.

(*a*) Replace the operator (+) with (.) and (.) with (+) given in the expression.

(*b*) Complement each of the terms or variables in the expression.

DeMorgan's theorems are applicable to any number of variables. For three variables A, B, and C, the equations are

$$(A.B.C)' = A' + B' + C' \qquad \text{and}$$
$$(A + B + C)' = A'.B'.C'$$

3.5.3 Other Important Theorems

Theorem 1(*a*): A + A = A

A + A = (A + A).1	by postulate 2(*b*)
= (A + A) . (A + A')	by postulate 5
= A + A.A'	
= A + 0	by postulate 4
= A	by postulate 2(*a*)

Theorem 1(*b*): A . A = A

A . A = (A . A) + 0	by postulate 2(*a*)
= (A . A) + (A . A')	by postulate 5
= A (A + A')	
= A . 1	by postulate 4
= A	by postulate 2(*b*)

Theorem 2(*a*): A + 1 = 1

Theorem 2(*b*): A . 0 = 0

Theorem 3(*a*): A + A.B = A

A + A.B = A . 1 + A.B	by postulate 2(*b*)
= A (1 + B)	by postulate 4(*a*)
= A . 1	by postulate 2(*a*)
= A	by postulate 2(*b*)

Theorem 3(*b*): A (A + B) = A   by duality

The following is the complete list of postulates and theorems useful for two-valued Boolean algebra.

Postulate 2	(*a*) A + 0 = A	(*b*) A.1 = A
Postulate 5	(*a*) A + A′ = 1	(*b*) A.A′ = 0
Theorem 1	(*a*) A + A = A	(*b*) A.A = A
Theorem 2	(*a*) A + 1 = 1	(*b*) A.0 = 0
Theorem 3, Involution	(A′)′ = A	
Theorem 3, Commutative	(*a*) A + B = B + A	(*b*) A.B = B.A
Theorem 4, Associative	(*a*) A + (B + C) = (A + B) + C	(*b*) A.(B.C) = (A.B).C
Theorem 4, Distributive	(*a*) A(B + C) = A.B + A.C	(*b*) A + B.C = (A + B).(A + C)
Theorem 5, DeMorgan	(*a*) (A + B)′ = A′.B′	(*b*) (A.B)′ = A′ + B′
Theorem 6, Absorption	(*a*) A + A.B = A	(*b*) A.(A + B) = A

Figure 3.6

3.6 VENN DIAGRAM

A Venn diagram is a helpful illustration to visualize the relationship among the variables of a Boolean expression. The diagram consists of a rectangle as shown in Figure 3.7, inside two overlapping circles are drawn, which represent two variables. Each circle is labeled by a variable. We consider that the area inside the circle belongs to the named variable and area outside circle does not belong to that variable. For example, if the variables are A and B, then A = 1 for inside the circle A, A = 0 for outside of circle A, and B = 1 for inside the circle B, B = 0 for outside of circle B. Now for two overlapping circles as in Figure 3.7, four distinct areas are available inside the rectangle area belonging to A only or AB′, area belonging to B only or A′B, area belonging to both A and B *i.e.*, A.B and area belonging to neither A or B *i.e.*, A′B′.

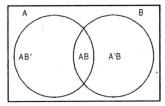

Figure 3.7

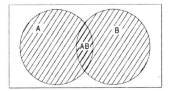

Figure 3.8

A Venn diagram may be used to illustrate the postulates of Boolean algebra or to explain the validity of theorems. For example, Figure 3.8 shows that the area belonging to AB is inside the circle A, and therefore A + AB = A.

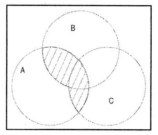

Figure 3.9

Figure 3.9 demonstrates the distributive law A(B + C) = AB + AC. In this figure three variables A, B, and C are used. It is possible to demonstrate eight distinct areas available for three variables in a Venn diagram. For this particular example, the distributive law is explained by showing the area intersecting the circle A with area enclosed by B or C is the same area belonging to AB or AC.

3.7 BOOLEAN FUNCTIONS

Binary variables have two values, either 0 or 1. A Boolean function is an expression formed with binary variables, the two binary operators AND and OR, one unary operator NOT, parentheses and equal sign. The value of a function may be 0 or 1, depending on the values of variables present in the Boolean function or expression. For example, if a Boolean function is expressed algebraically as

$$F = AB'C$$

then the value of F will be 1, when A = 1, B = 0, and C = 1. For other values of A, B, C the value of F is 0.

Boolean functions can also be represented by truth tables. A *truth table* is the tabular form of the values of a Boolean function according to the all possible values of its variables. For an n number of variables, 2^n combinations of 1s and 0s are listed and one column represents function values according to the different combinations. For example, for three variables the Boolean function F = AB + C truth table can be written as below in Figure 3.10.

A	B	C	F
0	0	0	0
0	0	1	1
0	1	0	0
0	1	1	1
1	0	0	0
1	0	1	1
1	1	0	1
1	1	1	1

Figure 3.10

A Boolean function from an algebraic expression can be realized to a logic diagram composed of logic gates. Figure 3.11 is an example of a logic diagram realized by the basic gates like AND, OR, and NOT gates. In subsequent chapters, more logic diagrams with various gates will be shown.

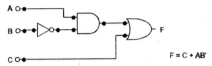

Figure 3.11

3.8 SIMPLIFICATION OF BOOLEAN EXPRESSIONS

When a Boolean expression is implemented with logic gates, each literal in the function is designated as input to the gate. The literal may be a primed or unprimed variable. Minimization of the number of literals and the number of terms leads to less complex circuits as well as less number of gates, which should be a designer's aim. There are several methods to minimize the Boolean function. In this chapter, simplification or minimization of complex algebraic expressions will be shown with the help of postulates and theorems of Boolean algebra.

Example 3.1. *Simplify the Boolean function F=AB+ BC + B'C.*

Solution. F = AB + BC + B'C

= AB + C(B + B')

= AB + C

Example 3.2. *Simplify the Boolean function F= A + A'B.*

Solution. F = A+ A'B

= (A + A') (A + B)

= A + B

Example 3.3. *Simplify the Boolean function F= A'B'C + A'BC + AB'.*

Solution. F = A'B'C + A'BC + AB'

= A'C (B'+B) + AB'

= A'C + AB'

Example 3.4. *Simplify the Boolean function F = AB + (AC)' + AB'C(AB + C).*

Solution. F = AB + (AC)' + AB'C(AB + C)

= AB + A' + C'+ AB'C.AB + AB'C.C

= AB + A' + C' + 0 + AB'C   (B.B' = 0 and C.C = C)

= ABC + ABC' + A' + C' + AB'C   (AB = AB(C + C') = ABC + ABC')

= AC(B + B') + C'(AB + 1) + A'

= AC + C'+A'   (B + B' = 1 and AB + 1 = 1)

= AC + (AC)'

= 1

Example 3.5. *Simplify the Boolean function F = ((XY′ + XYZ)′ + X(Y + XY′))′.*

Solution.
$$
\begin{aligned}
F &= ((XY' + XYZ)' + X(Y + XY'))' \\
  &= ((X(Y' + YZ))' + XY + XY')' \\
  &= ((X(Y'Z + Y' + YZ))' + X(Y + Y'))' \qquad (Y' = Y'(Z + 1) = Y'Z + Y') \\
  &= (X(Y' + Z))' + X)' \\
  &= (X' + (Y' + Z)' + X)' \\
  &= (1+ YZ')' \\
  &= 1' \\
  &= 0
\end{aligned}
$$

Example 3.6. *Simplify the Boolean function F = XYZ + XY′Z + XYZ′.*

Solution.
$$
\begin{aligned}
F &= XYZ + XY'Z + XYZ' \\
  &= XZ\ (Y + Y') + XY\ (Z + Z') \\
  &= XZ + XY \\
  &= X\ (Y + Z)
\end{aligned}
$$

3.9 CANONICAL AND STANDARD FORMS

Logical functions are generally expressed in terms of different combinations of logical variables with their true forms as well as the complement forms. Binary logic values obtained by the logical functions and logic variables are in binary form. An arbitrary logic function can be expressed in the following forms.

(*i*) Sum of the Products (SOP)

(*ii*) Product of the Sums (POS)

Product Term. In Boolean algebra, the logical product of several variables on which a function depends is considered to be a product term. In other words, the AND function is referred to as a product term or standard product. The variables in a product term can be either in true form or in complemented form. For example, ABC′ is a product term.

Sum Term. An OR function is referred to as a sum term. The logical sum of several variables on which a function depends is considered to be a sum term. Variables in a sum term can also be either in true form or in complemented form. For example, A + B + C′ is a sum term.

Sum of Products (SOP). The logical sum of two or more logical product terms is referred to as a sum of products expression. It is basically an OR operation on AND operated variables. For example, Y = AB + BC + AC or Y = A′B + BC + AC′ are sum of products expressions.

Product of Sums (POS). Similarly, the logical product of two or more logical sum terms is called a product of sums expression. It is an AND operation on OR operated variables. For example, Y = (A + B + C)(A + B′ + C)(A + B + C′) or Y = (A + B + C)(A′ + B′ + C′) are product of sums expressions.

Standard form. The standard form of the Boolean function is when it is expressed in sum of the products or product of the sums fashion. The examples stated above, like Y = AB + BC + AC or Y = (A + B + C)(A + B′ + C)(A + B + C′) are the standard forms.

However, Boolean functions are also sometimes expressed in nonstandard forms like $F = (AB + CD)(A'B' + C'D')$, which is neither a sum of products form nor a product of sums form. However, the same expression can be converted to a standard form with help of various Boolean properties, as

$$F = (AB + CD)(A'B' + C'D') = A'B'CD + ABC'D'$$

3.9.1 Minterm

A product term containing all n variables of the function in either true or complemented form is called the minterm. Each minterm is obtained by an AND operation of the variables in their true form or complemented form. For a two-variable function, four different combinations are possible, such as, A'B', A'B, AB', and AB. These product terms are called the fundamental products or standard products or minterms. In the minterm, a variable will possess the value 1 if it is in true or uncomplemented form, whereas, it contains the value 0 if it is in complemented form. For three variables function, eight minterms are possible as listed in the following table in Figure 3.12.

A	B	C	Minterm
0	0	0	A'B'C'
0	0	1	A'B'C
0	1	0	A'BC'
0	1	1	A'BC
1	0	0	AB'C'
1	0	1	AB'C
1	1	0	ABC'
1	1	1	ABC

Figure 3.12

So, if the number of variables is n, then the possible number of minterms is 2^n. The main property of a minterm is that it possesses the value of 1 for only one combination of n input variables and the rest of the $2^n - 1$ combinations have the logic value of 0. This means, for the above three variables example, if A = 0, B = 1, C = 1 *i.e.*, for input combination of 011, there is only one combination A'BC that has the value 1, the rest of the seven combinations have the value 0.

Canonical Sum of Product Expression. When a Boolean function is expressed as the logical sum of all the minterms from the rows of a truth table, for which the value of the function is 1, it is referred to as the *canonical sum of product expression*. The same can be expressed in a compact form by listing the corresponding decimal-equivalent codes of the minterms containing a function value of 1. For example, if the canonical sum of product form of a three-variable logic function F has the minterms A'BC, AB'C, and ABC', this can be expressed as the sum of the decimal codes corresponding to these minterms as below.

$$F (A,B,C) = (3,5,6)$$
$$= m_3 + m_5 + m_6$$
$$= A'BC + AB'C + ABC'$$

where Σ (3,5,6) represents the summation of minterms corresponding to decimal codes 3, 5, and 6.

The canonical sum of products form of a logic function can be obtained by using the following procedure.

1. Check each term in the given logic function. Retain if it is a minterm, continue to examine the next term in the same manner.
2. Examine for the variables that are missing in each product which is not a minterm. If the missing variable in the minterm is X, multiply that minterm with (X+X').
3. Multiply all the products and discard the redundant terms.

Here are some examples to explain the above procedure.

Example 3.7. *Obtain the canonical sum of product form of the following function.*

$$F (A, B) = A + B$$

Solution. The given function contains two variables A and B. The variable B is missing from the first term of the expression and the variable A is missing from the second term of the expression. Therefore, the first term is to be multiplied by (B + B') and the second term is to be multiplied by (A + A') as demonstrated below.

$$F (A, B) = A + B$$
$$= A.1 + B.1$$
$$= A (B + B') + B (A + A')$$
$$= AB + AB' + AB + A'B$$
$$= AB + AB' + A'B \qquad \text{(as AB + AB = AB)}$$

Hence the canonical sum of the product expression of the given function is

$$F (A, B) = AB + AB' + A'B.$$

Example 3.8. *Obtain the canonical sum of product form of the following function.*

$$F (A, B, C) = A + BC$$

Solution. Here neither the first term nor the second term is minterm. The given function contains three variables A, B, and C. The variables B and C are missing from the first term of the expression and the variable A is missing from the second term of the expression. Therefore, the first term is to be multiplied by (B + B') and (C + C'). The second term is to be multiplied by (A + A'). This is demonstrated below.

$$F (A, B, C) = A + BC$$
$$= A (B + B') (C + C') + BC (A + A')$$
$$= (AB + AB') (C + C') + ABC + A'BC$$
$$= ABC + AB'C + ABC' + AB'C' + ABC + A'BC$$
$$= ABC + AB'C + ABC' + AB'C' + A'BC \text{ (as ABC + ABC = ABC)}$$

Hence the canonical sum of the product expression of the given function is

$$F (A, B) = ABC + AB'C + ABC' + AB'C' + A'BC.$$

Example 3.9. *Obtain the canonical sum of product form of the following function.*

$$F (A, B, C, D) = AB + ACD$$

Solution. $F (A, B, C, D) = AB + ACD$
$$= AB (C + C') (D + D') + ACD (B + B')$$
$$= (ABC + ABC') (D + D') + ABCD + AB'CD$$
$$= ABCD + ABCD' + ABC'D + ABC'D' + ABCD + AB'CD$$
$$= ABCD + ABCD' + ABC'D + ABC'D' + AB'CD$$

Hence above is the canonical sum of the product expression of the given function.

3.9.2 Maxterm

A sum term containing all n variables of the function in either true or complemented form is called the maxterm. Each maxterm is obtained by an OR operation of the variables in their true form or complemented form. Four different combinations are possible for a two-variable function, such as, $A' + B'$, $A' + B$, $A + B'$, and $A + B$. These sum terms are called the standard sums or maxterms. Note that, in the maxterm, a variable will possess the value 0, if it is in true or uncomplemented form, whereas, it contains the value 1, if it is in complemented form. Like minterms, for a three-variable function, eight maxterms are also possible as listed in the following table in Figure 3.13.

A	B	C	Maxterm
0	0	0	$A + B + C$
0	0	1	$A + B + C'$
0	1	0	$A + B' + C$
0	1	1	$A + B' + C'$
1	0	0	$A' + B + C$
1	0	1	$A' + B + C'$
1	1	0	$A' + B' + C$
1	1	1	$A' + B' + C'$

Figure 3.13

So, if the number of variables is n, then the possible number of maxterms is 2^n. The main property of a maxterm is that it possesses the value of 0 for only one combination of n input variables and the rest of the $2^n - 1$ combinations have the logic value of 1. This means, for the above three variables example, if $A = 1$, $B = 1$, $C = 0$ *i.e.*, for input combination of 110, there is only one combination $A' + B' + C$ that has the value 0, the rest of the seven combinations have the value 1.

Canonical Product of Sum Expression. When a Boolean function is expressed as the logical product of all the maxterms from the rows of a truth table, for which the value of the function is 0, it is referred to as the *canonical product of sum expression*. The same can be expressed in a compact form by listing the corresponding decimal equivalent codes of the maxterms containing a function value of 0. For example, if the canonical product of sums form of a three-variable logic function F has the maxterms $A + B + C$, $A + B' + C$, and $A' + B + C'$, this can be expressed as the product of the decimal codes corresponding to these maxterms as below,

$$F (A,B,C) = \Pi (0,2,5)$$
$$= M_0 M_2 M_5$$
$$= (A + B + C) (A + B' + C) (A' + B + C')$$

where $\Pi (0,2,5)$ represents the product of maxterms corresponding to decimal codes 0, 2, and 5.

The canonical product of sums form of a logic function can be obtained by using the following procedure.

1. Check each term in the given logic function. Retain it if it is a maxterm, continue to examine the next term in the same manner.

2. Examine for the variables that are missing in each sum term that is not a maxterm. If the missing variable in the maxterm is X, multiply that maxterm with (X.X′).

3. Expand the expression using the properties and postulates as described earlier and discard the redundant terms.

Some examples are given here to explain the above procedure.

Example 3.10. *Obtain the canonical product of the sum form of the following function.*

$\quad$ F (A, B, C) = (A + B′) (B + C) (A + C′)

Solution. In the above three-variable expression, C is missing from the first term, A is missing from the second term, and B is missing from the third term. Therefore, CC′ is to be added with first term, AA′ is to be added with the second, and BB′ is to be added with the third term. This is shown below.

$\quad$ F (A, B, C) = (A + B′) (B + C) (A + C′)

$\qquad$ = (A + B′ + 0) (B + C + 0) (A + C′ + 0)

$\qquad$ = (A + B′ + CC′) (B + C + AA′) (A + C′ + BB′)

$\qquad$ = (A + B′ + C) (A + B′ + C′) (A + B + C) (A′ + B + C) (A + B + C′) (A + B′ + C′)

$\qquad$ [using the distributive property, as X + YZ = (X + Y)(X + Z)]

$\qquad$ = (A + B′ + C) (A + B′ + C′) (A + B + C) (A′ + B + C) (A + B + C′)

$\qquad$ [as (A + B′ + C′) (A + B′ + C′) = A + B′ + C′]

Hence the canonical product of the sum expression for the given function is

$\quad$ F (A, B, C) = (A + B′ + C) (A + B′ + C′) (A + B + C) (A′ + B + C) (A + B + C′)

Example 3.11. *Obtain the canonical product of the sum form of the following function.*

$\quad$ F (A, B, C) = A + B′C

Solution. In the above three-variable expression, the function is given at sum of the product form. First, the function needs to be changed to product of the sum form by applying the distributive law as shown below.

$\quad$ F (A, B, C) = A + B′C

$\qquad$ = (A + B′) (A + C)

Now, in the above expression, C is missing from the first term and B is missing from the second term. Hence CC′ is to be added with the first term and BB′ is to be added with the second term as shown below.

$\quad$ F (A, B, C) = (A + B′) (A + C)

$\qquad$ = (A + B′ + CC′) (A + C + BB′)

$\qquad$ = (A + B′ + C) (A + B′ + C′) (A + B + C) (A + B′ + C)

$\qquad$ [using the distributive property, as X + YZ = (X + Y) (X + Z)]

$\qquad$ = (A + B′ + C) (A + B′ + C′) (A + B + C)

$\qquad$ [as (A + B′ + C) (A + B′ + C) = A + B′ + C]

Hence the canonical product of the sum expression for the given function is

$\quad$ F (A, B, C) = (A + B′ + C) (A + B′ + C′) (A + B + C).

3.9.3 Deriving a Sum of Products (SOP) Expression from a Truth Table

The sum of products (SOP) expression of a Boolean function can be obtained from its truth table summing or performing OR operation of the product terms corresponding to the combinations containing a function value of 1. In the product terms the input variables appear either in true (uncomplemented) form if it contains the value 1, or in complemented form if it possesses the value 0.

Now, consider the following truth table in Figure 3.14, for a three-input function Y. Here the output Y value is 1 for the input conditions of 010, 100, 101, and 110, and their corresponding product terms are A'BC', AB'C', AB'C, and ABC' respectively.

Inputs			Output	Product terms	Sum terms
A	B	C	Y		
0	0	0	0		A + B + C
0	0	1	0		A + B + C'
0	1	0	1	A'BC'	
0	1	1	0		A + B' + C'
1	0	0	1	AB'C'	
1	0	1	1	AB'C	
1	1	0	1	ABC'	
1	1	1	0		A' + B' + C'

Figure 3.14

The final sum of products expression (SOP) for the output Y is derived by summing or performing an OR operation of the four product terms as shown below.

$$Y = A'BC' + AB'C' + AB'C + ABC'$$

In general, the procedure of deriving the output expression in SOP form from a truth table can be summarized as below.

1. Form a product term for each input combination in the table, containing an output value of 1.
2. Each product term consists of its input variables in either true form or complemented form. If the input variable is 0, it appears in complemented form and if the input variable is 1, it appears in true form.
3. To obtain the final SOP expression of the output, all the product terms are OR operated.

3.9.4 Deriving a Product of Sums (POS) Expression from a Truth Table

As explained above, the product of sums (POS) expression of a Boolean function can also be obtained from its truth table by a similar procedure. Here, an AND operation is performed on the sum terms corresponding to the combinations containing a function value of 0. In the sum terms the input variables appear either in true (uncomplemented) form if it contains the value 0, or in complemented form if it possesses the value 1.

Now, consider the same truth table as shown in Figure 3.14, for a three-input function Y. Here the output Y value is 0 for the input conditions of 000, 001, 011, and 111, and their corresponding product terms are A + B + C, A + B + C', A + B' + C', and A' + B' + C' respectively.

So now, the final product of sums expression (POS) for the output Y is derived by performing an AND operation of the four sum terms as shown below.

$$Y = (A + B + C) (A + B + C') (A + B' + C') (A' + B' + C')$$

In general, the procedure of deriving the output expression in POS form from a truth table can be summarized as below.

1. Form a sum term for each input combination in the table, containing an output value of 0.

2. Each product term consists of its input variables in either true form or complemented form. If the input variable is 1, it appears in complemented form and if the input variable is 0, it appears in true form.

3. To obtain the final POS expression of the output, all the sum terms are AND operated.

3.9.5 Conversion between Canonical Forms

From the above example, it may be noted that the complement of a function expressed as the sum of products (SOP) equals to the sum of products or sum of the minterms which are missing from the original function. This is because the original function is expressed by those minterms that make the function equal to 1, while its complement is 1 for those minterms whose values are 0. According to the truth table given in Figure 3.14:

$$F (A,B,C) = \Sigma (2,4,5,6)$$
$$= m_2 + m_4 + m_5 + m_6$$
$$= A'BC' + AB'C' + AB'C + ABC'.$$

This has the complement that can be expressed as

$$F' (A,B,C) = (0,1,3,7)$$
$$= m_0 + m_1 + m_3 + m_7$$

Now, if we take complement of F' by DeMorgan's theorem, we obtain F as

$$F (A,B,C) = (m_0 + m_1 + m_3 + m_7)'$$
$$= m_0'm_1'm_3'm_7$$
$$= M_0M_1M_3M_7$$
$$= \Pi(0,1,3,7)$$
$$= (A + B + C)(A + B + C') (A + B' + C') (A' + B' + C').$$

The last conversion follows from the definition of minterms and maxterms as shown in the tables in Figures 3.12 and 3.13. It can be clearly noted that the following relation holds true

$$m'_j = M_j.$$

That is, the maxterm with subscript j is a complement of the minterm with the same subscript j, and vice versa.

This example demonstrates the conversion between a function expressed in sum of products (SOP) and its equivalent in product of maxterms. A similar example can show the conversion between the product of sums (POS) and its equivalent sum of minterms. In general, to convert from one canonical form to other canonical form, it is required to interchange the symbols Σ and π, and list the numbers which are missing from the original form.

Note that, to find the missing terms, the total 2^n number of minterms or maxterms must be realized, where n is the number of variables in the function.

3.10 OTHER LOGIC OPERATORS

When the binary operators AND and OR are applied on two variables A and B, they form two Boolean Functions A.B and A+B respectively. However, 16 possible Boolean function can be generated using two variables, the binary operators AND and OR, and one unary operator NOT or INVERT or complement. These functions, with an accompanying name and a comment that explains each function in brief, are listed in the table in Figure 3.15.

Boolean Functions	Operator Symbol	Name	Comments
$F_0 = 0$		Null	Binary constant 0
$F_1 = AB$	A . B	AND	A and B
$F_2 = AB'$	A / B	Inhibition	A but not B
$F_3 = A$		Transfer	A
$F_4 = A'B$	B / A	Inhibition	B but not A
$F_5 = B$		Transfer	B
$F_6 = AB' + A'B$	A ⊕ B	Exclusive-OR	A or B but not both
$F_7 = A + B$	A + B	OR	A or B
$F_8 = (A+B)'$	A ↓ B	NOR	Not OR
$F_9 = AB + A'B'$	A B	Equivalence*	A equals B
$F_{10} = B'$	B'	Complement	Not B
$F_{11} = A + B'$	A ⊂ B	Implication	If B then A
$F_{12} = A'$	A'	Complement	Not A
$F_{13} = A' + B$	A ⊃ B	Implication	If A then B
$F_{14} = (AB)'$	A ↑ B	NAND	Not AND
$F_{15} = 1$		Identity	Binary constant 1

*Equivalence is also termed as equality, coincidence, and exclusive-NOR.

Figure 3.15

Although these functions can be represented in terms of AND, OR, and NOT operation, special operator symbols are assigned to some of the functions.

The 16 functions as listed in the table can be subdivided into three categories.

1. Two functions produce a constant 0 or 1.

2. Four functions with unary operations complement and transfer.

3. Ten functions with binary operators defining eight different operations—AND, OR, NAND, NOR, exclusive-OR, equivalence, inhibition, and implication.

3.11 DIGITAL LOGIC GATES

As Boolean functions are expressed in terms of AND, OR, and NOT operations, it is easier to implement the Boolean functions with these basic types of gates. However, for all practical purposes, it is possible to construct other types of logic gates. The following factors are to be considered for construction of other types of gates.

Name	Graphic Symbol	Algebraic Function	Truth Table		
			A	B	F
AND		$F = AB$	0	0	0
			0	1	0
			1	0	0
			1	1	1
			A	B	F
OR		$F = A + B$	0	0	0
			0	1	1
			1	0	1
			1	1	1
Inverter or NOT		$F = A'$		A	F
				0	1
				1	0
Buffer		$F = A$		A	F
				0	0
				1	1
			A	B	F
NAND		$F = (AB)'$	0	0	1
			0	1	1
			1	0	1
			1	1	0
			A	B	F
NOR		$F = (A + B)'$	0	0	1
			0	1	0
			1	0	0
			1	1	0
			A	B	F
Exclusive-OR (XOR)		$F = AB' + A'B$ $= A \oplus B$	0	0	0
			0	1	1
			1	0	1
			1	1	0
Equivalence Or Exclusive-NOR (XNOR)		$F = AB + A'B'$ $= A \ B$	A	B	F
			0	0	1
			0	1	0
			1	0	0
			1	1	1

Figure 3.16

1. The feasibility and economy of producing the gate with physical parameters.
2. The possibility of extending to more than two inputs.
3. The basic properties of the binary operator such as commutability and associability.
4. The ability of the gate to implement the Boolean functions alone or in conjunction with other gates.

Out of the 16 functions described in the table in Figure 3.15, we have seen that two are equal to constant, and four others are repeated twice. Two functions—inhibition and implication, are impractical to use as standard gates due to lack of commutative or associative properties. So, there are eight functions—Transfer (or buffer), Complement, AND, OR, NAND, NOR, Exclusive-OR (XOR), and Equivalence (XNOR) that may be considered to be standard gates in digital design.

The graphic symbols and truth tables of eight logic gates are shown in Figure 3.16. The transfer or buffer and complement or inverter or NOT gates are unary gates, *i.e.*, they have single input, while other logic gates have two or more inputs.

3.11.1 Extension to Multiple Inputs

A gate can be extended to have multiple inputs if its binary operation is commutative and associative. AND and OR gates are both commutative and associative.

For the AND function, $AB = BA$ -commutative

and

$(AB)C = A(BC) = ABC.$ -associative

For the OR function, $A + B = B + A$ -commutative

and

$(A + B) + C = A + (B + C).$ -associative

These indicate that the gate inputs can be interchanged and these functions can be extended to three or more variables very simply as shown in Figures 3.17(a) and 3.17(b).

Figure 3.17(a)

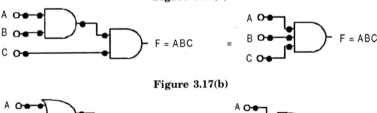

Figure 3.17(b)

The NAND and NOR functions are the complements of AND and OR functions respectively. They are commutative, but not associative. So these functions can not be extended to multiple input variables very simply. However, these gates can be extended to multiple inputs with slightly modified functions as shown in Figures 3.18(a) and 3.18(b) below.

For NAND function, $(AB)' = (BA)'.$ -commutative

But, $((AB)'C)' \neq (A(BC)')'.$ -does not follow associative property.

As $((AB)' \ C)' = (AB) + C'$ and

$(A(BC)')' = A' + BC$.

Similarly, for NOR function, $((A + B)' + C)' \neq (A + (B + C)')'$.

As, $((A + B)' + C)' = (A + B) C' = AC' + BC'$.

And $(A + (B + C)')' = A'(B + C) = A'B + A'C$.

Figure 3.18(a)

Figure 3.18(b)

The Exclusive-OR gates and equivalence gates both possess commutative and associative properties, and they can be extended to multiple input variables. For a multiple-input Ex-OR (XOR) gate output is low when even numbers of 1s are applied to the inputs, and when the number of 1s is odd the output is logic 0. Equivalence gate or XNOR gate is equivalent to XOR gate followed by NOT gate and hence its logic behavior is opposite to the XOR gate. However, multiple-input exclusive-OR and equivalence gates are uncommon in practice. Figures 3.19(a) and 3.19(b) describe the extension to multiple-input exclusive-OR and equivalence gates.

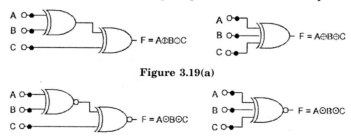

Figure 3.19(a)

Figure 3.19(b)

3.11.2 Universal Gates

NAND gates and NOR gates are called *universal gates* or *universal building blocks*, as any type of gates or logic functions can be implemented by these gates. Figures 3.20(a)-(e) show how various logic functions can be realized by NAND gates and Figures 3.21(a)-(d) show the realization of various logic gates by NOR gates.

NOT function: $F = A'$

Figure 3.20(a)

AND function: $F = AB$

Figure 3.20(b)

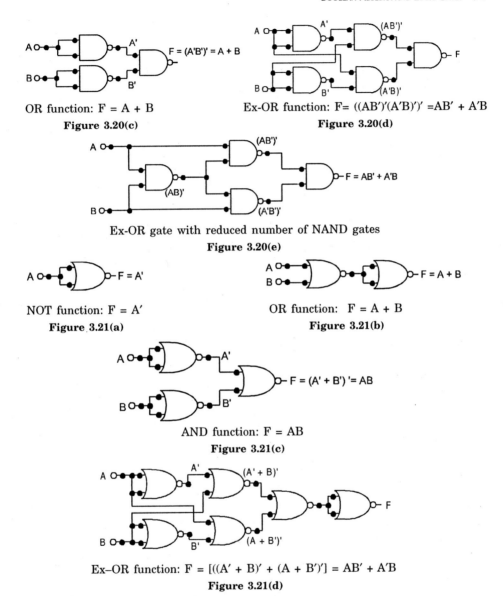

OR function: F = A + B

Figure 3.20(c)

Ex-OR function: F= ((AB')'(A'B)')' =AB' + A'B

Figure 3.20(d)

Ex-OR gate with reduced number of NAND gates

Figure 3.20(e)

NOT function: F = A′

Figure 3.21(a)

OR function: F = A + B

Figure 3.21(b)

AND function: F = AB

Figure 3.21(c)

Ex–OR function: F = [((A′ + B)′ + (A + B′)′] = AB′ + A′B

Figure 3.21(d)

3.11.3 Realization of Logic Functions by Nand Gates

Since any gate can be realized by the universal gates, *i.e.*, NAND gates or NOR gates, as shown above, any logic function can be realized by the universal gates. Universal gates are easier to fabricate with electronic components. The advantage of using the universal gates for implementation of logic functions is that it reduces the number of varieties of gates. As an example, if the logic function F = AB + CD is to be implemented, it requires two AND gates and an OR gate, that means two different types of ICs (Integrated Circuits) are required.

Whereas, the same logic function can be developed by two NAND gates or one single IC (generally one NAND IC contains four gates of similar function). So the logic functions can be implemented by only a single type of gate and thus reduces power consumption as well as the inventory and cost of inventory in practical situations in industry.

To achieve the realization of logic functions by NAND gates, the first step is to express the function in SOP form (sum of products) and simply replace the gates with NAND gates. In other words, logic functions with first level AND gates and second level OR gates can be replaced by NAND-NAND realization. The concept can be understood by the diagram in Figures 3.22(a)-(c), considering the logic expression F = AB + CD.

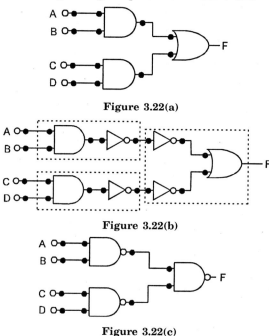

Figure 3.22(a)

Figure 3.22(b)

Figure 3.22(c)

Figure 3.22(a) shows the normal AND-OR realization of the function F = AB + CD. In Figure 3.22(b), two INVERTER gates are introduced at the outputs of AND gates. When two INVERTERs are cascaded, the function remains the same as complement to complement of a function is its true form. Now an AND gate followed by an INVERTER is a NAND gate, as explained in 3.10.1, and an OR gate preceded by INVERTERs can be replaced by a NAND, as from Figures 3.20(a) and 3.21(c). These are shown by the dashed lines in Figure 3.22(b). Thus the function F = AB + CD can be realized by NAND gates as in Figure 3.22(c).

The same can be explained using Boolean algebra and DeMorgan's theorems.

$$F = AB + CD$$
$$= ((AB + CD)')' \quad \text{- complement to complement operation}$$
$$= ((AB)' (CD)')' \quad \text{- applying DeMorgan's theorem}$$

Note that the derived expression is in terms of NAND function only.

A convenient way to implement a logic circuit with NAND gates is to obtain the simplified Boolean function in terms of AND, OR, and NOT and convert the functions to NAND logic as explained above. The conversion of the algebraic expression from AND, OR, and NOT operations is usually quite complicated because it involves a large number of applications of DeMorgan's theorem. This difficulty is avoided by the use of circuit manipulations as explained by the Figures 3.22(*a*), 3.22(*b*), and 3.22(*c*).

The implementation of Boolean functions with NAND gates by circuit manipulation or block diagram manipulation is simple and straightforward. The method requires two other logic diagrams to be drawn prior to obtaining the NAND logic diagram. Simple rules for circuit manipulation are outlined below.

1. From the given algebraic expression, draw the logic diagram with AND, OR, and NOT gates. Assume that both normal and complement inputs are available.

2. Draw a second logic diagram with NAND logic, as given in the Figures 3.20(a)-(e), substituted for each AND, OR, and NOT gate.

3. Remove any two cascaded inverters from the diagram, since double inversion does not perform a logic function. Remove inverters connected to single external inputs and complement the corresponding input variable. The new logic diagram is the required NAND gate implementation.

The procedure can be illustrated with the example as follows.

Example 3.12. *Realize the following function by NAND gates only, F = B(A + CD) + AC′.*

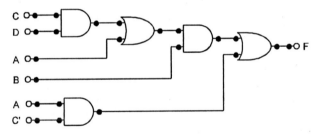

Figure 3.23(a) Realization by AND, OR, and NOT gates.

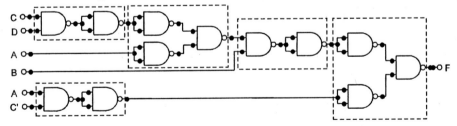

Figure 3.23(b) AND and OR gates are replaced by equivalent NAND gates.

The AND-OR implementation of the function is shown in Figure 3.23(*a*). Now each of the ANDs is replaced by a NAND gate followed by an INVERTER, and each of the OR gates is replaced by INVERTERS followed by a NAND gate. This logic diagram is shown in Figure 3.23(*b*). In the next step, two cascaded INVERTERS are removed to obtain the logic diagram in Figure 3.23(*c*), which is a required NAND realization of the given function.

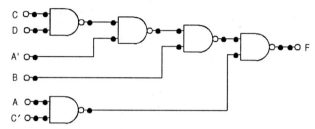

Figure 3.23(c) NAND gate realization after two cascaded inverters are removed.

It may be noticed that the number of NAND gates required to implement the Boolean function is equal to the number of AND-OR gates, provided both normal and complement inputs are available. If only the normal inputs are available, INVERTERs must be introduced to generate the complemented inputs.

3.11.4 Realization of Logic Functions by NOR Gates

Similarly, any logic function can be developed by using only NOR gates. To achieve the realization of logic functions by NOR gates only, the first step is to express the function at POS form (products of sums) and replace the AND gates and OR gates with NOR gates. Logic functions with first level OR gates and second level AND gates can be replaced by NOR-NOR realization. This can be demonstrated by the diagram in Figures 3.24(a)-(c), considering the logic expression F = (A + B) (C + D).

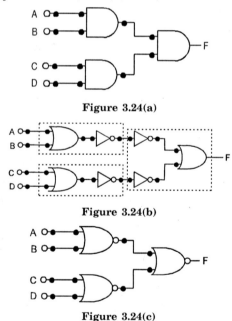

Figure 3.24(a)

Figure 3.24(b)

Figure 3.24(c)

Figure 3.24(*a*) shows the normal OR-AND realization of the function F = (A + B) (C +D). In Figure 3.24(*b*), two INVERTER gates are introduced at the outputs of OR gates. Two cascaded INVERTERs bring back the function to its original true form. Now an OR gate followed by an INVERTER is a NOR gate, as explained in 3.10.1, and an AND gate preceded by INVERTERs can be replaced by a NOR, as shown in Figures 3.21(a) and 3.20(c). The blocks formed by dashed lines at Figure 3.24(b) represent NOR gates. Thus the function $F = (A + B) (C + D)$ can be realized by NOR gates as in Figure 3.24(c).

This concept can also be explained by Boolean algebra and DeMorgan's theorem.

$$F = (A + B) (C + D)$$
$$= (((A + B) (C + D))')' \qquad \text{- complement to complement operation}$$
$$= ((A + B)' + (C + D)')' \qquad \text{- applying DeMorgan's theorem}$$

The derived expression is in terms of NOR function only.

Similar to realization with NAND gates of the Boolean functions, circuit manipulation techniques may be adopted to implement the Boolean functions with NOR gates. Here also, a simple procedure is followed to realize the function with NOR gates, which is illustrated below.

1. From the given algebraic expression, draw the logic diagram with AND, OR, and NOT gates. Assume that both normal and complement inputs are available.

2. Draw a second logic diagram with NOR logic, as given in Figures 3.21(a)-(d), substituted for each AND, OR, and NOT gate.

3. Remove pairs of cascaded inverters from the diagram, since double inversion does not perform a logic function. Remove inverters connected to single external inputs and complement the corresponding input variable. The new logic diagram is the required NOR gate implementation.

The procedure can be demonstrated with the example that follows.

Example 3.13. *Realize the following function by NOR gates only, $F = A(B + CD) + BC'$.*

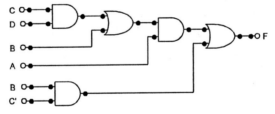

Figure 3.25(a) Circuit realization by AND-OR gates.

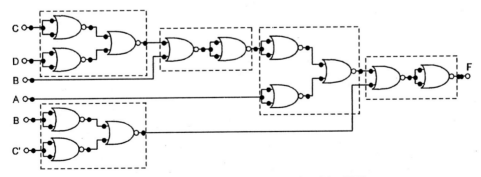

Figure 3.25(b) AND and OR gates are replaced by NOR gates.

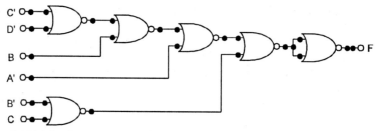

Figure 3.25(c) Implementation by NOR gates after two cascaded inverters are removed.

First the function is realized with AND-OR gates as shown in Figure 3.25(a). At the next step, each of the AND gates are replaced by INVERTERS followed by a NOR gate, and OR gates are substituted by NOR gates followed by INVERTERs as illustrated in Figure 3.25(b). Finally, the logic diagram is redrawn after removing two cascaded INVERTERs in Figure 3.25(c), which represents the NOR gate implementation of the given function.

The number of NOR gates for the Boolean function is equal to the number of AND-OR gates plus one additional INVERTER at the output. In general, the number of NOR gates required to implement a Boolean function equals the number of AND-OR gates, except for an occasional INVERTER. This is true if both normal and complemented inputs are provided, because the conversion requires certain complemented input.

3.11.5 Two-level Implementation of Logic Networks

The maximum number of gates cascaded in series between an input and output is called the *level of gates*. For example, a sum of products (SOP) expression can be implemented using a two-level gate network, *i.e.*, AND gates at the first-level and an OR gate at the second level. Similarly, a product of sums (POS) expression can be implemented by a two-level gate network, as OR gates at the first level and an AND gate at the second level. It is important to note that INVERTERS are not considered to decide the level of gate network.

Apart from the realization of Boolean functions using AND gates and OR gates, the NAND gates and NOR gates are most often found in the implementation of logic circuits as they are universal type by nature. Some of the NAND and NOR gates allow the possibility of a wire connection between the outputs of two gates to provide a specific logic function. This type of logic is called *wired logic*. (This will be discussed in detail in Chapter 11: Logic Family.) When two NAND gates are wired together as shown in Figure 3.26(a), they perform the wired-AND logic function. AND drawn with lines going through the center of the gate is symbolized as a wired-AND logic function. The wired-AND gate is not a physical gate, but only a symbol to designate the function obtained from the indicated wired connections. The logic function implemented by the circuit of Figure 3.26(a) is

$$F = (WX)'.(YZ)' = (WX + YZ)'.$$

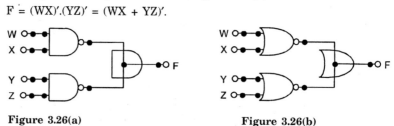

Figure 3.26(a)   Figure 3.26(b)

The above function is referred to as an AND-OR-INVERT function. Similarly, some specially constructed NOR gates outputs can be tied together to form the wired-OR function as shown in Figure 3.26(*b*). The logic function implemented by Figure 3.26(*b*) is

$$F = (W + X)' + (Y + Z)' = [(W + X). (Y + Z)]'.$$

This function is called an OR-AND-INVERT function.

The wired logic gate does not produce a physical second level gate since it is just the wire connection. However, according to the logic function concerned, wired logic is considered a two-level implementation.

Degenerate and Nondegenerate Forms

It may be noted that, although there may be 16 combinations of two-level implementation of gates possible, four types of gates AND, OR, NAND, and NOR are considered. Eight of these combinations are similar in nature. As an example, an AND gate at first level with an AND gate at second level is practically performing a single operation. Similarly, an OR gate followed by an OR gate performs a single operation. These types of combinations are called *degenerate forms*.

The other eight combinations produce sum of the products or product of sums functions and they are called *nondegenerate forms*. The eight nondegenerate forms are below.

AND-OR	OR-AND	NOR-NOR	NAND-NAND
AND-NOR	OR-NAND	NOR-OR	NAND-AND

In each form the first gate represents the first level and second gate is for second level.

3.11.6 Multilevel Gating Networks

The number of levels can be increased by factoring the sum of products expression for an AND-OR network, or by multiplying out some terms in the product of sums expression for an OR-AND network. If a switching network is implemented using gates in more than two levels, then it is called a *multilevel gate network*. Some examples are given here to illustrate the multilevel gate network.

Example 3.14. *Realize the function* $F = BC' + A'B + D$ *with a multilevel network.*

Solution. The function can be realized in a two-level AND-OR network as shown is Figure 3.27(*a*). However, by factoring some part of the function, it can be rewritten as $F = B (A' + C') + D$ and implemented as a multilevel gate network in Figure 3.27(*b*).

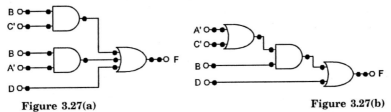

Figure 3.27(a)   **Figure 3.27(b)**

The logic circuit in Figure 3.27(*a*) consists of two 2-input AND gates, a 3-input OR gate, and five literals or inputs, whereas the logic circuit in Figure 3.27(*b*) is a three-level representation of the same function containing two 2-input OR gates, a 2-input AND gate, and four literals. Thus it reduces the number of gate inputs by one.

Example 3.15. *Realize the function Y = BD′E + BF + C′D′E + C′F + A with a multilevel network.*

By the straightforward method, the function can be realized in a two-level AND-OR network as shown in Figure 3.28(*a*). However, the expression may be factored into a different form as below.

$$
\begin{aligned}
Y &= BD'E + BF + C'D'E + C'F + A \\
&= B\,(D'E + F) + C'\,(D'E + F) + A \\
&= (D'E + F)\,(B + C') + A
\end{aligned}
$$

The same function can be realized as a multilevel gate network as shown in Figure 3.28(*b*).

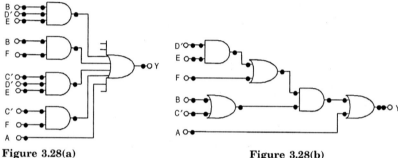

Figure 3.28(a)   **Figure 3.28(b)**

The logic diagram in Figure 3.28(*a*) is a normal two-level AND-OR network consisting of two 3-input AND gates, two 2-input AND gates, one 5-input OR gate (a 5-input OR gate is not normally available in practice and an 8-input OR gate is to be used in place of that), and eleven literals or inputs. However, equivalent multilevel network is realized in Figure 3.28(*b*), which contains two 2-input AND gates, three 2-input OR gates, and six inputs. Hence the multilevel network reduces the number of literals as well as the variety of gate types.

Hence, from the above examples, we observe that the multilevel network has distinct advantages over the two-level network, which may be summarized as below.

1. Multilevel networks use less number of literals or inputs, thus reducing the number of wires for connection.

2. Sometimes the multilevel network reduces the number of gates.

3. It reduces the variety type of gates and hence the number of ICs (integrated circuits).

4. Multilevel gate networks can be very easily converted to universal gates realization by the procedure described in sections 3.10.3 and 3.10.4 of this chapter. In that case the switching network can be implemented by less variety of the logic gates.

However, the biggest disadvantage of the multilevel network is that it increases the propagation delay. The propagation delay is the inherent characteristics of any logic gate, and it increases with the increase of number of levels. So a designer must consider these factors while designing a switching network and its application.

3.11.7 Some Examples of Realization of Logic Functions

Example 3.16. *Realize the function F = B′C′ + A′C′ + A′B′ by (i) basic gates, (ii) NAND gates only, (iii) NOR gates only.*

Solution. (*i*) The function is realized basic gates as in Figure 3.29.

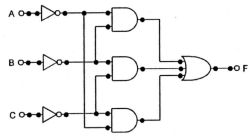

Figure 3.29

(*ii*) For the NAND realization, at the first step, each of the gates are converted to NAND gates as in Figure 3.30(*a*). Figure 3.30(*b*) demonstrates the all NAND realization.

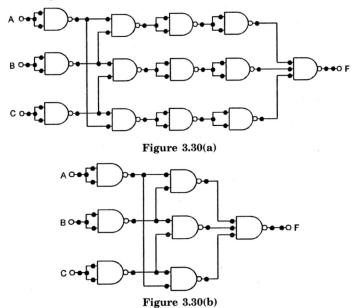

Figure 3.30(a)

Figure 3.30(b)

(*iii*) Figure 3.31(*a*) represents the conversion of each gate to a NOR gate and Figure 3.31(*b*) is the logic diagram for the given function realized with NOR gates only.

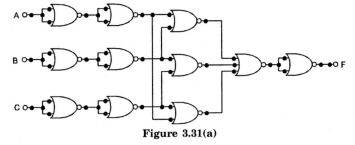

Figure 3.31(a)

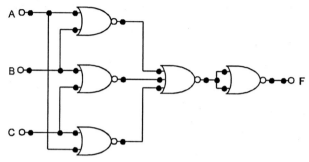

Figure 3.31(b)

Example 3.17. *Realize the function F = (A + B)(A' + C)(B + D) by (i) basic gates, (ii) NAND gates only, (iii) NOR gates only.*

Solution. (*i*) The function is realized basic gates as in Figure 3.32.

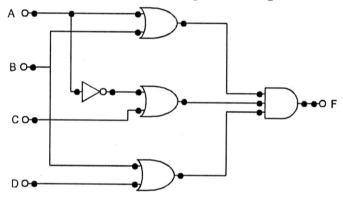

Figure 3.32

(*ii*) Realization by NAND gates only is demonstrated in Figure 3.33.

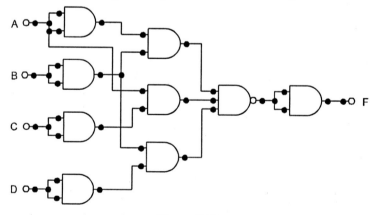

Figure 3.33

(*iii*) The given function has been implemented with NOR gates only in Figure 3.34.

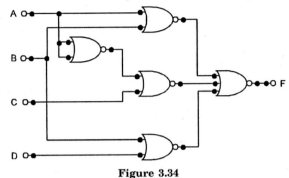

Figure 3.34

Example 3.18. *Realize the function F = (AB)′ + A + (B + C)′ NAND gates only.*

Solution. Figure 3.35 is the NAND gate implementation of the given function.

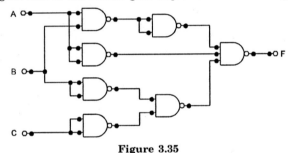

Figure 3.35

Example 3.19. (*a*) *Realize the function F = A + BCD′ using NAND gates only.*

(*b*) *Realize the function F = (A + C)(A + D′) (A + B + C′) using NOR gates only.*

Solution. (*a*)   F = A + BCD′ = [(A + BCD′)′]′

$$= [A'(BCD')']'$$

Figure 3.36 is the NAND gate implementation of the given function.

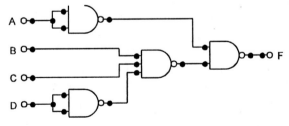

Figure 3.36

(*b*)   F = (A + C) (A + D′) (A + B + C′)

$$= [\{(A + C) (A + D') (A + B + C')\}']'$$

$$= [(A + C)' + (A + D')' + (A + B + C')']'$$

The logic diagram of the given function is implemented in Figure 3.37 with NOR gates only.

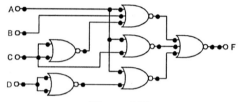

Figure 3.37

Example 3.20. *(a) Realize the function* $F = (A + C)\ (B' + D')\ (A' + B' + C')$ *with multilevel NAND gates. Use 2-input NAND gates only.*

(b) Realize the function in Figure 3.20(a) with multilevel NOR gates. Use 2-input NOR gates only.

Solution. The function is first realized by basic gates as in Figure 3.38(a) and it is implemented with all 2-input gates as in Figure 3.38(b).

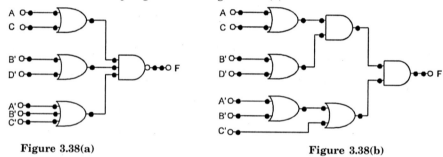

Figure 3.38(a)   **Figure 3.38(b)**

(a) All the basic gates are converted to NAND gates as shown in Figure 3.39(a). At the next step, cascaded pairs of INVERTERs gates are removed and also the INVERTERs at the inputs are eliminated assuming complement inputs are available. Figure 3.39(b) is the NAND realization of the given function.

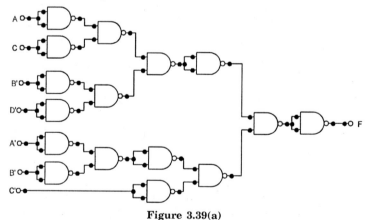

Figure 3.39(a)

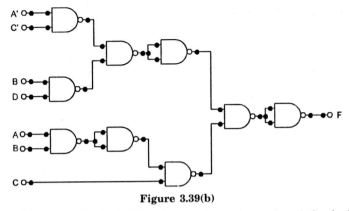

Figure 3.39(b)

(b) For realization with all NOR gates, each of the gates of the logic diagram of Figure 3.37(b) is converted to NOR gates as shown in Figure 3.40(a). Then cascaded pairs of INVERTERs are removed and the final logic diagram with all NOR gates is realized as in Figure 3.40(b).

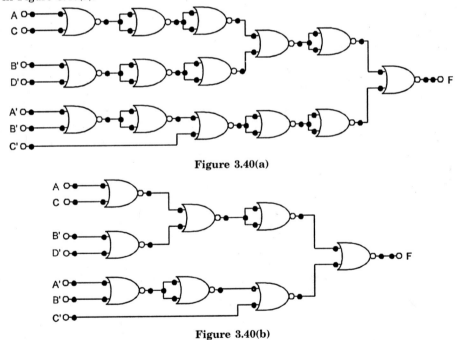

Figure 3.40(a)

Figure 3.40(b)

3.12 POSITIVE AND NEGATIVE LOGIC

The binary signals at the inputs or outputs of any gate may be one of two values, except during transitions. One signal value represents logic 1, and the other is logic 0. For a *positive logic* system, the most positive voltage level represents logic 1 state or HIGH level

(H) and the lowest voltage level represents logic 0 state or LOW level (L). For a *negative logic* system, the most positive voltage level represents logic 0 state and the lowest voltage level represents logic 1 state. For example, if the voltage levels are –1 volt and –10 volt in a positive logic system, then –1 volt represents logic 1 and –10 volt represents logic 0. In a negative logic system, logic 1 state is represented by –10 volt and logic 0 is represented by –1 volt. Figure 3.41(*a*) represents the positive logic system choosing the highest voltage level as logic 1 and the lowest voltage level as logic 0. Whereas Figure 3.41(*b*) represents the negative logic system assigning the highest voltage level as logic 0 and the lowest voltage level as logic 1.

Figure 3.41(a)   **Figure 3.41(b)**

The effect of changing one logic system to an other logic system is equivalent to complementing the logic function. The simple method of converting from one logic system to an other is to change all 0s of a truth table with 1s and all 1s with 0s. The resulting logic function is determined accordingly. For example, if 0s and 1s are interchanged in the truth table, the positive logic OR function converts to a negative logic AND function. Similarly, a positive logic NOR function turns to a negative logic NAND function.

The logic gates are commercially available in integrated circuit (IC) form, and according to the construction of basic structure and fabrication process they are classified into various groups termed as logic families. Parameters and characteristics are different for different logic families. In each family, there is a range of voltage values that the circuit will recognize as HIGH or LOW level. The table in Figure 3.42 describes the ranges of voltage levels for some of the widely used logic families.

IC family types	Supply Voltage (V)	High-level voltage (V)		Low-level voltage (V)	
		Range	*Typical*	*Range*	*Typical*
TTL	$V_{CC} = 5$	2.4 to 5	3.5	0 to 0.4	0.2
ECL	$V_{EE} = -5.2$	–0.95 to –0.7	–0.8	–1.9 to –1.6	–1.8
CMOS	$V_{DD} = 3$ to 10	V_{DD}	V_{DD}	0 to 0.5	0
Positive logic		Logic 1		Logic 0	
Negative logic		Logic 0		Logic 1	

Figure 3.42

However, there is no real advantage of either logic system over the other and the choice of using a positive logic system or negative logic system solely depends on the logic designer. In practice, a positive logic system is followed mostly.

3.13 CONCLUDING REMARKS

The basic digital principles, postulates, Boolean algebra and its simplification rules and implementation with logic gates have been discussed in this chapter. Logic gates are the electronic circuits constructed with basic electronic components such as resistors, diodes, transistors, etc., and fabricated in one chip referred to as integrated circuit or IC with the interconnections among the components within the chip. According to the construction and fabrication process of the basic structure of the logic gates, they are classified into different logic families, the parameters and characteristics of which are different for one family to an other. The governing characteristics are propagation delay, operating voltage level, fan out, power dissipation, etc., and they play an important part in the logic design. These will be discussed in detail in Chapter 11.

REVIEW QUESTIONS

3.1 State the methods used to simplify the Boolean equations.

3.2 State and explain the basic Boolean logic operations.

3.3 What are the applications of Boolean algebra?

3.4 Define truth table.

3.5 How is the AND multiplication different from the ordinary multiplication?

3.6 How does OR addition differ from the ordinary addition method?

3.7 What are the basic laws of Boolean algebra?

3.8 State and prove Absorption and Simplification theorems.

3.9 State and prove Associative and Distributive theorems.

3.10 What is meant by duality in Boolean algebra?

3.11 State DeMorgan's theorem.

3.12 State and explain the DeMorgan's theorem that converts a sum into a product and vice versa. Draw the equivalent logic circuits using basic gates.

3.13 Explain the terms—(*a*) input variable, (*b*) minterm, (*c*) maxterm.

3.14 Prove DeMorgan's theorem for a 4-variable function.

3.15 What is the truth table and logic symbol of a three-input OR gate?

3.16 Write the expression for a 4-input AND gate. Construct the complete truth table showing the output for all possible cases.

3.17 Does any three-input INVERTER exist?

3.18 Define NAND and NOR gates with their truth tables.

3.19 What is a logic gate? Explain logic designation.

3.20 Discuss the operation of Ex-OR and Ex-NOR gates with truth tables and logic diagram.

3.21 Explain the term 'universal gate.' Name the universal gates.

3.22 Explain how basic gates can be realized by NAND gates.

3.23 Explain how basic gates can be realized by NOR gates.

3.24 Construct a two-input XOR gate using NAND gates. Construct the same with NOR gates.

3.25 Realize an INVERTER with two-input XOR gate only.

3.26 Realize the logic expression for $A \oplus B \oplus C \oplus D$.

3.27 Draw a logic circuit for the function $F = (A + B)(B + C)(A + C)$, using NOR gates only.

3.28 How can an AND-OR network be converted to all NAND network?

3.29 How can an AND-OR network be converted to an all-NOR network?

3.30 What are the advantages and disadvantages of a multilevel gate network?

3.31 For the function $F = AB'C' + AB$, find the logic value of F under the conditions—

    (a) A = 1, B = 0, C = 1;   (b) A = 0, B = 1, C = 1;

    (c) A = 0, B = 0, C = 0

3.32 Simplify the following expressions:

    (a) $AB'C' + A'B'C' + A'BC' + A'B'C$

    (b) $ABC + A'BC + AB'C + ABC' + AB'C' + A'BC' + A'B'C'$

    (c) $A(A + B + C) (A' + B + C) (A + B' + C) (A + B + C')$

    (d) $(A + B + C) (A + B' + C') (A + B + C') (A + B' + C)$

3.33 Draw truth tables for the following expressions:

    (a) $F = AC + AB$ (b) $F = AB (B + C + D')$

    (c) $Y = A (B' + C')$ (d) $Y = (A + B + C) AB'$

    (e) $F = ABC (C + D')$ (f) $F = AB + BA + C (A + B)$

3.34 Reduce the Boolean expressions given below:

    (a) $A + A' + B + C$ (b) $AB + BB + C + B'$

    (c) $ABC (ABC + 1)$ (d) $AB + B + A + C$

    (e) $AAB + ABB + BCC$ (f) $A (A' + B)$

    (g) $AB (B + C)$ (h) $ABB (ABC + BC)$

    (i) $(AB + C) (AB + D)$ (j) $AB'C + A'B'C$

    (k) $AB'C + A'BC + ABC$ (l) $(A'B) AB + AB$

    (m) $(AB' + AC') (BC + BC') (ABC)$ (n) $A + B'C (A + B'C)$

    (o) $A [(ABC)' + AB'C]$ (p) $[(ABC)' + A'B' + BC]$

    (q) $A [B + C(AB + AC)']$ (r) $(M + N) (M' + P) (N' + P)$

3.35   Find the complements of the following expressions:

    (a) $A + BC + AB$ (b) $(A + B)(B + C)(A + C)$

    (c) $AB + BC + CD$ (d) $AB (C'D + B'C)$

    (e) $A (B + C) (C' + D')$

3.36 Apply DeMorgan's theorem to each of the following expressions:

    (a) $(AB' + C + D')'$ (b) $[AB (CD + EF)]'$

    (c) $(A + B' + C + D')' + (ABCD')'$ (d) $(AB + CD)'$

    (e) $[(A' + B + C + D')' + (AB'C'D)]'$ (f) $[(AB)' (CD + E'F) ((AB)' + (CD)')]'$

    (g) $(AB)' + (CD)'$ (h) $(A + B') (C' + D)$

3.37 Simplify the following Boolean expressions using Boolean technique:

    (a) $AB + A (B + C) + B (B + C)$ (b) $AB(C + BD') (AB)'$

    (c) $A + AB + AB'C$ (d) $(A' + B)C + ABC$

    (e) $AB'C (BD + CDE) + AC'$ (f) $BD + B (D + E) + D' (D + F)$

(g) A'B'C + (A + B + C')' + A'B'C'D'

(h) (B + BC) (B + B'C) (B + D)

(i) ABCD + AB (CD)' + (AB)'CD

(j) ABC [AB + C' (BC + AC)]

(k) A + A'B + (A + B)' C + (A + B + C + D)

(l) AB' + AC + BCD + D'

(m) A + A'B' + BCD' + BD'

(n) AB'C + (B' + C') (B' + D') + (A + C + D)'

3.38 Prove the following using Boolean theorems:

(a) (A + C)(A + D)(B + C)(B + D) = AB + CD

(b) (A' + B' + D') (A' + B + D') (B + C + D) (A + C') (A + C' + D) = A'C'D + ACD'' + BC'D'

3.39 (a) Find the Boolean expression for F, when F is 1 only if A is 1 and B is 1, or if A is 0 and B is 0.

(b) Find the Boolean expression for F, when F is 1 only if A, B, C are all 1s, or if one of the variables is 0.

3.40 (a) Convert Y = ABCD + A'BC + B'C' into a sum of minterms by algebraic method.

(b) Convert Y = AB + B'CD into a product of maxterms by algebraic method.

3.41 Find the canonical sum of products and product of sums expression for the function
$F = X_1X_2X_3 + X_1X_3X_4 + X_1X_2X_4$.

3.42 (a) Express the function Y = (1,3,5,7) as a product of maxterms.

(b) Express the complement of the function as a sum of the minterms.

(c) Express the complement of the function as a product of maxterms.

3.43 Simplify the function F = (0,2,3,6,8,10,11,14,15) and implement it with

(a) AND-OR network,

(b) OR-AND network,

(c) NAND-NAND network, and

(d) NOR-NOR network.

3.44 Realize the following function using a multilevel NAND-NAND network and NOR-NOR network:
$F = A'B + B (C + D) + EF' (B' + D')$.

3.45 Seven switches operate a lamp in the following way; if switches 1, 3, 5, and 7 are closed and switch 2 is opened, or if switches 2, 4, and 6 are closed and switch 3 is opened, or if all seven switches are closed the lamp will glow. Use basic gates to show how the switches are to be connected.

3.46 A corporation having 100 shares entitles the owner of each share to cast one vote at the share-holders' meeting. Assume that A has 60 shares, B has 30 shares, C has 20 shares, and D has 10 shares. A two-third majority is required to pass a resolution in a share-holders' meeting. Each of these four men has a switch which he closes to vote YES and opens to vote NO for his percentage of shares. When the resolution passed, one output LED is ON. Derive a truth table for the output function and give the sum of product equation for it.

3.47 Prove that (X + Y) ⊕ (X + Z) = X' (Y ⊕ Z).

□ □ □

4

SIMPLIFICATION AND MINIMIZATION OF BOOLEAN FUNCTIONS

4.1 INTRODUCTION

The complexity of digital logic gates to implement a Boolean function is directly related to the complexity of algebraic expression. Also, an increase in the number of variables results in an increase of complexity. Although the truth table representation of a Boolean function is unique, its algebraic expression may be of many different forms. Boolean functions may be simplified or minimized by algebraic means as described in Chapter 3. However, this minimization procedure is not unique because it lacks specific rules to predict the succeeding step in the manipulative process. The map method, first proposed by Veitch and slightly improvised by Karnaugh, provides a simple, straightforward procedure for the simplification of Boolean functions. The method is called Veitch diagram or Karnaugh map, which may be regarded either as a pictorial representation of a truth table or as an extension of the Venn diagram.

The Karnaugh map provides a systematic method for simplification and manipulation of a Boolean expression. The map is a diagram consisting of squares. For n variables on a Karnaugh map there are 2^n numbers of squares. Each square or cell represents one of the minterms. Since any Boolean function can be expressed as a sum of minterms, it is possible to recognize a Boolean function graphically in the map from the area enclosed by those squares whose minterms appear in the function. It is also possible to derive alternative algebraic expressions or simplify the expression with a minimum number of variables or literals and sum of products or product of sums terms, by analyzing various patterns. In fact, the map represents a visual diagram of all possible ways a function can be expressed in a standard form and the simplest algebraic expression consisting of a sum of products or product of sums can be selected. Note that the expression is not necessarily unique.

4.2 TWO-VARIABLE KARNAUGH MAPS

A two-variable Karnaugh map is shown in Figure 4.1. Since a two-variable system can form four minterms, the map consists of four cells—one for each minterm. The map has been redrawn in Figure 4.1(*b*) to show the relationship between the squares and the two variables A and B. Note that, in the first row, the variable A is complemented, in the second row A is uncomplemented, in the first column variable B is complemented and in the second column B is uncomplemented.

The two-variable Karnaugh map is a useful way to represent any of the 16 Boolean functions of two variables as described in section 3.7, if the squares are marked with 1 whose minterms belong to a certain function. As an example, the function AB has been shown in Figure 4.2(*a*). Since the function AB is equal to the minterm m_3, a 1 is placed in the cell corresponding to m_3. Similarly, the function A + B has three minterms of A′B, AB′, and AB, as

$$A + B = A (B + B') + B (A + A') = AB + AB' + AB + A'B = A'B + AB' + AB.$$

So the squares corresponding to A′B, AB′, and AB are marked with 1 as shown in Figure 4.2(*b*).

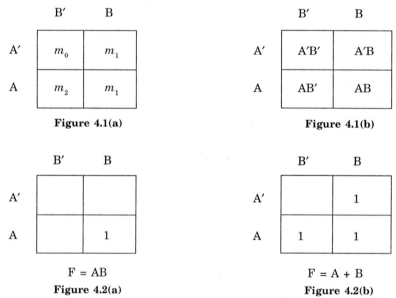

Figure 4.1(a)

Figure 4.1(b)

F = AB

Figure 4.2(a)

F = A + B

Figure 4.2(b)

4.3 THREE-VARIABLE KARNAUGH MAPS

Since, there are eight minterms for three variables, the map consists of eight cells or squares, which is shown in Figure 4.3(*a*). It may be noticed that the minterms are arranged, not according to the binary sequence, but according to the sequence similar to the reflected code, which means, between two consecutive rows or columns, only one single variable changes its logic value from 0 to 1 or from 1 to 0. Figure 4.3(*b*) shows the relationship between the squares and the variables. Two rows are assigned to A′ and A, and four columns to B′C′, B′C, BC, and BC′. The minterm m_3, for example, is assigned in the square corresponding to row 0 and column 11, thus making the binary number 011. Another way of analyzing the square m_3, is to consider it to be in the row A′ and column BC, as m_3 = A′BC. Note that, each of the variables has four squares where its logic value is 0 and four squares with logic value 1.

	B'C'	B'C	BC	BC'
A'	m_0	m_1	m_3	m_2
A	m_4	m_5	m_7	m_6

Figure 4.3 (a)

	B'C'	B'C	BC	BC'
A'	A'B'C'	A'B'C	A'BC	A'BC'
A	AB'C'	AB'C	ABC	ABC'

Figure 4.3 (b)

	A'	A
B'C'	m_0	m_4
B'C	m_1	m_5
BC	m_3	m_7
BC'	m_2	m_6

Figure 4.4 (b)

	A'B'	A'B	AB	AB'
C'	m_0	m_2	m_6	m_4
C	m_1	m_3	m_7	m_5

Figure 4.4 (a)

The three-variable Karnaugh Map can be constructed in other ways, too. Figure 4.4(a) shows if variable C is assigned to the rows and variables A and B are assigned along the columns. Figure 4.4(b) demonstrates where variable A is along columns and variables B and C are along the rows. Corresponding minterms are shown in the figures.

To understand the usefulness of the map for simplifying the Boolean functions, we must observe the basic properties of the adjacent squares. Any two adjacent squares in the Karnaugh map differ by only one variable, which is complemented in one square and uncomplemented in one of the adjacent squares. For example, in Figure 4.3(a), m_1 and m_3 are placed at adjacent squares, where variable B is complemented at m_1 while it is uncomplemented at m_3. From the postulates of Boolean algebra, the sum of two minterms can be simplified to a single AND term consisting of less number of literals. As in the case of m_1 and m_3, $m_1 + m_3$ can be reduced to the term below.

$$m_1 + m_3 = AB'C + ABC = AC (B' + B) = AC$$

So it can be observed, the variable which has been changed at the adjacent squares can be removed, if the minterms of those squares are ORed together.

Example 4.1. *Simplify the Boolean function*

$$F = A'BC + A'BC' + AB'C' + AB'C.$$

Solution. First, a three-variable Karnaugh map is drawn and 1s are placed at the squares according to the minterms of the function as shown in Figure 4.5. Now two 1s of adjacent squares are grouped together. As in the figure, A'BC and A'BC' are grouped together at the first row, and AB'C' and AB'C are grouped together. From the first row, the reduced term of ABC + A'BC' is A'B, as C is the variable which changes its form. Similarly from the second row, AB'C'+ AB'C can be simplified to AB'. Now, as further simplification is not possible for this particular Boolean function, the simplified sum of the product of the function can be written as,

$$F = A'B + AB'.$$

	B'C'	B'C	BC	BC'
A'			1	1
A	1	1		

Figure 4.5

Example 4.2. *Simplify the expression F = A'BC + AB'C' + ABC + ABC'.*

Solution. The Karnaugh map for this function is shown in Figure 4.6. There are four squares marked with 1s, each for one of the minterms of the function.

	B'C'	B'C	BC	BC'
A'			1	
A	1		1	1

Figure 4.6

In the third column, two adjacent squares are grouped together to produce the simplified term BC. The other two 1s are placed at the first column and last column of the same second row. Note that these 1s or minterms can be combined to produce a reduced term. Here the B variable is changing its form, from uncomplemented to complemented. After combining these two minterms, we get the reduced term AC'. This can be confirmed by applying the Boolean algebra, AB'C' + ABC' = AC' (B + B') = AC'.

Therefore, the final simplified expression can be written as,

$$F = BC + AC'.$$

As in the previous examples, it is shown that two adjacent squares consisting of 1s can be combined to form reduced terms. Similarly, it is possible to combine four adjacent squares consisting of 1s, in the process of simplification of Boolean functions. Let us consider the next example.

Example 4.3. *Simplify the expression F = A'B'C + A'BC + A'BC' + AB'C + ABC.*

Solution. The Karnaugh map is shown in Figure 4.7. The four adjacent squares comprising the minterms A'B'C, A'BC, AB'C, and ABC can be combined. Here, it may observed that two of the variables A and B are changing their forms form uncomplemented to complemented. Therefore, these variables can be removed to form the reduced expression to C.

	B'C'	B'C	BC	BC'
A'		1	1	1
A		1	1	

Figure 4.7

Again, two adjacent squares comprising the minterms A'BC and A'BC' can be combined to produce the reduced term A'B. So the final simplified expression of the given function is

F = C + A'B.

Note that squares that are already considered in one group, can be combined with other group or groups.

Example 4.4. *Simplify the expression F (A, B, C) = Σ (0, 2, 4, 5, 6).*

	B'C'	B'C	BC	BC'
A'	1			1
A	1	1		1

Figure 4.8

The Karnaugh map is shown in Figure 4.8. Here, the minterms are given by their decimal-equivalent numbers. The squares according to those minterms are filled with 1s. A'B'C', ABC', AB'C', and ABC' are grouped to produce the reduced term of C' and, AB'C' and AB'C are grouped to produce the term AB'. So the final simplified expression may be written as

F = C' + AB'.

Note that four squares of the first column and last column may be combined just like the two squares combination explained in Example 4.2.

4.4 FOUR-VARIABLE KARNAUGH MAPS

Similar to the method used for two-variable and three-variable Karnaugh maps, four-variable Karnaugh maps may be constructed with 16 squares consisting of 16 minterms as shown in Figure 4.9(a). The same is redrawn in Figure 4.9(b) to show the relationship with the four binary variables. The rows and columns are numbered in a reflected code sequence, where only one variable is changing its form between two adjacent squares. The minterm of a particular square can be obtained by combining the row and column. As an example, the minterm of the second row and third column is A'BCD *i.e.*, m_7.

	C'D'	C'D	CD	CD'
A'B'	m_0	m_1	m_3	m_2
A'B	m_4	m_5	m_7	m_6
AB	m_{12}	m_{13}	m_{15}	m_{14}
AB'	m_8	m_9	m_{11}	m_{10}

Figure 4.9(a)

	C'D'	C'D	CD	CD'
A'B'	A'B'C'D'	A'B'C'D	A'B'CD	A'B'CD'
A'B	A'BC'D'	A'BC'D	A'BCD	A'BCD''
AB	ABC'D'	ABC'D	ABCD	ABCD'
AB'	AB'C'D'	AB'C'D	AB'CD	AB'CD'

Figure 4.9(b)

Different four-variable Karnaugh maps can be redrawn, if the variables are assigned an other way. Figure 4.10(a) and 4.10(b) also demonstrate the location of minterms for four-variable Karnaugh maps when variables A and B are assigned along the columns and variables C and D are assigned along the rows.

	A′B′	A′B	AB	AB′
C′D′	m_0	m_4	m_{12}	m_8
C′D	m_1	m_5	m_{13}	m_9
CD	m_3	m_7	m_{15}	m_{11}
CD′	m_2	m_6	m_{14}	m_{10}

Figure 4.10(a)

	A′B′	A′B	AB	AB′
C′D′	A′B′C′D′	A′BC′D′	ABC′D′	AB′C′D′
C′D	A′B′C′D	A′BC′D	ABC′D	AB′C′D
CD	A′B′CD	A′BCD	ABCD	AB′CD
CD′	A′B′CD′	A′BCD′	ABCD′	AB′CD′

Figure 4.10(b)

The minimization of four-variable Boolean functions using Karnaugh maps is similar to the method used to minimize three-variable functions. Two, four, or eight adjacent squares can be combined to reduce the number of literals in a function. The squares of the top and bottom rows as well as leftmost and rightmost columns may be combined. For example, m_0 and m_2 can be combined, as can m_4 and m_6, m_{12} and m_{14}, m_8 and m_{10}, m_0 and m_8, m_1 and m_9, m_3 and m_{11}, and m_2 and m_{10}. Similarly, the four squares of the corners *i.e.*, the minterms m_0, m_2, m_8, and m_{10} can also be combined.

When two adjacent squares are combined, it is called a pair and represents a term with three literals.

Four adjacent squares, when combined, are called a quad and its number of literals is two.

If eight adjacent squares are combined, it is called an octet and represents a term with one literal.

If, in the case all sixteen squares can be combined, the function will be reduced to 1.

Example 4.5. *Simplify the expression F (A, B, C, D) = $m_1 + m_5 + m_{10} + m_{11} + m_{12} + m_{13} + m_{15}$.*

Solution. The Karnaugh map for the above expression is shown in Figure 4.11.

	C′D′	C′D	CD	CD′
A′B′		1		
A′B		1		
AB	1	1	1	
AB′			1	1

Figure 4.11

From the figure, it can be seen that four pairs can be formed. The simplified expression may be written as, F = A′C′D + ABC′ + ACD + AB′C.

Note that the reduced expression is not a unique one, because if pairs are formed in different ways as shown in Figure 4.12, the simplified expression will be different. But both expressions are logically correct.

The simplified expression of the given function as per the Karnaugh map of Figure 4.12 is

F = A′C′D + ABC′ + ABD + AB′C.

	C′D′	C′D	CD	CD′
A′B′		1		
A′B		1		
AB	1	1	1	
AB′			1	1

Figure 4.12

Example 4.6. *Simplify the expression* $F(A, B, C, D) = m_7 + m_9 + m_{10} + m_{11} + m_{12} + m_{13} + m_{14} + m_{15}$.

Solution. The Karnaugh map for the above expression is shown in Figure 4.13.

	C′D′	C′D	CD	CD′
A′B′				
A′B			1	
AB	1	1	1	1
AB′		1	1	1

Figure 4.13

Three quads and one pair are formed as shown in the figure.

The simplified expression of the given function is,

F = AB + AC + AD + BCD.

Example 4.7. *Plot the logical expression F(A, B, C, D) = ABCD + AB′C′D′ + AB′C + AB on a four-variable Karnaugh map. Obtain the simplified expression.*

Solution. To form a Karnaugh map for a logical expression, the function is to be expanded to either canonical SOP form or canonical POS form. The canonical SOP form for the above expression can be obtained as follows.

F (A, B, C, D) = ABCD + AB′C′D′ + AB′C + AB
  = ABCD + AB′C′D′ + AB′C (D + D′) + AB (C + C′) (D + D′)
  = ABCD + AB′C′D′ + AB′CD + AB′CD′ + (ABC + ABC′) (D + D′)
  = ABCD + AB′C′D′ + AB′CD + AB′CD′ + ABCD + ABC′D +
    ABCD′ + ABC′D′
  = ABCD + AB′C′D′ + AB′CD + AB′CD′ + ABC′D + ABCD′ + ABC′D′
  = Σ (8, 10, 11, 12, 13, 14, 15)

The Karnaugh map for the above expression is shown in Figure 4.14.

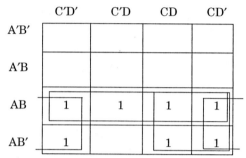

	C'D'	C'D	CD	CD'
A'B'				
A'B				
AB	1	1	1	1
AB'	1		1	1

Figure 4.14

Three quads (one of them is a roll-over type formed with first column and fourth column) are formed. The simplified expression is

$$F = AB + AC + AD'.$$

Example 4.8. *Simplify the expression F (W,X,Y,Z) = Σ (0, 1, 2, 4, 5, 6, 8, 9, 12, 13, 14).*

Solution. The Karnaugh map for the above function is shown in Figure 4.15.

One octet and two quads are formed. The simplified expression is

$$F = Y' + W'Z' + XZ'.$$

	Y'Z'	Y'Z	YZ	YZ'
W'X'	1	1		1
W'X	1	1		1
WX	1	1		1
WX'	1	1		

Figure 4.15

Example 4.9. *Simplify the expression F (W, X, Y, Z) = W'X'Y' + X'YZ' + W'XYZ' + WX'Y'.*

Solution. To obtain the minterms for the above expression, it needs to be expanded to the canonical SOP form as below.

$$F (W,X,Y,Z) = W'X'Y' + X'YZ' + W'XYZ' + WX'Y'$$
$$= W'X'Y' (Z + Z') + X'YZ'(W + W') + W'XYZ' + WX'Y'(Z + Z')$$
$$= W'X'Y'Z + W'X'Y'Z' + WX'YZ' + W'X'YZ' + W'XYZ' + WX'Y'Z + WX'Y'Z'$$

The Karnaugh map for the above function is shown in Figure 4.16.

One pair and two quads are formed (one quad consists of the four squares of the corners). The simplified expression is

$$F = X'Y' + X'Z' + W'YZ'.$$

	Y'Z'	Y'Z	YZ	YZ'
W'X'	1	1		1
W'X				1
WX				
WX'	1	1		1

Figure 4.16

Note that, to form the Karnaugh map above like an expression, it is not always necessary to expand the Boolean expression as described above. For the term W'X'Y', the squares W'X'Y'Z and W'X'Y'Z' are marked with 1s. For the term X'YZ'', the squares WX'YZ' and W'X'YZ' are marked with 1s. For the term WX'Y', the squares WX'Y'Z and WX'Y'Z' are marked with 1s. Lastly, the term W'XYZ' is the minterm itself, and is marked with 1. After forming the Karnaugh map, SOP expression can be realized as above.

Example 4.10. *Simplify the expression F (W,X,Y,Z) = Σ (3, 4, 5, 7, 9, 13, 14, 15).*

Solution. The Karnaugh map for the above function is shown in Figure 4.17.

Four pairs are formed. It may be noted that one quad can also be formed, but it is redundant as the squares contained by the quad are already covered by the pairs which are essential. The simplified expression may be written as

$$F = W'XY' + W'YZ + WY'Z + WXY.$$

	Y'Z'	Y'Z	YZ	YZ'
W'X'			1	
W'X	1	1	1	
WX		1	1	1
WX'		1		

Figure 4.17

Example 4.11. *Simplify the expression F (W,X,Y,Z) = Π (0, 1, 4, 5, 6, 8, 9, 12, 13, 14).*

Solution. The above expression is given in respect to the maxterms. In the Karnaugh map, 0s are to placed instead of 1s at the corresponding maxterm squares. The rest of the squares are filled with 1s.

The Karnaugh map for the above function is shown in Figure 4.18(*a*). There are two ways to achieve the minimized expression above. One way to is consider the 0s of the Karnaugh map. One octet and one quad has been formed with 0s. As we are considering the 0s, the simplified expression will be,

$$F' = Y' + XZ'.$$

Or, $F = (Y' + XZ')' = Y (X' + Z)$.

	Y'Z'	Y'Z	Y'Z'	YZ'
W'X'	0	0	1	1
W'X	0	0	1	0
WX	0	0	1	0
WX'	0	0	1	1

Figure 4.18(a)

The other way to achieve the minimized expression is to consider the 1s of the Karnaugh map as shown in Figure 4.18(b). Two quads are formed considering the 1s.

	Y'Z'	Y'Z	YZ	YZ'
W'X'	0	0	1	1
W'X	0	0	1	0
WX	0	0	1	0
WX'	0	0	1	1

Figure 4.18(b)

The minimized expression can be written as

$$F = YZ + X'Y$$
$$= Y(X' + Z).$$

Note that the final expressions are the same in both cases.

Example 4.12. *Obtain (a) the minimal sum of the products and (b) minimal product of the sums for the function F (W,X,Y,Z) = Σ (0, 1, 2, 5, 8, 9, 10).*

Solution.

	Y'Z'	Y'Z	YZ	YZ'
W'X'	1	1	0	1
W'X	0	1	0	0
WX	0	0	0	0
WX'	1	1	0	1

Figure 4.19(a)

(a) The Karnaugh map for the above function is shown in Figure 4.19(a). Two quads and a pair are formed considering the 1s of the Karnaugh map.

The SOP expression of the above is F = X′Y′ + X′Z′ + W′Y′Z.

(b) The Karnaugh map for the above function is shown in Figure 4.19(b). Three quads are formed considering the 0s of the Karnaugh map.

	Y′Z′	Y′Z	YZ	YZ′
W′X′	1	1	0	1
W′X	0	1	0	0
WX	0	0	0	0
WX′	1	1	0	1

Figure 4.19(b)

The POS expression of above funtion can be derived as,

$$F' = XZ' + WX + YZ.$$

Or, $$F = (X' + Z)\ (W' + X')\ (Y' + Z').$$

4.5 FIVE-VARIABLE KARNAUGH MAPS

Karnaugh maps with more than four variables are not simple to use. The number of cells or squares becomes excessively large and combining the adjacent squares becomes complex. The number of cells or squares is always equal to the number of minterms. A five-variable Karnaugh map contains 2^5 or 32 cells, which are used to simplify any five-variable logic function. Figures 4.20(a) and 4.20(b) demonstrate the five-variable Karnaugh map and its minterms.

	C′D′E′	C′D′E	C′DE	C′DE′	CDE′	CDE	CD′E	CD′E′
A′B′	m_0	m_1	m_3	m_2	m_6	m_7	m_5	m_4
A′B	m_8	m_9	m_{11}	m_{10}	m_{14}	m_{15}	m_{13}	m_{12}
AB	m_{24}	m_{25}	m_{27}	m_{26}	m_{30}	m_{31}	m_{29}	m_{28}
AB′	m_{16}	m_{17}	m_{19}	m_{18}	m_{22}	m_{23}	m_{21}	m_{20}

Figure 4.20(a)

	C′D′E′	C′D′E	C′DE	C′DE′	CDE′	CDE	CD′E	CD′E′
A′B′	A′B′C′D′E′	A′B′C′D′E	A′B′C′DE	A′B′C′DE′	A′B′CDE′	A′B′CDE	A′B′CD′E	A′B′CD′E′
A′B	A′BC′D′E′	A′BC′D′E	A′BC′DE	A′BC′DE′	A′BCDE′	A′BCDE	A′BCD′E	A′BCD′E′
AB	ABC′D′E′	ABC′D′E	ABC′DE	ABC′DE′	ABCDE′	ABCDE	ABCD′E	ABCD′E′
AB′	AB′C′D′E′	AB′C′D′E	AB′C′DE	AB′C′DE′	AB′CDE′	AB′CDE	AB′CD′E	AB′CD′E′

Figure 4.20(b)

Figures 4.21, 4.22, and 4.23 also demonstrate five-variable Karnaugh maps, if the variables are assigned in different ways. The five-variable Karnaugh maps have properties similar to the two-, three-, or four-variable Karnaugh maps described earlier, *i.e.*, adjacent squares can be grouped together. In addition to those, while making groups or combinations, in Figures 4.20 and 4.21, the 1st column with 4th column, 2nd column with 7th column, and 3rd column with 6th column can be combined together, as there is only one variable which is changing its form for those columns. Similarly, according to Figures 4.22 and 4.23, the 1st row with 4th row, 2nd row with 7th row, and 3rd row with 6th row can be combined together to get the terms of reduced literals.

	A'B'C'	A'B'C	A'BC	A'BC'	ABC'	ABC	AB'C	AB'C'
C'D'	m_0	m_4	m_{12}	m_8	m_{24}	m_{28}	m_{20}	m_{16}
C'D	m_1	m_5	m_{13}	m_9	m_{25}	m_{29}	m_{21}	m_{17}
CD	m_3	m_7	m_{15}	m_{11}	m_{27}	m_{31}	m_{23}	m_{19}
CD'	m_2	m_6	m_{14}	m_{10}	m_{26}	m_{30}	m_{22}	m_{18}

Figure 4.21

	D'E'	D'E	DE	DE'
A'B'C'	m_0	m_1	m_3	m_2
A'B'C	m_4	m_5	m_7	m_6
A'BC	m_{11}	m_{12}	m_{15}	m_{14}
A'BC'	m_8	m_9	m_7	m_{10}
ABC'	m_{24}	m_{25}	m_{27}	m_{26}
ABC	m_{28}	m_{29}	m_{31}	m_{30}
AB'C	m_{20}	m_{21}	m_{23}	m_{22}
AB'C'	m_{16}	m_{17}	m_{19}	m_{18}

Figure 4.22

	A'B'	A'B	AB	AB'
C'D'E'	m_0	m_8	m_{24}	m_{16}
C'D'E	m_1	m_9	m_{25}	m_{17}
C'DE	m_3	m_{11}	m_{27}	m_{19}
C'DE'	m_2	m_{10}	m_{26}	m_{18}
CDE'	m_6	m_{14}	m_{30}	m_{22}
CDE	m_7	m_{15}	m_{31}	m_{23}
CD'E	m_5	m_{13}	m_{29}	m_{21}
CD'E'	m_4	m_{12}	m_{28}	m_{20}

Figure 4.23

4.6   SIX-VARIABLE KARNAUGH MAPS

Six-variable Karnaugh maps consist of 2^6 or 64 squares or cells. Similar to the method described above, six-variable Karnaugh maps are formed with 64 minterms as demonstrated in Figure 4.24(*a*). Figure 4.24(*b*) also represents six-variable Karnaugh maps when the variables are assigned differently.

Apart from the properties described for two-, three- and four-variable Karnaugh maps that adjacent squares can be grouped together, similar to five-variable maps, the 1st column with 4th column, 2nd column with 7th column, 3rd column with 6th column, 1st row with 4th row, 2nd row with 7th row, and 3rd row with 6th row can be combined together to get the terms of reduced literals.

	D'E'F'	D'E'F	D'EF	D'EF'	DEF'	DEF	DE'F	DE'F'
A'B'C'	m_0	m_1	m_3	m_2	m_6	m_7	m_5	m_4
A'B'C	m_8	m_9	m_{11}	m_{10}	m_{14}	m_{15}	m_{13}	m_{12}
A'BC	m_{24}	m_{25}	m_{27}	m_{26}	m_{30}	m_{31}	m_{29}	m_{28}
A'BC'	m_{16}	m_{17}	m_{19}	m_{18}	m_{22}	m_{23}	m_{21}	m_{20}
ABC'	m_{48}	m_{49}	m_{51}	m_{50}	m_{54}	m_{55}	m_{53}	m_{52}
ABC	m_{56}	m_{57}	m_{59}	m_{58}	m_{62}	m_{63}	m_{61}	m_{60}
AB'C	m_{40}	m_{41}	m_{43}	m_{42}	m_{46}	m_{47}	m_{45}	m_{44}
AB'C'	m_{32}	m_{33}	m_{35}	m_{34}	m_{38}	m_{39}	m_{37}	m_{36}

Figure 4.24(a)

	A'B'C'	A'B'C	A'BC	A'BC'	ABC'	ABC	AB'C	AB'C'
D'E'F'	m_0	m_8	m_{24}	m_{16}	m_{48}	m_{56}	m_{40}	m_{32}
D'E'F	m_1	m_9	m_{25}	m_{17}	m_{49}	m_{57}	m_{41}	m_{33}
D'EF	m_3	m_{11}	m_{27}	m_{19}	m_{51}	m_{59}	m_{43}	m_{35}
D'EF'	m_2	m_{10}	m_{26}	m_{18}	m_{50}	m_{58}	m_{42}	m_{34}
DEF'	m_6	m_{14}	m_{30}	m_{22}	m_{54}	m_{62}	m_{46}	m_{38}
DEF	m_7	m_{15}	m_{31}	m_{23}	m_{55}	m_{63}	m_{47}	m_{39}
DE'F	m_5	m_{13}	m_{29}	m_{21}	m_{53}	m_{61}	m_{45}	m_{37}
DE'F'	m_4	m_{12}	m_{28}	m_{20}	m_{52}	m_{60}	m_{44}	m_{36}

Figure 4.24(b)

Example 4.13. *Obtain the minimal sum of the products for the function*

$F (A, B, C, D, E) = \Sigma\ (0, 2, 5, 7, 9, 11, 13, 15, 16, 18, 21, 23, 25, 27, 29, 31).$

Solution. The five-variable Karnaugh map for the above function is shown in Figure 4.25.

	C′D′E′	C′D′E	C′DE	C′DE′	CDE′	CDE	CD′E	CD′E′
A′B′	1			1		1	1	
A′B		1	1			1	1	
AB		1	1			1	1	
AB′	1			1		1	1	

Figure 4.25

An octet at the 6th and 7th column with 1st to 4th rows, one octet at the 2nd, 3rd, 6th, and 7th columns with 2nd and 3rd rows, and one quad at the 1st and 4th rows with 1st and 4th columns are formed. The minimized expression can be written as,

$$F = CE + BE + B′C′E′.$$

Example 4.14. *Obtain the minimal sum of the products for the function*

$$F\ (A,\ B,\ C,\ D,\ E) = \Sigma\ (0,\ 2,\ 4,\ 6,\ 9,\ 11,\ 13,\ 15,\ 17,\ 21,\ 25,\ 27,\ 29,\ 31).$$

Solution. The five-variable Karnaugh map for the function is shown in Figure 4.26.

	C′D′E′	C′D′E	C′DE	C′DE′	CDE′	CDE	CD′E	CD′E′
A′B′	1			1	1			1
A′B		1	1			1	1	
AB		1	1			1	1	
AB′		1					1	

Figure 4.26

An octet at the 2nd, 3rd, 6th, and 7th columns with 2nd and 3rd rows, one quad at the 1st row with 1st, 4th, 5th, and 8th columns, and one quad at 3rd and 4th rows with 2nd and 6th columns are formed. The minimized expression can be written as,

$$F = BE + A′B′E′ + AD′E.$$

4.7 DON'T-CARE COMBINATIONS

In certain digital systems, some input combinations never occur during the process of a normal operation because those input conditions are guaranteed never to occur. Such input combinations are called *Don't-Care Combinations*. The function output may be either 1 or 0 and these functions are called incompletely specified functions. These input combinations can be plotted on the Karnaugh map for further simplification of the function. The don't-care combinations are represented by *d* or *x* or Φ.

When an incompletely specified function, *i.e.*, a function with don't-care combinations is simplified to obtain minimal SOP expression, the value 1 can be assigned to the selected don't care combinations. This is done to form groups like pairs, quadoctet, etc., for further simplification. In each case, choice depends only on need to achieve simplification. Similarly, selected don't care combinations may be assumed as 0s to form groups of 0s for obtaining the POS expression.

Example 4.15. *Obtain the minimal sum of the products for the function*

$$F\ (A,\ B,\ C,\ D) = \Sigma\ (1,3,7,11,15) + \Phi(0,2,5).$$

The Karnaugh map for the above function is shown in Figure 4.27.

	C'D'	C'D	CD	CD'
A'B'	X	1	1	X
A'B		X	1	
AB			1	
AB'			1	

Figure 4.27

In the Karnaugh map of Figure 4.27, the minterm m_0 and m_2 *i.e.*, A'B'C'D' and A'B'CD', are the don't care terms which have been assumed as 1s, while making a quad. The simplified SOP expression of above function can be written as

$$F = A'B' + CD.$$

4.8 THE TABULATION METHOD

The Karnaugh map method is a very useful and convenient tool for simplification of Boolean functions as long as the number of variables does not exceed four (at the most six). But if the number of variables increases, the visualization and selection of patterns of adjacent cells in the Karnaugh map becomes complicated and difficult. The tabular method, also known as the *Quine-McCluskey method*, overcomes this difficulty. It is a specific step-by-step procedure to achieve guaranteed, simplified standard form of expression for a function.

The following steps are followed for simplification by the tabular or Quine-McCluskey method.

1. An exhaustive search is done to find the terms that may be included in the simplified functions. These terms are called *prime implicants*.

2. Form the set of prime implicants, essential prime implicants are determined by preparing a prime implicants chart.

3. The minterms that are not covered by the essential prime implicants, are taken into consideration by selecting some more prime implications to obtain an optimized Boolean expression.

4.8.1 Determination of Prime Implicants

The prime implicants are obtained by the following procedure:

1. Each minterm of the function is expressed by its binary representation.

2. The minterms are arranged according to increasing index (index is defined as the number of 1s in a minterm). Each set of minterms possessing the same index are separated by lines.

3. Now each of the minterms is compared with the minterms of a higher index. For each pair of terms that can combine, the new terms are formed. If two minterms are differed by only one variable, that variable is replaced by a '-' (dash) to form the new term with one less number of literals. A line is drawn in when all the minterms of one set is compared with all the minterms of a higher index.

4. The same process is repeated for all the groups of minterms. A new list of terms is obtained after the first stage of elimination is completed.

5. At the next stage of elimination two terms from the new list with the '-' of the same position differing by only one variable are compared and again another new term is formed with a less number of literals.

6. The process is to be continued until no new match is possible.

7. All the terms that remain unchecked *i.e.*, where no match is found during the process, are considered to be the prime implicants.

4.8.2 Prime Implicant Chart

1. After obtaining the prime implicants, a chart or table is prepared where rows are represented by the prime implicants and the columns are represented by the minterms of the function.

2. Crosses are placed in each row to show the composition of the minterms that makes the prime implicants.

3. A completed prime implicant table is to be inspected for the columns containing only a single cross. Prime implicants that cover the minterms with a single cross are called the essential prime implicants.

The above process to find the prime implicants and preparation of the chart can be illustrated by the following examples.

Example 4.16. *Obtain the minimal sum of the products for the function*

$$F\ (A,\ B,\ C,\ D) = \Sigma\ (1,\ 4,\ 6,\ 7,\ 8,\ 9,\ 10,\ 11,\ 15).$$

Solution. The table in Figure 4.28 shows the step-by-step procedure the Quine-McCluskey method uses to obtain the simplified expression of the above function.

Column I consists of the decimal equivalent of the function or the minterms and column II is the corresponding binary representation. They are grouped according to their index *i.e.*, number of 1s in the binary equivalents. In column III, two minterms are grouped if they are differed by only a single variable and equivalent terms are written with a '-' in the place where the variable changes its logic value. As an example, minterms 1 (0001) and 9 (1001) are grouped and written as 1,9 (– 001) and so on for the others. Also, the terms of column II, which are considered to form the group in column III, are marked with '√'.

I	II					III				IV		
Decimal equivalent	Binary equivalent						ABCD				ABCD	
	A	B	C	D								
1	0	0	0	1	√	1,9	−001			8,9,10,11	10− −	
4	0	1	0	0	√	4,6	01−0			8,10,9,11	10− −	
8	1	0	0	0	√	8,9	100−	√				
						8,10	10−0	√				
6	0	1	1	0	√	6,7	011−					
9	1	0	0	1	√	9,11	10−1	√				
10	1	0	1	0	√	10,11	101−	√				
7	0	1	1	1	√	7,15	−111					
11	1	0	1	1	√	11,15	1−11					
15	1	1	1	1	√							

Figure 4.28

The terms which are not marked with '√' are the Prime implicants. To express the prime implicants algebraically, variables are to be considered as true form in place of 1s, as complemented form in place of 0s, and no variable if '-' appears. Here the prime implicants are B′C′D, A′BD′, A′BC, BCD, ACD (from column III), and AB′ (from column IV). So the Boolean expression of the given function can be written as

$$F = AB' + B'C'D + A'BD' + A'BC + BCD + ACD.$$

But the above expression may not be of minimized form, as all the prime implicants may not be necessary. To find out the essential prime implicants, the following steps are carried out. A table or chart consisting of prime implicants and the decimal equivalent of minterms as given in the expression, as in Figure 4.29 is prepared.

Prime Implicants	1	4	6	7	8	9	10	11	15
√ AB′					X	X	X	X	
√ B′C′D	X					X			
√ A′BD′		X	X						
A′BC			X	X					
BCD				X					X
ACD								X	X
	√	√	√		√	√	√	√	

Figure 4.29

In the table, the prime implicants are listed in the 1ˢᵗ column and Xs are placed against the corresponding minterms. The completed prime implicant table is now inspected for the columns containing only a single X. As in Figure 4.29, the minterm 1 is represented by only a single prime implicant B′C′D, and only a single X in that column, it should be marked as well as the corresponding column should be marked. Similarly, the prime implicants AB′ and AB′D′ are marked. These are the essential prime implicants as they are absolutely necessary to form the minimized Boolen expression. Now all the other minterms corresponding to these prime implicants are marked at the end of the columns *i.e.*, the minterms 1, 4, 6, 8, 9, 10, and 11 are marked. Note that the terms A′BC, BCD, and ACD are not marked. So they are not the essential prime implicants. However, the minterms 7 and 15 are still unmarked and both of them are covered by the term BCD and are included in the Boolean expression. Therefore, the simplified Boolen expression of the given function can be written as

$$F = AB' + B'C'D + A'BD' + BCD.$$

The simplified expressions derived in the preceeding example are in the sum of products form. The Quine-McClusky method can also be adopted to derive the simplified expression in product of sums form. In the Karnaugh map method the complement of the function was considered by taking 0s from the initial list of the minterns. Similarly the tabulation method or Quine-McClusky method may be carried out by considering the 0s of the function to derive the sum of products form. Finally, by making the complement again, we obtain the simplified expression in the form of product of sums.

A function with don't-care conditions can be simplified by the tabulation method with slight modification. The don't-care conditions are to be included in the list of minterms while determining the prime implicants. This allows the derivation of prime implicants with the least number of literals. But the don't-care conditions are excluded in the list of minterms when the prime implicants table is prepared, because these terms do not have to be covered by the selected prime implicants.

4.9 MORE EXAMPLES

Example 4.17. *Obtain the minimal sum of the products for the function F (A,B,C,D) = Σ (1, 2, 3, 7, 8, 9, 10, 11, 14, 15) by the Quine-McClusky method.*

Solution. The first step is to find out the prime implicants as described by the table in Figure 4.30.

The prime implicants are B′D, B′C, AB′, CD, and AC. The prime implicant table is prepared as in Figure 4.31.

I	II					III			IV	
Decimal equivalent	*Binary equivalent*									
	A	*B*	*C*	*D*		*ABCD*			*ABCD*	
1	0	0	0	1	√	1,3	00–1	√	1,3,9,11	–0–1
2	0	0	1	0	√	1,9	–001	√	2,3,10,11	–01–
8	1	0	0	0	√	2,3	001–	√	8,9,10,11	10– –
						2,10	–010	√	3,7,11,15	– –11

						8,9	100–	√	10,11,14,15 1–1–
						8,10	10–0	√	
3	0	0	1	1	√	3,7	0–11	√	
9	1	0	0	1	√	3,11	–011	√	
10	1	0	1	0	√	9,11	10–1	√	
						10,11	101–	√	
						10,14	1–10	√	
7	0	1	1	1	√	7,15	–111	√	
11	1	0	1	1	√	11,15	1–11	√	
14	1	1	1	0	√	14,15	111–	√	
15	1	1	1	1	√				

Figure 4.30

Prime Implicants	1	2	3	7	8	9	10	11	14	15
√ B′D	X		X			X		X		
√ B′C		X	X				X	X		
√ AB′					X	X	X	X		
√ CD			X	X				X		X
√ AC							X	X	X	X
	√	√	√	√	√	√	√	√	√	√

Figure 4.31

All the prime implicants are essential. So the simplified Boolean expression of the given function is

$$F = B'D + B'C + AB' + CD + AC.$$

Example 4.18. *Using the Quine-McClusky method, obtain the minimal sum of the products expression for the function $F(A, B, C, D) = \Sigma (1, 3, 4, 5, 9, 10, 11) + \Phi (6, 8)$.*

Solution. The prime implicants are obtained from the table in Figure 4.32.

I	II					III			IV	
Decimal equivalent	Binary equivalent						ABCD			ABCD
	A	B	C	D						
1	0	0	0	1	√	1,3	00–1	√	1,3,9,11	–0 –1
4	0	1	0	0	√	1,5	0–01		8,9,10,11	10– –
8	1	0	0	0	√	1,9	–001	√		
						4,5	010–			
						4,6	01–0			
						8,9	100–	√		
						8,10	10–0	√		
3	0	0	1	1	√	3,11	–011	√		
5	0	1	0	1	√	9,11	10–1	√		
6	0	1	1	0	√	10,11	101–	√		
9	1	0	0	1	√					
10	1	0	1	0	√					
11	1	0	1	1	√					

Figure 4.32

The prime implicants are A'C'D, A'BC', A'BD', B'D, and AB'. The prime implicant table is prepared as in Figure 4.33.

Prime Implicants	1	3	4	5	9	10	11
A'C'D	X			X			
A'BC'			X	X			
A'BD'			X				
√ B'D	X	X			X		X
√ AB'					X	X	X
	√	√			√	√	√

Figure 4.33

From the table, we obtain the essential prime implicants B′D and AB′. The minterms 4 and 5 are not marked in the table. The term A′BC′ is considered, which covers both the minterms 4 and 5. So the simplified Boolean expression for the given function is

$$F = A'BC' + B'D + AB'.$$

Example 4.19. *Using the Karnaugh map method obtain the minimal sum of the products and product of sums expressions for the function*

$$F(A,B,C,D) = \Sigma \ (1, \ 3, \ 4, \ 5, \ 6, \ 7, \ 9, \ 12, \ 13).$$

Solution. The Karnaugh map for the above function is in Figure 4.34. To obtain the SOP expression, 1s of the Karnaugh map are considered.

	C′D′	C′D	CD	CD′
A′B′		1	1	
A′B	1	1	1	1
AB	1	1		
AB′		1		

Figure 4.34

The simplified Boolean expression for the function is

$$F = A'B + BC' + C'D + A'D.$$

To derive the POS expression, the 0s of the Karnaugh map are considered as in Figure 4.35.

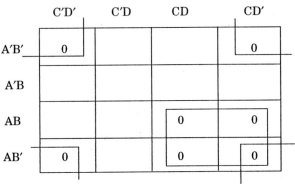

	C′D′	C′D	CD	CD′
A′B′	0			0
A′B				
AB			0	0
AB′	0		0	0

Figure 4.35

From the Karnaugh map we obtain F′ = AC + B′D′.

So the POS expression for the above function is

$$F = (AC + B'D')' = (AC)'. (B'D')' = (A' + C'). (B + D).$$

Example 4.20. *Using the Karnaugh map method obtain the minimal sum of the products and product of sums expressions for the function*

$$F(A,B,C,D) = \Sigma \ (1, 5, 6, 7, 11, 12, 13, 15).$$

Solution. The Karnaugh map for the above function is in Figure 4.36. To obtain the SOP expression, 1s of the Karnaugh map are considered.

The simplified Boolean expression for the function is

$$F = A'C'D + A'BC + ABC' + ACD = B(A'C + AC') + D(A'C' + AC)$$

$$= B \ (A \oplus C) + D(A \oplus C)'.$$

	C'D'	C'D	CD	CD'
A'B'		1		
A'B		1	1	1
AB	1	1	1	
AB'			1	

Figure 4.36

To derive the POS expression, the 0s of the Karnaugh map are considered as in Figure 4.37.

	C'D'	C'D	CD	CD'
A'B'	0		0	0
A'B	0			
AB				0
AB'	0	0		0

Figure 4.37

From the Karnaugh map we obtain

$$F' = A'C'D' + A'B'C + ABC' + ACD'.$$

So the POS expression for the above function is

$$F = (A + C + D) (A + B + C') (A' + B + C) (A' + C' + D).$$

Example 4.21. *Using the Karnaugh map method obtain the minimal sum of the products expression for the function*

$$F(A,B,C,D) = \Sigma \ (0, 2, 3, 6, 7) + d \ (8, 10, 11, 15).$$

Solution. The Karnaugh map for the above function is in Figure 4.38.

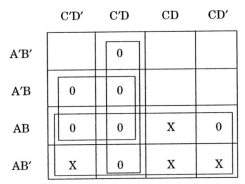

Figure 4.38

The simplified Boolean expression for the function is

$$F = A'C + B'D'.$$

Example 4.22. *Using the Karnaugh map method obtain the minimal product of the sums expression for the function given in example 4.21.*

Solution. To derive the POS expression, the 0s of the Karnaugh map are considered as in Figure 4.39.

	C'D'	C'D	CD	CD'
A'B'		0		
A'B	0	0		
AB	0	0	X	0
AB'	X	0	X	X

Figure 4.39

The simplified Boolean expression for the function is

$$F' = A + C'D + BC'.$$

So $$F = A' (C + D') (B' + C).$$

Example 4.23. *Using the Karnaugh map method obtain the minimal sum of the products expression for the function*

$$F(A,B,C,D) = \Sigma\ (1,\ 5,\ 7,\ 13,\ 14,\ 15,\ 17,\ 18,\ 21,\ 22,\ 25,\ 29) + d(6,\ 9,\ 19,\ 23,\ 30).$$

Solution. The Karnaugh map for the above function is in Figure 4.40.

	C'D'E'	C'D'E	C'DE	C'DE'	CDE'	CDE	CD'E	CD'E'
A'B'		1			X	1	1	
A'B		X			1	1	1	
AB		1			X		1	
AB'		1	X	1	1	X	1	

Figure 4.40

The simplified Boolean expression for the function is

$$F = D'E + A'CD + AB'D.$$

Example 4.24. *Using the Quine-McClusky method obtain the minimal sum of the products expression for the function*

$$F(A,B,C,D,E) = \Sigma\ (0,\ 2,\ 3,\ 5,\ 7,\ 9,\ 11,\ 13,\ 14,\ 16,\ 18,\ 24,\ 26,\ 28,\ 30).$$

Solution. The prime implicants are obtained from the table in Figure 4.41.

I	II						III			IV	
Decimal	*Binary equivalent*										
equivalent	*A*	*B*	*C*	*D*	*E*			*ABCDE*			*ABCDE*
0	0	0	0	0	0	√	0,2	000–0	√	0,2,16,18	–00–0
2	0	0	0	1	0	√	0,16	–0000	√	16,18,24,26	1–0–0
16	1	0	0	0	0	√				24,26,28,30	11––0
3	0	0	0	1	1	√	2,3	0001–			
5	0	0	1	0	1	√	2,18	–0010	√		
9	0	1	0	0	1	√	16,18	100–0	√		
18	1	0	0	1	0	√	16,24	1–000	√		
24	1	1	0	0	0	√					
7	0	0	1	1	1	√	3,7	00–11			
11	0	1	0	1	1	√	3,11	0–011			
13	0	1	1	0	1	√	5,7	001–1			
14	0	1	1	1	0	√	5,13	0–101			
26	1	1	0	1	0	√	9,13	01–01			

28	1 1 1 0 0 √	18,26	1–010	√	
		24,26	110–0	√	
		24,28	11–00	√	
30	1 1 1 1 0 √	14,30	–1110		
		26,30	11–10	√	
		28,30	111–0	√	

Figure 4.41

The prime implicant table is prepared as in Figure 4.42.

The essential prime implicants are B′C′E′, ABE′, A′C′DE, A′BD′E, and BCDE′, as each of them represent at least one minterm which is not represented by any of the other prime implicants. The term A′B′CE may be considered to include minterms 5 and 7.

Prime Implicants	0	2	3	5	7	9	11	13	14	16	18	24	26	28	30
√ B′C′E′	X	X								X	X				
AC′E′										X	X	X	X		
√ ABE′												X	X	X	X
A′B′C′D		X	X												
A′B′DE			X		X										
√ A′C′DE			X			X									
A′B′CE				X	X										
A′CD′E				X				X							
√ A′BD′E						X		X							
√ BCDE′									X						X
	√	√	√			√	√	√	√	√	√	√	√	√	√

Figure 4.42

The simplified expression of the function is

$$F = B'C'E + ABE' + A'C'DE + A'B'CE + A'BD'E + BCDE'.$$

4.10 VARIABLE-ENTERED KARNAUGH MAPS

There is another method of simplification of Boolean functions which is not widely used, but certainly has some importance from an academic point of view. Earlier in this section we have already discussed five-variable and six-variable Karnaugh maps, which are a little complex and difficult while making pairs, quads, or octets. *Variable-entered Karnaugh maps* may be used in cases where the number of variables exceeds four. It is the useful extension of normal Karnaugh maps as discussed earlier. In variable-entered Karnaugh maps, one or more Boolean variables can be used as map entries along with 1s, 0s, and don't-cares. The variables associated with the entries in these maps are called *map-entered variables*.

The significance of variable-entered maps is that they provide for map compression. Normally an nth order Karnaugh map is associated with the Boolean functions of n number of variables. However, if one of the Boolean variables is used as a map-entered variable, the Karnaugh map will be reduced to an order of $n-1$. In general, if the number of map-entered variables is m of the total number of variables n, the Karnaugh map of the order of $n-m$ will suffice to associate in the necessary simplification process. ($n > m$).

A useful application for variable-entered maps arises in the problems that have infrequently appearing variables. In such situations it is convenient to have the functions of the infrequently appearing variables as the entries within the map, allowing a high-order Boolean function to be represented by a low-order map.

4.10.1 Contruction of Variable-entered MAPS

To understand the construction of variable-entered Karnaugh maps, consider the generic truth table in Figure 4.43, where the functional value for row i is denoted by F_i. From the truth table the Karnaugh map is constructed in Figure 4.44. The entries within the cells are the F_i's, which, in turn, correspond to the 0s, 1s, and don't-cares that normally appear in the last column of the truth table.

Alternatively, the generic minterm canonical formula for the truth table of Figure 4.43 can be written as

$$F (A,B,C) = F_0 \cdot A'B'C' + F_1 \cdot A'B'C + F_2 \cdot A'BC' + F_3 \cdot A'BC + F_4 \cdot AB'C' + F_5 \cdot AB'C + F_6 \cdot ABC' + F_7 \cdot ABC.$$

Using Boolean algebra, this expression can be modified as

$$F(A,B,C) = A'B' (F_0 \cdot C' + F_1 \cdot C) + A'B (F_2 \cdot C' + F_3 \cdot C) + AB' (F_4 \cdot C' + F_5 \cdot C) + AB (F_6 \cdot C' + F_7 \cdot C).$$

A	B	C	F_i
0	0	0	F_0
0	0	1	F_1
0	1	0	F_2
0	1	1	F_3
1	0	0	F_4
1	0	1	F_5
1	1	0	F_6
1	1	1	F_7

Figure 4.43

	B'C'	B'C	BC	BC'
A'	F_0	F_1	F_3	F_2
A	F_4	F_5	F_7	F_6

Figure 4.44

Since this equation consists of four combinations of A and B variables in their complemented and uncomplemented form, a map can be prepared from the equation by using A and B variables as the row and column labels and the terms within the parentheses as cell entries. This is illustrated in Figure 4.45. Now too may noticed that the Karnaugh map of order three has been reduced to the order of two. Hence, the map compression is achieved. In this example, A and B are called map variables, and C is treated as a map-entered variable as it is appearing in the cell entries.

	B′	B
A′	$F_0.C' + F_1.C$	$F_2.C' + F_3.C$
A	$F_4.C' + F_5.C$	$F_6.C' + F_7.C$

Figure 4.45

The above expression may also be written as

$F (A,B,C) = A'C' (F_0.B' + F_2.B) + A'C (F_1.B'++ F_3.B) + AC' (F_4. B' + F_6.B) + AC (F_5.B' + F_7.B)$.

Here A and C are the map variables and B may be used as a map-entered variable, if the Karnaugh map, according to this expression, is to be formed.

Also, the same equation may be expressed as

$F (A,B,C) = B'C'(F_0.A' + F_4.A) + B'C(F_1.A' + F_5.A) + BC'(F_2.A' + F_6.A) + BC(F_3.A' + F_7.A)$.

Where B and C are the map variables, and A is the map-entered variable for its Karnaugh map.

It may be noted that the Karnaugh map for the above expression can be further compressed in respect of its order, if the expression is rewritten as below.

$F (A,B,C) = A'(F_0.B'C' + F_1.B'C + F_2.BC' + F_3.BC) + A(F_4. B'C'+ F_5.B'C + F_6.BC' + F_7.BC)$

In this case, A is the map variable, and B and C are the map-entered variable. The Karnaugh map can be constructed as in Figure 4.46.

A′	A
$F_0.B'C'+F_1.B'C + F_2.BC'+ F_3.BC$	$F_4.B'C'+ F_5.B'C + F_6.BC'+ F_7.BC$

Figure 4.46

Thus higher order Karnaugh maps can be reduced to lower order maps using one or more variables within the cells. However, the degree of difficulty in interpreting compressed maps lies in the complexities of the entered function. Alternatively, variable-entered maps may be derived by partitioning of the truth table. The truth table of Figure 4.43 may be reconstructed like Figure 4.47, where rows are paired such that they correspond to equal values of A and B. The two possible values of the C variable appear within each pair and the last column of the truth table consists of single variable functions corresponding to C. Since within each of the partitions A and B possess the same value, the partitioned truth table can now be used to form the variable-entered map. This means that for each combination of A and B variables, the cell entries become the function of the C variable.

We have so far discussed the map entries as single-variable functions. The map entries functions may be generalized as $F_i.V' + F_j.V$ where F_i and F_j are the functions F_0, F_1, ..., F_7. Now, with the assumption of completely specified Boolean function and truth table,

A	B	C	F_i	F_i
0	0	0	F_0	$F_0.C' + F_1.C$
0	0	1	F_1	
0	1	0	F_2	$F_2.C' + F_3.C$
0	1	1	F_3	
1	0	0	F_4	$F_4.C' + F_5.C$
1	0	1	F_5	
1	1	0	F_6	$F_6.C' + F_7.C$
1	1	1	F_7	

Figure 4.47

F_i	F_j	$F_i.V' + F_j.V$	Map entry
0	0	$0 + 0 = 0$	0
0	1	$0 + V = V$	V
1	0	$V' + 0 = V'$	V'
1	1	$V' + V = 1$	1

Figure 4.48

i.e., there are no don't-care condition, it may be noted that values of F_i and F_j are restricted to only 0s and 1s. The table in Figure 4.48 illustrates the four possible value assignments to F_i and F_j and corresponding map entries in respect to V, where V is generalized as a map-entered variable. In addition to these, the don't-care conditions are to be considered as map entries wherever necessary.

A	B	C	F
0	0	0	1
0	0	1	1
0	1	0	1
0	1	1	0
1	0	0	0
1	0	1	1
1	1	0	0
1	1	1	0

Figure 4.49

	B'	B
A'	1	C'
A	C	0

Figure 4.50

Now let us consider a practical example of a Boolean function according to the truth table of Figure 4.49. The Boolean expression of the function is

$$F(A,B,C) = A'B'C' + A'B'C + A'BC' + AB'C.$$

The expression may be modified as

$$F(A,B,C) = A'B'(C' + C) + A'BC' + AB'C$$
$$= A'B'(1) + A'B(C') + AB'(C).$$

The expressions within the parentheses correspond to the single variable map entries. With C as the map-entered variable, the Karnaugh map is constructed as in Figure 4.50.

Another example may be cited here for the Boolean expression as below.

$$F(X,Y,A,B,C) = XA'B'C' + A'B'C + A'BC + YABC' + ABC$$

In this expression, variables X and Y appear infrequently and may be considered as map-entered variables. The Karnaugh map is constructed as in Figure 4.51.

	B'C'	B'C	BC	BC'
A'	X	1	1	0
A	0	0	1	Y

Figure 4.51

4.10.2 Formation of Minimal Sums and Products with Map-entered Variables

Just like normal Karnaugh maps, minimal sum terms or product terms can be obtained from variable-entered Karnaugh maps by grouping and subgrouping of cells. However, while obtaining a minimal sum, it is necessary to form groups involving the map-entered variable in addition to 1s. Similarly, the map-entered variables are to be considered for formation of group in addition to 0s, while obtaining minimal products. We know from the Boolean algebra that $V + V' = 1$, where V is the generalized notation of a map-entered variable. Therefore a 1 in a cell can be grouped with V as well as V'. This may be clarified by considering the Karnaugh map of Figure 4.47. The 1 at cell A'B' may be represented as $C + C'$ (here the map-entered variable is C). The map can be reconstructed as in Figure 4.52(a). Now the 1 can be grouped with C as well as C', if they are placed at adjacent cells. This has been demonstrated in Figure 4.52(b). The minimal terms are obtained as B'C (where two Cs are grouped) and A'C' (for the group of two Cs).

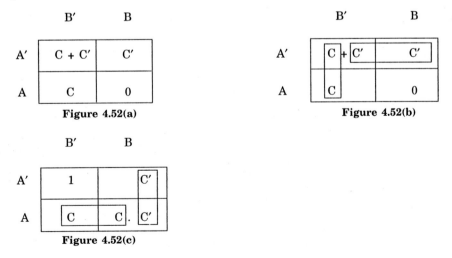

Figure 4.52(a)   Figure 4.52(b)

Figure 4.52(c)

Therefore, the final minimal sum of the products expression is

F (A,B,C) = B′C + A′C′.

The same technique may be applied to derive the minimal product of the sums expression, where groups or subgroups are formed with 0s as from Boolean algebra V.V′=0. This has been shown in Figure 4.52(c). Therefore, the product of the sums expression of function is

F′ (A,B,C) = (B + C′). (A + C).

This may be summarized to a two-step procedure for obtaining minimal sums for completely specified Boolean functions from a variable-entered map as stated below.

1. Consider each map entry having literal V and V′. Form the groups with the maximum number of elements involving the literal V using cells containing 1s as don't-care cells and V′ as 0-cells. Next form the groups involving the literal V′ using cells containing 1s as don't-care cells and the cells containing literal V as 0-cells.

2. After formation of groups involving V and V′, form the groups of cells containing 1s but not completely covered in step 1. One approach for doing this is to let all the cells containing the literals V and V′ become 0 and all 1s that were completely covered at step 1 become don't-care cells. Another way for the not completely covered 1-cells is to use V-cells and V′-cells from step 1 that ensures that all the 1-cells are completely covered.

Example 4.25. *Let us consider the Boolean function that has the Karnaugh map as per figure 4.53(a).*

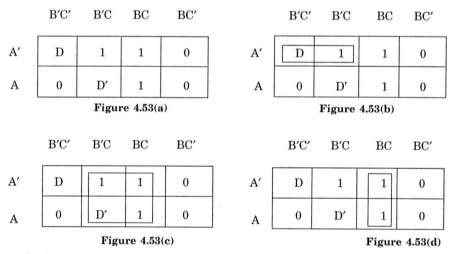

Figure 4.53(a)   Figure 4.53(b)

Figure 4.53(c)   Figure 4.53(d)

At the first step, the cell with D is grouped with adjacent 1 as in Figure 4.53(b). For this the minimal term is obtained as A′B′D. Next D′ is grouped with adjacent 1s as a quad, which is demonstrated in Figure 4.53(c). The minimal term is obtained as CD′. Now, note that the 1-cell at A′B′C is grouped with D as well as D′. Therefore, it can be stated that this 1-cell is completely covered. But 1s at A′BC and ABC are considered in the grouping of D′ only, so these 1-cells are not completely covered and they are grouped again separately as shown in Figure 4.53(d). The minimal term obtained is BC. So, considering all the minimal terms obtained, the final expression is derived as

$$F(A,B,C,D) = A'B'D + CD' + BC.$$

In the above example, the minimal terms are determined in three steps. This is done for illustration purposes and in practice minimal terms may be determined in a single-step only.

Example 4.26. *Find the sum of the product expression for the Boolean function whose variable-entered Karnaugh map is shown in Figure 4.54(a).*

	B'C'	B'C	BC	BC'
A'	D'	D'	D'	D'
A	D'	1	D	D'

Figure 4.54(a)

	B'C'	B'C	BC	BC'
A'	D'	D'	D'	D'
A	D'	1	D	D'

Figure 4.54(b)

Solution. The required sum of the product expression is

$$F (A,B,C,D) = A'D' + C'D' + B'D' + ACD.$$

The 1-cell has been grouped with D as well as D' to make it completely covered.

4.10.3 Variable-entered MAPS with Don't-care Conditions

So far variable-entered mapping has been discussed for the Boolean functions that are completely specified. However, incompletely specified Boolean functions *i.e.*, those having don't-care conditions, commonly occur in logic design process. It is possible to generalize the construction and reading of variable-entered maps with don't-care conditions.

Let us assume again that map entries in a variable-entered map correspond to single-variable functions. Previously it was shown that the map entries functions may be generalized as $F_i.V' + F_j.V$ where F_i and F_j are the functions F_0, F_1,....etc. F_i and F_j may have the values of 0, 1, or don't-care. A table in Figure 4.55 lists the nine possible assignments to F_i and F_j, the evaluation of $F_i.V' + F_j.V$ for each case, and the corresponding entries for a variable-entered map. The don't-cares are denoted by X in the expressions.

F_i	F_j	$F_i.V' + F_j.V$	Map entry
0	0	0.V'+0.V = 0+0 = 0	0
0	1	0.V'+1.V = 0+1 = 1	1
0	X	0.V'+X.V = 0+X.V = X.V	V, 0
1	0	1.V'+0.V = V'+0 = V'	V'
1	1	1.V'+1.V = V'+V = 1	1
1	X	1.V'+X.V = V'+X.V	V', 1
X	0	X.V'+0.V = X.V'+0 = X.V'	V', 0
X	1	X.V'+1.V = X.V'+V	V, 1
X	X	X.V'+X.V	X

Figure 4.55

It may be noted from the table in Figure 4.55 that there are some double entries in the map, which signifies that these cells can have both values. The first part of the double entry is referred to as the literal part and the second part is the constant part. The process of reading a variable-entered map with don't-care conditions is more complex, since the double entry cells in the map provide flexibility. Again, two-step process may be adopted for the formation of groups and subgroups to obtain the minimal terms.

1. Form an optimal collection or group for all entries that consist of only a single literal, *i.e.*, V or V', using 1s, Xs and double entries having a 1 constant part as don't-cares. In addition, double entries having a 0 constant part can be used as don't-cares for the formation of groups that agree with the literal part of that double entry.

2. Form a step 2 as follows,

   (*a*) Replace the single literal entries, *i.e.*, V and V' by 0.

   (*b*) Retain the single 0 and X entries.

   (*c*) Replace each single 1 entry with an X, if it was completely covered in step 1, otherwise retain a single 1 entry.

   (*d*) Replace the double entries having a 0 as a constant part, *i.e.*, V,0 and V',0 as 0.

   (*e*) Replace each double entry having a 1 as a constant part by X, if the cell was used in step 1 to form at least one group agreeing with the literal part, otherwise replace the double entry having a 1 constant part by a 1. (It should be noted that the second case corresponds to the cell not being covered at all or only used in association with the complement of the literal part of a double entry).

The resulting step 2 map only has 0, 1, and X entries. Minimal terms are to be obtained considering the X-cells, also. This may be illustrated by considering the practical example that follows.

Example 4.27. *Consider the Boolean function*

$$F (A,B,C,D) = \Sigma (3, 5, 6, 7, 8, 9, 10) + \Phi (4, 11, 12, 14, 15).$$

The corresponding truth table is presented in Figure 4.56. Variables A, B, and C are considered as map variables, and D is considered as a map-entered variable. In the truth table, columns are provided for the function $F_i .V' + F_j.V$ and map-entry values. The variable-entered Karnaugh map is shown in Figure 4.57(*a*). Figure 4.57(*b*) demonstrates how map entries are grouped at step 1. Here, one D entry is grouped with 1, X, and (D',1), as (D',1) may be interpreted as don't-care while grouping with D. Thus the minimal term is formed as CD.

At the next step, replace (D',1) at the cell position AB'C with 1 as this was once grouped with D. Replace (D,1) of cell position A'BC' with 1 and (D',0) of cell position ABC' with 0. The Karnaugh map is reconstructed with changes as shown in Figure 4.57(*c*). Minimal terms are obtained from this map as AB' and A'B.

So the final expression for the given Boolean function can be written as

$$F (A,B,C,D) = CD + AB' + A'B.$$

A	B	C	D	F	$F_i.V' + F_j.V$	Map entry
0	0	0	0	0	0+0	0
0	0	0	1	0		
0	0	1	0	0	0+D	D
0	0	1	1	1		
0	1	0	0	X	X.D'+D	D,1
0	1	0	1	1		
0	1	1	0	1	D'+D	1
0	1	1	1	1		
1	0	0	0	1	D'+D	1
1	0	0	1	1		
1	0	1	0	1	D'+X.D	D',1
1	0	1	1	X		
1	1	0	0	X	X.D'+0	D',0
1	1	0	1	0		
1	1	1	0	X	X.D'+X.D	X
1	1	1	1	X		

Figure 4.56

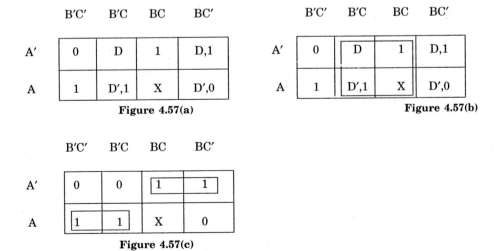

Figure 4.57(a)

Figure 4.57(b)

Figure 4.57(c)

Example 4.28. *Find the Boolean expression for the following function using variable-entered map technique.*

$$F\ (A,B,C,D) = \Sigma\ (0,\ 4,\ 5,\ 6,\ 13,\ 14,\ 15) + \Phi\ (2,\ 7,\ 8,\ 9)$$

Solution. Let A, B, and C be map variables and D be a map-entered variable. The truth table with map entry is prepared as in Figure 4.58(*a*). The Karnaugh map according to the truth table is formed in Figure 4.58(*b*). At first step, group formation consisting of single literal is done. The minimal terms thus obtained are A′D′ and BD. (D′, 1 may be considered as don't-care while grouped with D, as once it has been considered as D′.)

At the next step, the Karnaugh map is reconstructed in Figure 4.58(*c*). Here the single literal entries D and D′ are replaced by 0s as they are already used. (D′,0) at cell position A′B′C is replaced by 0. (D′,1) at cell position A′BC is replaced by X as this has been grouped with both D as well as D′ and similarly 1 at cell position A′BC replaced by X as this also has been grouped with D as well as D′. The minimal term obtained from this map is BC.

Therefore, the final expression of the given Boolean function may be derived as

$$F\ (A,B,C,D) = A′D′ + BD + BC.$$

A	B	C	D	F	$F_i .V′ + F_j V$	Map entry
0	0	0	0	1	D′+0	D′
0	0	0	1	0		
0	0	1	0	X	X.D′+0	D′,0
0	0	1	1	0		
0	1	0	0	1	D′+D	1
0	1	0	1	1		
0	1	1	0	1	D′+X.D	D′,1
0	1	1	1	X		
1	0	0	0	X	X.D′+X.D	X
1	0	0	1	X		
1	0	1	0	0	0+0	0
1	0	1	1	0		
1	1	0	0	0	0+D	D
1	1	0	1	1		
1	1	1	0	1	D′+D	1
1	1	1	1	1		

Figure 4.58(a)

	B'C'	B'C	BC	BC'
A'	D'	D',0	D',1	1
A	X	0	1	D

Figure 4.58(b)

	B'C'	B'C	BC	BC'
A'	0	0	X	X
A	X	0	1	0

Figure 4.58(c)

The above examples demonstrate how a single variable is used as a map-entered variable. However, more than one variable may used as map-entered variables. But these increase the complexities and difficulties in the simplification process.

4.11 CONCLUDING REMARKS

Two methods of Boolean function simplification are shown in this chapter. The criterion for simplification is to minimize the number of literals in the sum of products or product of sums forms. Both the map and the tabulation methods are restricted in their capabilities as they are useful for simplifying the Boolean functions expressed in the standard forms. Although this is a disadvantage, it is not very critical, as most of the applications prefer the standard forms over any other form. The gate implementation of standard expressions requires no more than two levels of gates. Expressions not in the standard forms are implemented with more than two levels.

It should be observed that the reflected code sequence chosen for the maps is not unique. It is possible to draw the maps assigning different code sequences to rows and columns keeping reflected code sequence in mind. This is already shown in this chapter. Also, the simplified expression for a function may not be unique, if pairs, quads, etc., are considered differently. The map method is preferable because of its simplicity when the number of variables is restricted to four, at the most five. For more than five variables, grouping of binary sequences leads to confusion and error.

The tabulation method has the distinct advantage at this point, as a step-by-step procedure is followed to minimize the literals. Moreover, this formal procedure is suitable for computer mechanization. But the tabulation process always starts with the minterm list of the function. If the function is not in this form, it is to be converted and the list of minterms is to be prepared.

In this chapter, we have considered the simplifications of functions with many input variables and a single output. However, some digital circuits have more than one output. Such circuits with multiple outputs may sometimes have common terms among the various functions which can be utilized to form common gates during the implementation. This results in further simplification which is not found in the simplification process if done separately. There exists an extension of the tabulation process for multiple-output functions. However, the method is too specialized and very tedious for human manipulation.

REVIEW QUESTIONS

4.1 What are the don't-care conditions?

4.2 Explain the terms (a) prime implicant, and (b) essential prime implicant.

4.3 What are the advantages of the tabulation method?

4.4 Draw a Karnaugh map for a four-variable Ex-OR function and derive its expression.

4.5 How does a Karnaugh map differ from a truth table?

4.6 What kind of network is developed by sum of the products?

4.7 Using a Karnaugh map, simplify the following functions and implement them with basic gates.

(a) F (A, B, C, D) = Σ (0, 2, 3, 6, 7, 8, 10, 11, 12, 15)

(b) F (A, B, C, D) = Σ (0, 2, 3, 5, 7, 8, 13) + d (1, 6, 12)

(c) F (A, B, C, D) = Σ (1, 7, 9, 10, 12, 13, 14, 15) + d (4, 5, 8)

(d) F (A, B, C, D) = π (0, 8, 10, 11, 14) + d (6)

(e) F (A, B, C, D) = π (2, 8, 11, 15) + d (3, 12, 14)

(f) F (W, X, Y, Z) = π (0, 2, 6, 11, 13, 15) + d (1, 9, 10, 14)

4.8 Prepare a Karnaugh map for the following functions.

(a) F = ABC + A'BC + B'C'

(b) F = A + B + C'

(c) Y = AB + B'CD

4.9 Using the Karnaugh map method, simplify the following functions, obtain their sum of the products form, and product of the sums form. Realize them with basic gates.

(a) F (W, X, Y, Z) = Σ (1, 3, 4, 5, 6, 7, 9, 12, 13)

(b) F (W, X, Y, Z) = Σ (1, 5, 6, 7, 11, 12, 13, 15)

4.10 Determine the don't-care conditions for the Boolean expression BE + B'DE', which is the simplified version of the expression A'BE + BCDE + BC'D'E + A'B'DE' + B'C'DE'.

4.11 Obtain the sum of the products expressions for the following functions and implement them with NAND gates as well as NOR gates.

(a) F = Σ (1, 4, 7, 8, 9, 11) + d (0, 3, 5)

(b) F = Σ (0, 2, 3, 5, 6, 7, 8, 9) + φ(10, 11, 12, 13, 14, 15)

4.12 A combinational switching network has four inputs A, B, C, and D, and one output Z. The output is to be 0, if the input combination is a valid Excess-3 coded decimal digit. If any other combinations of inputs appear, the output is to be 1. Implement the network using basic gates.

4.13 Design a circuit similar to problem 4.7 for BCD digits.

4.14 Using the Quine-McCluskey method obtain all the prime implicants, essential prime implicants, and minimized Boolean expression for the following functions.

(a) F (A, B, C, D, E) = Σ (4, 5, 6, 7, 9, 10, 14, 19, 26, 30, 31)

(b) F (A, B, C, D) = Σ (7, 9, 12, 13, 14, 15) + d (4, 11)

(c) F (A, B, C, D, E) = Σ (1, 3, 6, 10, 11, 12, 14, 15, 17, 19, 20, 22, 24, 29, 30)

(d) F (A, B, C, D) = Σ (4, 5, 8, 9, 12, 13) + d (0, 3, 7, 10, 11)

4.15 Determine the simplified expression for each of following functions using variable-entered maps where A, B, and C are map variables.

(a) F (A, B, C, D) = Σ (2, 3, 5, 12, 14) + d (0, 4, 8, 10, 11)

(b) F (A, B, C, D) = Σ (1, 5, 6, 7, 9, 11, 12, 13) + d (0, 3, 4)

(c) F (A, B, C, D) = Σ (1, 5, 7, 10, 11) + d (2, 3, 6, 13)

(d) F (A, B, C, D) = Σ (5, 6, 7, 12, 13, 14) + d (3, 8, 9)

Chapter **5** *COMBINATIONAL LOGIC CIRCUITS*

5.1 INTRODUCTION

The digital system consists of two types of circuits, namely

(*i*) Combinational circuits and

(*ii*) Sequential circuits

A combinational circuit consists of logic gates, where outputs are at any instant and are determined only by the present combination of inputs without regard to previous inputs or previous state of outputs. A combinational circuit performs a specific information-processing operation assigned logically by a set of Boolean functions. Sequential circuits contain logic gates as well as memory cells. Their outputs depend on the present inputs and also on the states of memory elements. Since the outputs of sequential circuits depend not only on the present inputs but also on past inputs, the circuit behavior must be specified by a time sequence of inputs and memory states. The sequential circuits will be discussed later in the chapter.

In the previous chapters we have discussed binary numbers, codes, Boolean algebra and simplification of Boolean function, logic gates, and economical gate implementation. Binary numbers and codes bear discrete quantities of information and the binary variables are the representation of electric voltages or some other signals. In this chapter, formulation and analysis of various systematic design of combinational circuits and application of information-processing hardware will be discussed.

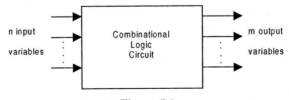

n input variables | Combinational Logic Circuit | m output variables

Figure 5.1

A combinational circuit consists of input variables, logic gates, and output variables. The logic gates accept signals from inputs and output signals are generated according to the logic

125

circuits employed in it. Binary information from the given data transforms to desired output data in this process. Both input and output are obviously the binary signals, *i.e.*, both the input and output signals are of two possible states, logic 1 and logic 0. Figure 5.1 shows a block diagram of a combinational logic circuit. There are n number of input variables coming from an electric source and m number of output signals go to an external destination. The source and/or destination may consist of memory elements or sequential logic circuit or shift registers, located either in the vicinity of the combinational logic circuit or in a remote external location. But the external circuit does not interfere in the behavior of the combinational circuit.

For n number of input variables to a combinational circuit, 2^n possible combinations of binary input states are possible. For each possible combination, there is one and only one possible output combination. A combinational logic circuit can be described by m Boolean functions and each output can be expressed in terms of n input variables.

5.2 DESIGN PROCEDURE

Any combinational circuit can be designed by the following steps of design procedure.

1. The problem is stated.
2. Identify the input variables and output functions.
3. The input and output variables are assigned letter symbols.
4. The truth table is prepared that completely defines the relationship between the input variables and output functions.
5. The simplified Boolean expression is obtained by any method of minimization—algebraic method, Karnaugh map method, or tabulation method.
6. A logic diagram is realized from the simplified expression using logic gates.

It is very important that the design problem or the verbal specifications be interpreted correctly to prepare the truth table. Sometimes the designer must use his intuition and experience to arrive at the correct interpretation. Word specification are very seldom exact and complete. Any wrong interpretation results in incorrect truth table and combinational circuit.

Varieties of simplification methods are available to derive the output Boolean functions from the truth table, such as the algebraic method, the Karnaugh map, and the tabulation method. However, one must consider different aspects, limitations, restrictions, and criteria for a particular design application to arrive at suitable algebraic expression. A practical design approach should consider constraints like—(1) minimum number of gates, (2) minimum number of outputs, (3) minimum propagation time of the signal through a circuit, (4) minimum number of interconnections, and (5) limitations of the driving capabilities of each logic gate. Since the importance of each constraint is dictated by the particular application, it is difficult to make all these criteria satisfied simultaneously, and also difficult to make a general statement on the process of achieving an acceptable simplification. However, in most cases, first the simplified Boolean expression at standard form is derived and then other constraints are taken care of as far as possible for a particular application.

5.3 ADDERS

Various information-processing jobs are carried out by digital computers. Arithmetic operations are among the basic functions of a digital computer. Addition of two binary digits is the most basic arithmetic operation. The simple addition consists of four possible elementary operations, which are 0+0 = 0, 0+1 = 1, 1+0 = 1, and 1+1 = 10. The first three operations produce a sum of one digit, but the fourth operation produces a sum consisting of two digits. The higher significant bit of this result is called the *carry*. A combinational circuit that performs the addition of two bits as described above is called a *half-adder*. When the augend and addend numbers contain more significant digits, the carry obtained from the addition of two bits is added to the next higher-order pair of significant bits. Here the addition operation involves three bits—the augend bit, addend bit, and the carry bit and produces a sum result as well as carry. The combinational circuit performing this type of addition operation is called a *full-adder*. In circuit development two half-adders can be employed to form a full-adder.

5.3.1 Design of Half-adders

As described above, a half-adder has two inputs and two outputs. Let the input variables augend and addend be designated as A and B, and output functions be designated as S for sum and C for carry. The truth table for the functions is below.

Input variables		Output variables	
A	*B*	*S*	*C*
0	0	0	0
0	1	1	0
1	0	1	0
1	1	0	1

Figure 5.2

From the truth table in Figure 5.2, it can be seen that the outputs S and C functions are similar to Exclusive-OR and AND functions respectively, as shown in Figure 3.5 in Chapter 3. The Boolean expressions are

$$S = A'B+AB' \quad \text{and}$$

$$C = AB.$$

Figure 5.3 shows the logic diagram to implement the half-adder circuit.

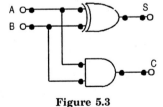

Figure 5.3

5.3.2 Design of Full-adders

A combinational circuit of full-adder performs the operation of addition of three bits—the augend, addend, and previous carry, and produces the outputs sum and carry. Let us designate the input variables augend as A, addend as B, and previous carry as X, and outputs sum as S and carry as C. As there are three input variables, eight different input combinations are possible. The truth table is shown in Figure 5.4 according to its functions.

Input variables			Outputs	
X	A	B	S	C
0	0	0	0	0
0	0	1	1	0
0	1	0	1	0
0	1	1	0	1
1	0	0	1	0
1	0	1	0	1
1	1	0	0	1
1	1	1	1	1

Figure 5.4

To derive the simplified Boolean expression from the truth table, the Karnaugh map method is adopted as in Figures 5.5(a)-(b).

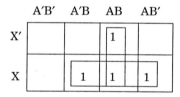

Figure 5.5(a) Map for function S. **Figure 5.5(b)** Map for function C.

The simplified Boolean expressions of the outputs are

$$S = X'A'B + X'AB' + XA'B' + XAB$$ and

$$C = AB + BX + AX.$$

The logic diagram for the above functions is shown in Figure 5.6. It is assumed complements of X, A, and B are available at the input source.

Note that one type of configuration of the combinational circuit diagram for full-adder is realized in Figure 5.6, with two-input and three-input AND gates, and three input and four-input OR gates. Other configurations can also be developed where number and type of gates are reduced. For this, the Boolean expressions of S and C are modified as followo.

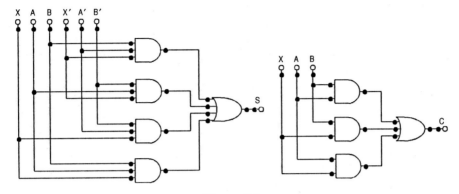

Figure 5.6

$$S = X'A'B + X'AB' + XA'B' + XAB$$
$$= X' (A'B + AB') + X (A'B' + AB)$$
$$= X' (A \oplus B) + X (A \oplus B)'$$
$$= X \oplus A \oplus B$$
$$C = AB + BX + AX = AB + X (A + B)$$
$$= AB + X (AB + AB' + AB + A'B)$$
$$= AB + X (AB + AB' + A'B)$$
$$= AB + XAB + X (AB' + A'B)$$
$$= AB + X (A \oplus B)$$

Logic diagram according to the modified expression is shown Figure 5.7.

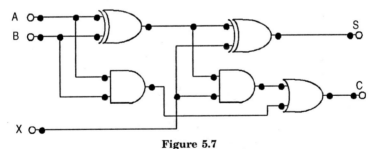

Figure 5.7

You may notice that the full-adder developed in Figure 5.7 consists of two 2-input AND gates, two 2-input XOR (Exclusive-OR) gates and one 2-input OR gate. This contains a reduced number of gates as well as type of gates as compared to Figure 5.6. Also, if compared with a half-adder circuit, the full-adder circuit can be formed with two half-adders and one OR gate.

5.4 SUBTRACTORS

Subtraction is the other basic function of arithmetic operations of information-processing tasks of digital computers. Similar to the addition function, subtraction of two binary digits consists of four possible elementary operations, which are 0–0 = 0, 0–1 = 1 with borrow of

1, 1–0 = 1, and 1–1 = 0. The first, third, and fourth operations produce a subtraction of one digit, but the second operation produces a difference bit as well as a *borrow* bit. The borrow bit is used for subtraction of the next higher significant bit. A combinational circuit that performs the subtraction of two bits as described above is called a *half-subtractor*. The digit from which another digit is subtracted is called the minuend and the digit which is to be subtracted is called the *subtrahend*. When the minuend and subtrahend numbers contain more significant digits, the borrow obtained from the subtraction of two bits is subtracted from the next higher-order pair of significant bits. Here the subtraction operation involves three bits—the minuend bit, subtrahend bit, and the borrow bit, and produces a different result as well as a borrow. The combinational circuit that performs this type of addition operation is called a *full-subtractor*. Similar to an adder circuit, a full-subtractor combinational circuit can be developed by using two half-subtractors.

5.4.1 Design of Half-subtractors

A half-subtractor has two inputs and two outputs. Let the input variables minuend and subtrahend be designated as X and Y respectively, and output functions be designated as D for difference and B for borrow. The truth table of the functions is as follows.

Input variables		Output variables	
X	Y	D	B
0	0	0	0
0	1	1	1
1	0	1	0
1	1	0	0

Figure 5.8

By considering the minterms of the truth table in Figure 5.8, the Boolean expressions of the outputs D and B functions can be written as

$$D = X'Y + XY' \qquad \text{and}$$
$$B = X'Y.$$

Figure 5.9 shows the logic diagram to realize the half-subtractor circuit.

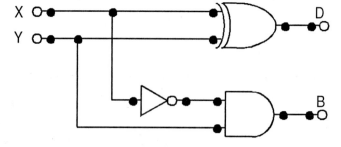

Figure 5.9

COMBINATIONAL LOGIC CIRCUITS 131

5.4.2 Design of Full-subtractors

A combinational circuit of full-subtractor performs the operation of subtraction of three bits—the minuend, subtrahend, and borrow generated from the subtraction operation of previous significant digits and produces the outputs difference and borrow. Let us designate the input variables minuend as X, subtrahend as Y, and previous borrow as Z, and outputs difference as D and borrow as B. Eight different input combinations are possible for three input variables. The truth table is shown in Figure 5.10(a) according to its functions.

Input variables			Outputs	
X	Y	Z	D	B
0	0	0	0	0
0	0	1	1	1
0	1	0	1	1
0	1	1	0	1
1	0	0	1	0
1	0	1	0	0
1	1	0	0	0
1	1	1	1	1

Figure 5.10(a)

Figure 5.10(b) Map for function D.   **Figure 5.10(c)** Map for function B.

Karnaugh maps are prepared to derive simplified Boolean expressions of D and B as in Figures 5.10(b) and 5.10(c), respectively.

The simplified Boolean expressions of the outputs are

$$S = X'Y'Z + X'YZ' + XY'Z' + XYZ$$ and

$$C = X'Z + X'Y + YZ.$$

The logic diagram for the above functions is shown in Figure 5.11.

Similar to a full-adder circuit, it should be noticed that the configuration of the combinational circuit diagram for full-subtractor as shown in Figure 5.11 contains two-input and three-input AND gates, and three-input and four-input OR gates. Other configurations can also be developed where number and type of gates are reduced. For this, the Boolean expressions of D and B are modified as follows.

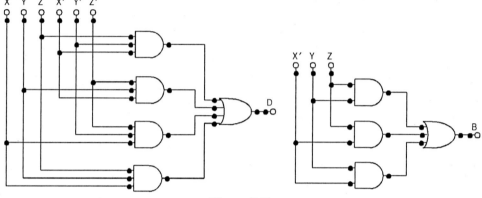

Figure 5.11

$$D = X'Y'Z + X'YZ' + XY'Z' + XYZ$$
$$= X'\ (Y'Z + YZ') + X\ (Y'Z' + YZ)$$
$$= X'\ (Y \oplus Z) + X\ (Y \oplus Z)'$$
$$= X \oplus Y \oplus Z$$

$$B = X'Z + X'Y + YZ\ = X'Y + Z\ (X' + Y)$$
$$= X'Y + Z(X'Y + X'Y' + XY + X'Y)$$
$$= X'Y + Z(X'Y + X'Y' + XY)$$
$$= X'Y + X'YZ + Z(X'Y' + XY)$$
$$= X'Y + Z(X \oplus Y)'$$

Logic diagram according to the modified expression is shown in Figure 5.12.

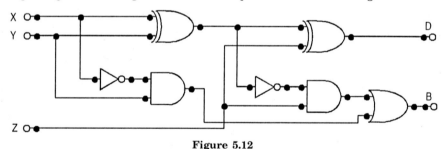

Figure 5.12

Note that the full-subtractor developed in Figure 5.12 consists of two 2-input AND gates, two 2-input XOR (Exclusive-OR) gates, two INVERTER gates, and one 2-input OR gate. This contains a reduced number of gates as well as type of gates as compared to Figure 5.12. Also, it may be observed, if compared with a half-subtractor circuit, the full-subtractor circuit can be developed with two half-subtractors and one OR gate.

5.5 CODE CONVERSION

We have seen in Chapter 2 that a large variety of codes are available for the same discrete elements of information, which results in the use of different codes for different digital systems. It is sometimes necessary to interface two digital blocks of different coding systems. A conversion circuit must be inserted between two such digital systems to use information of one digital system to other. Therefore, a code converter circuit makes two systems compatible when two systems use different binary codes.

To convert from one binary code A to binary code B, the input lines must provide the bit combination of elements as specified by A and the output lines must generate the corresponding bit combinations of code B. A combinational circuit consisting of logic gates performs this transformation operation. Some specific examples of code conversion techniques are illustrated in this chapter.

5.5.1 Binary-to-gray Converter

The bit combinations 4-bit binary code and its equivalent bit combinations of gray code are listed in the table in Figure 5.13. The four bits of binary numbers are designated as A, B, C, and D, and gray code bits are designated as W, X, Y, and Z. For transformation of binary numbers to gray, A, B, C, and D are considered as inputs and W, X, Y, and Z are considered as outputs. The Karnaugh maps are shown in Figures 5.14(a)-(d).

Binary				Gray			
A	B	C	D	W	X	Y	Z
0	0	0	0	0	0	0	0
0	0	0	1	0	0	0	1
0	0	1	0	0	0	1	1
0	0	1	1	0	0	1	0
0	1	0	0	0	1	1	0
0	1	0	1	0	1	1	1
0	1	1	0	0	1	0	1
0	1	1	1	0	1	0	0
1	0	0	0	1	1	0	0
1	0	0	1	1	1	0	1
1	0	1	0	1	1	1	1
1	0	1	1	1	1	1	0
1	1	0	0	1	0	1	0
1	1	0	1	1	0	1	1
1	1	1	0	1	0	0	1
1	1	1	1	1	0	0	0

Figure 5.13

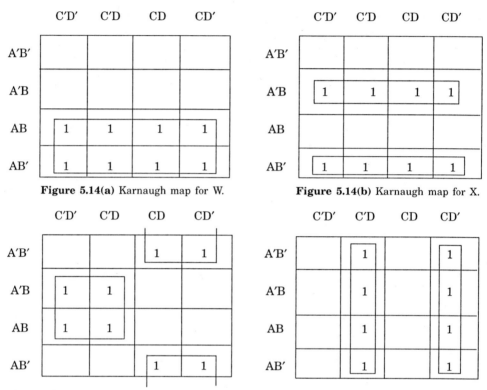

Figure 5.14(a) Karnaugh map for W.

Figure 5.14(b) Karnaugh map for X.

Figure 5.14(c) Karnaugh map for Y.

Figure 5.14(d) Karnaugh map for Z.

From the Karnaugh maps of Figure 5.14, we get

$$W = A, \qquad\qquad X = A'B + AB' = A \oplus B,$$
$$Y = BC' + B'C = B \oplus C, \quad \text{and} \quad Z = C'D + CD' = C \oplus D.$$

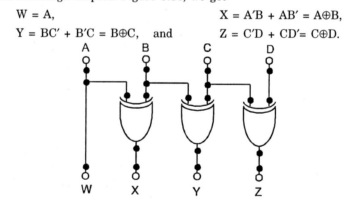

Figure 5.15

Figure 5.15 demonstrates the circuit diagram with logic gates.

5.5.2 Gray-to-binary Converter

Using the same conversion table as in Figure 5.13, the Karnaugh maps are formed in Figures 5.16(a)-(d). Here the inputs are considered as W, X, Y, and Z, whereas, outputs are A, B, C, and D.

	Y'Z'	Y'Z	YZ	YZ'
W'X'				
W'X				
WX	1	1	1	1
W'X'	1	1	1	1

Figure 5.16(a) Karnaugh map for A.

	Y'Z'	Y'Z	YZ	YZ'
W'X'				
W'X	1	1	1	1
WX				
WX'	1	1	1	1

Figure 5.16(b) Karnaugh map for B.

	Y'Z'	Y'Z	YZ	YZ'
W'X'			1	1
W'X	1	1		
WX			1	1
WX'	1	1		

Figure 5.16(c) Karnaugh map for C.

	Y'Z'	Y'Z	YZ	YZ'
W'X'		1		1
W'X	1		1	
WX		1		1
WX'	1		1	

Figure 5.16(d) Karnaugh map for D.

The Boolean expressions from Figure 5.16 are,

$A = W$

$B = W'X + WX' = W \oplus X$

$C = W'X'Y + W'XY' + WXY + WX'Y'$

$\quad = W'(X'Y + XY') + W(XY + X'Y')$

$\quad = W'(X \oplus Y) + W(X \oplus Y)'$

$\quad = W \oplus X \oplus Y \qquad\qquad$ or, $\qquad C = B \oplus Y$

$D = W'X'Y'Z + W'X'YZ' + W'XY'Z' + W'XYZ + WXY'Z + WXYZ' + WX'Y'Z' + WX'YZ$

$\quad = W'X'(Y'Z + YZ') + W'X(Y'Z' + YZ) + WX(Y'Z + YZ') + WX'(Y'Z' + YZ)$

$\quad = W'X'(Y \oplus Z) + W'X(Y \oplus Z)' + WX(Y \oplus Z) + WX'(Y \oplus Z)'$

$\quad = (W'X + WX')(Y \oplus Z)' + (W'X' + WX)(Y \oplus Z)$

$\quad = (W \oplus X)(Y \oplus Z)' + (W \oplus X)'(Y \oplus Z)$

$\quad = W \oplus X \oplus Y \oplus Z \qquad\qquad$ or, $\qquad D = C \oplus Z.$

From the Boolean expressions above, the circuit diagram of a gray-to-binary code converter is shown in Figure 5.17.

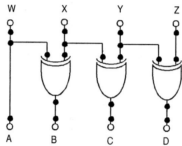

Figure 5.17

It may be noticed that a binary-to-gray converter and a gray-to-binary converter as illustrated above are four bits. However, these codes are not limited to four bits only. By similar process both the binary-to-gray and gray-to-binary code converter can be developed for a higher number of bits.

5.5.3 BCD-to-excess-3 Code Converter

The bit combinations of both the BCD (Binary Coded Decimal) and Excess-3 codes represent decimal digits from 0 to 9. Therefore each of the code systems contains four bits and so there must be four input variables and four output variables. Figure 5.18 provides the list of the bit combinations or truth table and equivalent decimal values. The symbols A, B, C, and D are designated as the bits of the BCD system, and W, X, Y, and Z are designated as the bits of the Excess-3 code system. It may be noted that though 16 combinations are possible from four bits, both code systems use only 10 combinations. The rest of the bit combinations never occur and are treated as don't-care conditions.

Decimal	BCD code				Excess-3 code			
Equivalent	A	B	C	D	W	X	Y	Z
0	0	0	0	0	0	0	1	1
1	0	0	0	1	0	1	0	0
2	0	0	1	0	0	1	0	1
3	0	0	1	1	0	1	1	0
4	0	1	0	0	0	1	1	1
5	0	1	0	1	1	0	0	0
6	0	1	1	0	1	0	0	1
7	0	1	1	1	1	0	1	0
8	1	0	0	0	1	0	1	1
9	1	0	0	1	1	1	0	0

Figure 5.18

For the BCD-to-Excess-3 converter, A, B, C, and D are the input variables and W, X, Y, and Z are the output variables. Karnaugh maps are shown in Figures 5.19(a)-(d) to derive each of the output variables. The simplified Boolean expressions of W, X, Y, and Z are given below.

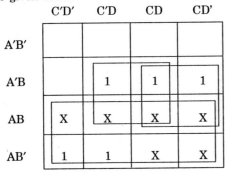

Figure 5.19(a) Karnaugh map for W.

Figure 5.19(b) Karnaugh map for X.

Figure 5.19(c) Karnaugh map for Y.

Figure 5.19(d) Karnaugh map for Z.

$$W = A + BC + BD$$
$$X = B'C + B'D + BC'D'$$
$$Y = CD + C'D'$$
$$Z = D'$$

According to the Boolean expression derived above, the logic diagram of a BCD-to-Excess-3 converter circuit is shown in Figure 5.20.

A good designer will always look forward to reduce the number and types of gates. It can be shown that reduction in the types and number of gates is possible to construct the BCD-to-Excess-3 code converter circuit if the above Boolean expressions are modified as follows.

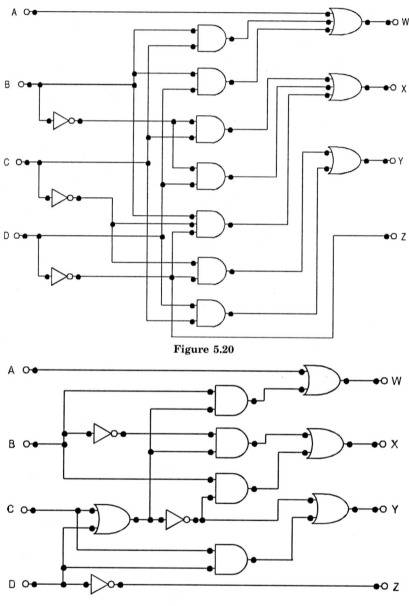

Figure 5.20

Figure 5.21

$$W = A + BC + BD = A + B(C + D)$$
$$X = B'C + B'D + BC'D' = B'(C + D) + BC'D' = B'(C + D) + B(C + D)'$$
$$Y = CD + C'D' = CD + (C + D)'$$
$$Z = D'$$

The BCD-to-Excess-3 converter circuit has been redrawn in Figure 5.21 according to the modified Boolean expressions above. Here, three-input AND gates and three-input OR gates are totally removed and the required number of gates has been reduced.

5.5.4 Excess-3-to-BCD Code Converter

To construct the Excess-3-to-BCD converter circuit, a similar truth table as in Figure 5.18 may be used. In this case, W, X, Y, and Z are considered as input variables and A, B, C, and D are termed as output variables. The required Karnaugh maps are prepared as per Figures 5.22(a)-(d).

	Y'Z'	Y'Z	YZ	YZ'
W'X'	X	X		X
W'X				
WX	1	X	X	X
WX'			1	

Figure 5.22(a) Karnaugh map for A.

	Y'Z'	Y'Z	YZ	YZ'
W'X'	X	X		X
W'X			1	
WX		X	X	X
WX'	1	1		1

Figure 5.22(b) Karnaugh map for B.

	Y'Z'	Y'Z	YZ	YZ'
W'X'	X	X		X
W'X		1		1
WX		X	X	X
WX'		1		1

Figure 5.22(c) Karnaugh map for C.

	Y'Z'	Y'Z	YZ	YZ'
W'X'	X	X		X
W'X	1			1
WX	1	X	X	X
WX'	1			X

Figure 5.22(d) Karnaugh map for D.

The Boolean expressions of the outputs are

$$A = WX + WYZ$$
$$B = X'Y' + X'Z' + XYZ$$
$$C = Y'Z + YZ'$$
$$D = Z'.$$

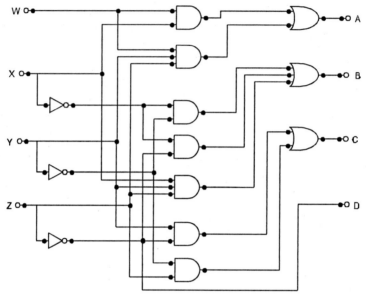

Figure 5.23

The logic diagram of an Excess-3-to-BCD converter is shown in Figure 5.23.

The alternative circuit diagram of Figure 5.24 can be made after the following modification on the above Boolean expressions.

$$A = WX + WYZ = W(X + YZ)$$
$$B = X'Y' + X'Z' + XYZ = X'(Y' + Z') + XYZ = X'(YZ)' + XYZ$$
$$C = Y'Z + YZ'$$
$$D = Z'$$

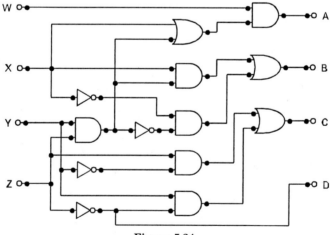

Figure 5.24

5.6 PARITY GENERATOR AND CHECKER

Parity is a very useful tool in information processing in digital computers to indicate any presence of error in bit information. External noise and loss of signal strength cause loss of data bit information while transporting data from one device to other device, located inside the computer or externally. To indicate any occurrence of error, an extra bit is included with the message according to the total number of 1s in a set of data, which is called *parity*. If the extra bit is considered 0 if the total number of 1s is even and 1 for odd quantities of 1s in a set of data, then it is called *even parity*. On the other hand, if the extra bit is 1 for even quantities of 1s and 0 for an odd number of 1s, then it is called *odd parity*.

5.6.1 Parity Generator

A parity generator is a combination logic system to generate the parity bit at the transmitting side. A table in Figure 5.25 illustrates even parity as well as odd parity for a message consisting of four bits.

Four bit Message $D_3D_2D_1D_0$	Even Parity (P_e)	Odd Parity (P_o)
0000	0	1
0001	1	0
0010	1	0
0011	0	1
0100	1	0
0101	0	1
0110	0	1
0111	1	0
1000	1	0
1001	0	1
1010	0	1
1011	1	0
1100	0	1
1101	1	0
1110	1	0
1111	0	1

Figure 5.25

If the message bit combination is designated as $D_3D_2D_1D_0$, and P_e , P_o are the even and odd parity respectively, then it is obvious from the table that the Boolean expressions of even parity and odd parity are

$$P_e = D_3 \oplus D_2 \oplus D_1 \oplus D_0 \quad \text{and}$$
$$P_o = (D_3 \oplus D_2 \oplus D_1 \oplus D_0)'.$$

These can be confirmed by Karnaugh maps, also (not shown here). The logic diagrams are shown in Figures 5.26(a)-(b).

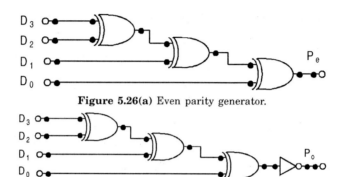

Figure 5.26(a) Even parity generator.

Figure 5.26(b) Odd parity generator.

The above illustration is given for a message with four bits of information. However, the logic diagrams can be expanded with more XOR gates for any number of bits.

5.6.2 Parity Checker

The message bits with the parity bit are transmitted to their destination, where they are applied to a parity checker circuit. The circuit that checks the parity at the receiver side is called the *parity checker*. The parity checker circuit produces a check bit and is very similar to the parity generator circuit. If the check bit is 1, then it is assumed that the received data is incorrect. The check bit will be 0 if the received data is correct.

4-bit message $D_3D_2D_1D_0$	Even Parity (P_e)	Even Parity Checker (C_e)
0000	0	0
0001	1	0
0010	1	0
0011	0	0
0100	1	0
0101	0	0
0110	0	0
0111	1	0
1000	1	0
1001	0	0
1010	0	0
1011	1	0
1100	0	0
1101	1	0
1110	1	0
1111	0	0

Figure 5.27(a) Even parity checker.

4-bit message $D_3D_2D_1D_0$	Odd Parity (P_o)	Odd Parity Checker (C_o)
0000	1	0
0001	0	0
0010	0	0
0011	1	0
0100	0	0
0101	1	0
0110	1	0
0111	0	0
1000	0	0
1001	1	0
1010	1	0
1011	0	0
1100	1	0
1101	0	0
1110	0	0
1111	1	0

Figure 5.27(b) Odd parity checker.

The tables in Figures 5.27(a)-(b) demonstrate the above. Note that the check bit is 0 for all the bit combinations of correct data. For incorrect data the parity check bit will be another logic value. Parity checker circuits are the same as parity generator circuits as shown in Figures 5.28(a)-(b).

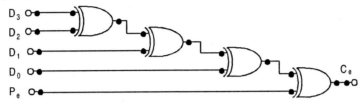

Figure 5.28(a) Even parity checker.

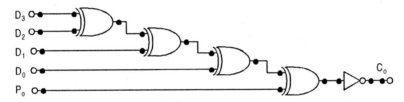

Figure 5.28(b) Odd parity checker.

5.7 SOME EXAMPLES OF COMBINATIONAL LOGIC CIRCUITS

Example 5.1. *Find the squares of 3-bit numbers.*

Solution. With three bits a maximum of eight combinations are possible with decimal equivalents of 0 to 7. By squaring of the decimal numbers the maximum decimal number produced is 49, which can be formed with six bits. Let us consider three input variables are X, Y, and Z, and six output variables are A, B, C, D, E, and F. A truth table is prepared as in Figure 5.29 and Karnaugh maps for each of the output variables are shown in Figures 5.30(a)-(f).

Input variables				Output variables						
Decimal Equivalent	X	Y	Z	Decimal Equivalent	A	B	C	D	E	F
0	0	0	0	0	0	0	0	0	0	0
1	0	0	1	1	0	0	0	0	0	1
2	0	1	0	4	0	0	0	1	0	0
3	0	1	1	9	0	0	1	0	0	1
4	1	0	0	16	0	1	0	0	0	0
5	1	0	1	25	0	1	1	0	0	1
6	1	1	0	36	1	0	0	1	0	0
7	1	1	1	49	1	1	0	0	0	1

Figure 5.29

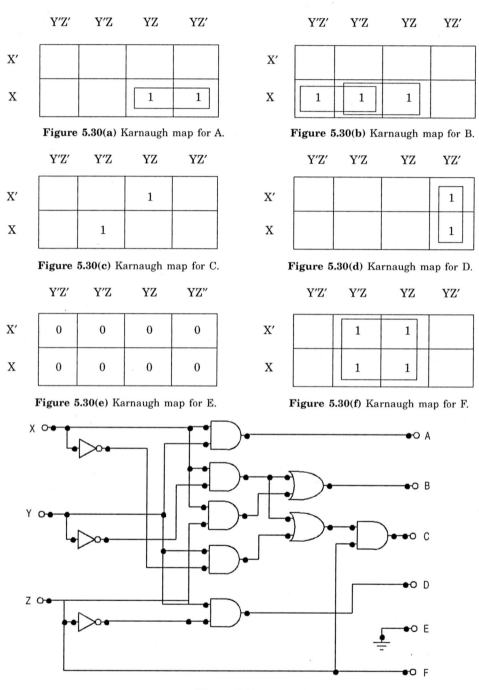

	Y'Z'	Y'Z	YZ	YZ'
X'				
X			1	1

Figure 5.30(a) Karnaugh map for A.

	Y'Z'	Y'Z	YZ	YZ'
X'				
X	1	1	1	

Figure 5.30(b) Karnaugh map for B.

	Y'Z'	Y'Z	YZ	YZ'
X'			1	
X		1		

Figure 5.30(c) Karnaugh map for C.

	Y'Z'	Y'Z	YZ	YZ'
X'				1
X				1

Figure 5.30(d) Karnaugh map for D.

	Y'Z'	Y'Z	YZ	YZ"
X'	0	0	0	0
X	0	0	0	0

Figure 5.30(e) Karnaugh map for E.

	Y'Z'	Y'Z	YZ	YZ'
X'		1	1	
X		1	1	

Figure 5.30(f) Karnaugh map for F.

Figure 5.31

The Boolean expressions of the output variables are

$A = XY$

$B = XY' + XZ$

$C = X'YZ + XY'Z = (X'Y + XY')Z$

$D = YZ'$

$E = 0$ and

$F = Z$.

The circuit diagram of the combinational network to obtain squares of three-bit numbers is shown in Figure 5.31.

Example 5.2. *Find the cubes of 3-bit numbers.*

Solution. Eight combinations are possible with 3-bit numbers and produce decimal equivalents of a maximum of 343 when cubes of them are calculated. These can be formed with nine bits. Let us consider the three input variables are X, Y, and Z, and the nine output variables are A, B, C, D, E, F, G, H, and I. A truth table is prepared as in Figure 5.32 and Karnaugh maps for each of the output variables are shown in Figures 5.34(a)-(i). The circuit diagram of this combinational network is shown in Figure 5.33.

The Boolean expressions of the output variables are

$A = XYZ$

$B = XYZ'$

$C = X$

$D = XY'Z$

$E = XY + YZ + XZ$

$F = X'Y + YZ' + XY'Z = (X' + Z')Y + XY'Z = (XZ)'Y + XZY'$

$G = XZ$

$H = YZ$

$I = Z$.

Input variables				Output variables									
Decimal Equivalents	X	Y	Z	Decimal Equivalents	A	B	C	D	E	F	G	H	I
0	0	0	0	0	0	0	0	0	0	0	0	0	0
1	0	0	1	1	0	0	0	0	0	0	0	0	1
2	0	1	0	8	0	0	0	0	0	1	0	0	0
3	0	1	1	27	0	0	0	0	1	1	0	1	1
4	1	0	0	64	0	0	1	0	0	0	0	0	0
5	1	0	1	125	0	0	1	1	1	1	1	0	1
6	1	1	0	216	0	1	1	0	1	1	0	0	0
7	1	1	1	343	1	0	1	0	1	0	1	1	1

Figure 5.32

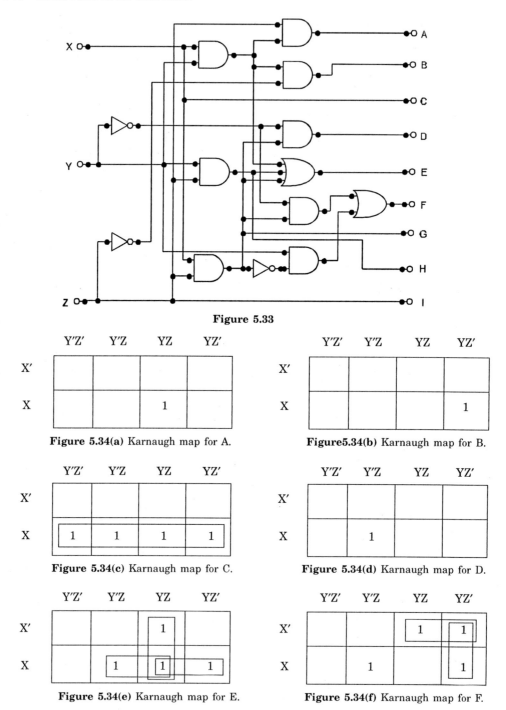

Figure 5.33

	Y′Z′	Y′Z	YZ	YZ′
X′				
X			1	

Figure 5.34(a) Karnaugh map for A.

	Y′Z′	Y′Z	YZ	YZ′
X′				
X				1

Figure5.34(b) Karnaugh map for B.

	Y′Z′	Y′Z	YZ	YZ′
X′				
X	1	1	1	1

Figure 5.34(c) Karnaugh map for C.

	Y′Z′	Y′Z	YZ	YZ′
X′				
X		1		

Figure 5.34(d) Karnaugh map for D.

	Y′Z′	Y′Z	YZ	YZ′
X′			1	
X		1	1	1

Figure 5.34(e) Karnaugh map for E.

	Y′Z′	Y′Z	YZ	YZ′
X′			1	1
X		1		1

Figure 5.34(f) Karnaugh map for F.

	Y'Z'	Y'Z	YZ	YZ'
X'				
X		1	1	

Figure 5.34(g) Karnaugh map for G.

	Y'Z'	Y'Z	YZ	YZ'
X'			1	
X			1	

Figure 5.34(h): Karnaugh map for H.

	Y'Z'	Y'Z	YZ	YZ'
X'		1	1	
X		1	1	

Figure 5.34(i) Karnaugh map for I.

Example 5.3. *Design a combinational circuit for converting 2421 code to BCD code.*

Solution. Both the 2421 code and BCD code are 4-bit codes and represent the decimal equivalents 0 to 9. To design the converter circuit for the above, first the truth table is prepared as in Figure 5.35 with the input variables W, X, Y, and Z of 2421 code, and the output variables A, B, C, and D. Karnaugh maps to obtain the simplified expressions of the output functions are shown in Figures 5.36(a)-(d). Unused combinations are considered as don't-care condition.

Decimal	Input varibles				Output variables			
Equivalent	2421 code				BCD code			
	W	X	Y	Z	A	B	C	D
0	0	0	0	0	0	0	0	0
1	0	0	0	1	0	0	0	1
2	0	0	1	0	0	0	1	0
3	0	0	1	1	0	0	1	1
4	0	1	0	0	0	1	0	0
5	1	0	1	1	0	1	0	1
6	1	1	0	0	0	1	1	0
7	1	1	0	1	0	1	1	1
8	1	1	1	0	1	0	0	0
9	1	1	1	1	1	0	0	1

Figure 5.35

	Y'Z'	Y'Z	YZ	YZ'
W'X'				
W'X		X	X	X
WX			1	1
WX'	X	X		X

Figure 5.36(a) Karnaugh map for A.

	Y'Z'	Y'Z	YZ	YZ'
W'X'				
W'X	1	X	X	X
WX	1	1		
WX'	X	X	1	X

Figure 5.36(b) Karnaugh map for B.

	Y'Z'	Y'Z	YZ	YZ'
W'X'			1	1
W'X		X	X	X
WX	1	1		
WX'	X	X		X

Figure 5.36(c) Karnaugh map for C.

	Y'Z'	Y'Z	YZ	YZ'
W'X'		1	1	
W'X		X	X	X
WX		1	1	
WX'	X	X	1	X

Figure 5.36(d) Karnaugh map for D.

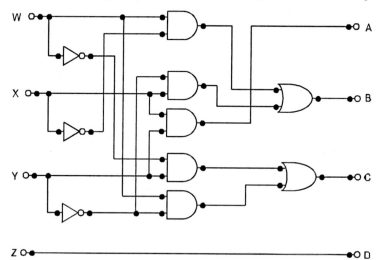

Figure 5.37

The Boolean expressions for the output functions are

$$A = XY \qquad\qquad B = XY'+WX'$$
$$C = W'Y + WY' \qquad D = Z.$$

The logic diagram of the required converter is shown in Figure 5.37.

Example 5.4. *Design a combinational circuit that converts 2421 code to 84-2-1 code, and also the converter circuit for 84-2-1 code to 2421 code.*

Solution. Both the codes represent binary codes for decimal digits 0 to 9. Let A, B, C, and D be represented as 2421 code variables and W, X, Y, and Z be variables for 84-2-1. The truth table is shown in Figure 5.38. The Karnaugh maps for W, X, Y, and Z in respect to A, B, C, and D are shown in Figure 5.39(a)-(d).

Decimal	2421 Code				84-2-1 code			
Digits	A	B	C	D	W	X	Y	Z
0	0	0	0	0	0	0	0	0
1	0	0	0	1	0	1	1	1
2	0	0	1	0	0	1	1	0
3	0	0	1	1	0	1	0	1
4	0	1	0	0	0	1	0	0
5	1	0	1	1	1	0	1	1
6	1	1	0	0	1	0	1	0
7	1	1	0	1	1	0	0	1
8	1	1	1	0	1	0	0	0
9	1	1	1	1	1	1	1	1

Figure 5.38

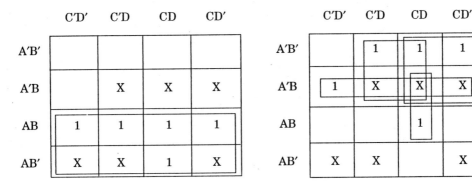

Figure 5.39(a) Karnaugh map for W.   **Figure 5.39(b)** Karnaugh map for X.

	C'D'	C'D	CD	CD'
A'B'		1		1
A'B		X	X	X
AB	1		1	
AB'	X	X	1	X

Figure 5.39(c) Karnaugh map for Y.

	C'D'	C'D	CD	CD'
A'B'		1	1	
A'B		X	X	X
AB		1	1	
AB'	X	X	1	X

Figure 5.39(d) Karnaugh map for Z.

The Boolean expressions for a 2421-to-84-2-1 code converter are

W = A

X = A'B + A'C + A'D + BCD = A'(B + C + D) + BCD

Y = AC'D' + ACD + A'C'D + A'CD'

Z = D.

The circuit diagram for a 2421-to-84-2-1 code converter is shown in Figure 5.40.

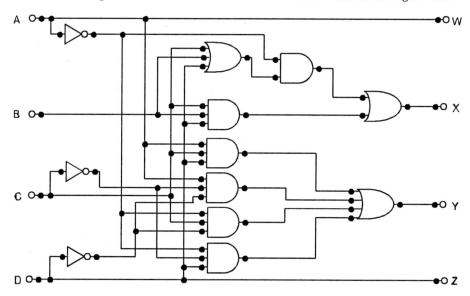

Figure 5.40

To design the 84-2-1-to-2421 code converter, the Karnaugh maps for the variables A, B, C, and D in respect to W, X, Y, and Z are shown in Figures 5.41(a)-(d).

	Y'Z'	Y'Z	YZ	YZ'
W'X'		X	X	X
W'X				
WX	X	X	1	X
WX'	1	1	1	1

Figure 5.41(a) Karnaugh map for A.

	Y'Z'	Y'Z	YZ	YZ'
W'X'		X	X	X
W'X	1			
WX	X	X	1	X
WX'	1	1		1

Figure 5.41(b) Karnaugh map for B.

	Y'Z'	Y'Z	YZ	YZ'
W'X'		X	X	X
W'X		1		1
WX	X	X	1	X
WX'	1		1	

Figure 5.41(c) Karnaugh map for C.

	Y'Z'	Y'Z	YZ	YZ'
W'X'		X	X	X
W'X		1	1	
WX	X	X	1	X
WX'		1	1	

Figure 5.41(d) Karnaugh map for D.

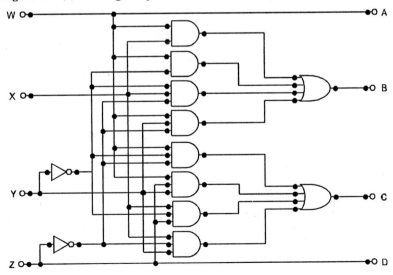

Figure 5.42

The Boolean expressions for an 84-2-1-to-2421 code converter are

A = W

B = WX + WY′ + XY′Z′ + WYZ′

C = WY′Z′ + WYZ + XY′Z + XYZ′

D = Z.

The combinational circuit for an 84-2-1-to-2421 code converter is shown in Figure 5.42.

Example 5.5. *Design a combinational circuit for a BCD-to-seven-segment decoder.*

Solution. Visual display is one of the most important parts of an electronic circuit. Often it is necessary to display the data in text form before the digits are displayed. Various types of display devices are commercially available. *Light Emitting Diode or LED* is one of the most widely used display devices and it is economical, low-power-consuming, and easily compatible in electronic circuits. They are available in various sizes, shapes, and colors. Here our concern is to display the decimal numbers 0 to 9 with the help of LEDs. Special display modules consisting of seven LEDs 'a, b, c, d, e, f, and g' of a certain shape and placed at a certain orientation as in Figure 5.43(a) are employed for this purpose. For its shape and as each of the LEDs can be controlled individually, this display is called the *seven segment display.*

Decimal digits 0 to 9 can be displayed by glowing some particular LED segments. As an example, digit '0' may be represented by glowing the segments a, b, c, d, e, and f as in Figure 5.43(b). Digit '1' may be represented by glowing b and c as in Figure 5.43(c). Other digits are also displayed by glowing certain segments as illustrated in Figures 5.43(d) to 5.43(k). In the figures, thick segments represent the glowing LEDs.

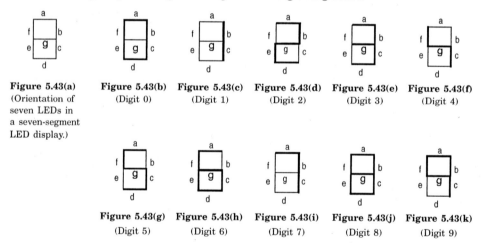

Figure 5.43(a)	Figure 5.43(b)	Figure 5.43(c)	Figure 5.43(d)	Figure 5.43(e)	Figure 5.43(f)
(Orientation of seven LEDs in a seven-segment LED display.)	(Digit 0)	(Digit 1)	(Digit 2)	(Digit 3)	(Digit 4)

Figure 5.43(g)	Figure 5.43(h)	Figure 5.43(i)	Figure 5.43(j)	Figure 5.43(k)
(Digit 5)	(Digit 6)	(Digit 7)	(Digit 8)	(Digit 9)

Two types of seven-segment display modules are available—*common cathode* type and *common anode* type, the equivalent electronic circuits are shown in Figures 5.44(a) and 5.44(b). From the equivalent circuit, it is clear that to glow a particular LED of common cathode type, logic 1 is to be applied at the anode of that LED as all the cathodes are grounded. Alternatively, logic 0 is to be applied to glow certain LEDs of common anode type, as all the anodes are connected to high-voltage Vcc.

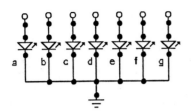

Figure 5.44(a) Common cathode LED.

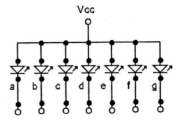

Figure 5.44(b) Common anode LED.

Decimal	Input Variables				Output Variables as Seven Segment Display						
Numbers	A	B	C	D	a	b	c	d	e	f	g
0	0	0	0	0	1	1	1	1	1	1	0
1	0	0	0	1	0	1	1	0	0	0	0
2	0	0	1	0	1	1	0	1	1	0	1
3	0	0	1	1	1	1	1	1	0	0	1
4	0	1	0	0	0	1	1	0	0	1	1
5	0	1	0	1	1	0	1	1	0	1	1
6	0	1	1	0	0	0	1	1	1	1	1
7	0	1	1	1	1	1	1	0	0	0	0
8	1	0	0	0	1	1	1	1	1	1	1
9	1	0	0	1	1	1	1	0	0	1	1

Figure 5.45 (For a common cathode display.)

Every decimal digit of 0 to 9 is represented by the BCD data, consisting of four input variables A, B, C, and D. A truth table can be made for each of the LED segments. A truth table for a common cathode display is shown in Figure 5.45. The Boolean expression for output variables a to g are obtained with the help of the Karnaugh maps as shown in Figures 5.46(a) to 5.46(g). The circuit diagram is developed as shown in Figure 5.47.

Note that the Boolean expressions of the outputs of a common anode type display are the complemented form of the respective outputs of a common cathode type.

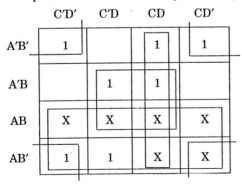

Figure 5.46(a) Karnaugh map for a.

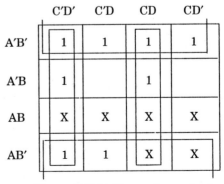

Figure 5.46(b) Karnaugh map for b.

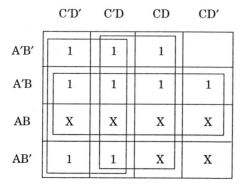

Figure 5.46(c) Karnaugh map for *c*.

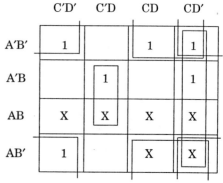

Figure 5.46(d) Karnaugh map for *d*.

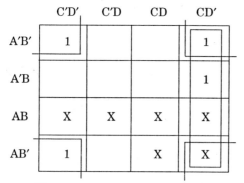

Figure 5.46(e) Karnaugh map for *e*.

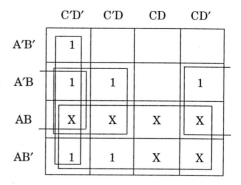

Figure 5.46(f) Karnaugh map for *f*.

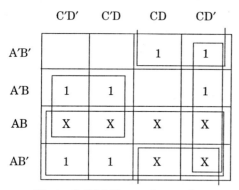

Figure 5.46(g) Karnaugh map for *g*.

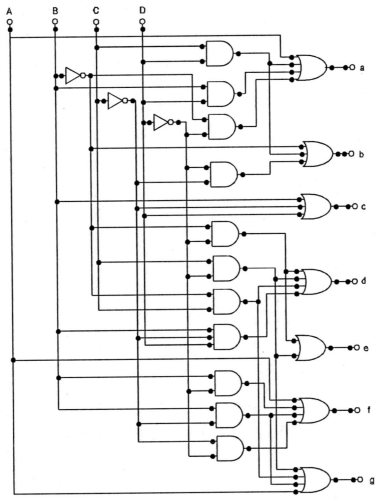

Figure 5.47

The Boolean expressions for a to g are given as

$$a = A + CD + BD + B'D'$$
$$b = B' + C'D' + CD$$
$$c = B + C' + D$$
$$d = B'D' + CD' + B'C + BC'D$$
$$e = B'D' + CD'$$
$$f = A + C'D' + BC' + BD'$$
$$g = A + BC' + CD' + B'C.$$

The BCD-to-seven-segment decoders are commercially available in a single IC package.

5.8 COMBINATIONAL LOGIC WITH MSI AND LSI

The purpose of simplification of Boolean functions is to obtain an algebraic expression with less number of literals and less numbers of logic gates. This results in low-cost circuit implementation. The design procedure for combinational circuits as described in the preceding sections is intended to minimize the number of logic gates to implement a given function. This classical procedure realizes the logic circuit with fewer gates with the assumption that the circuit with fewer gates will cost less. However, in practical design, with the arrival of a variety of integrated circuits (IC), this concept is always true.

Since one single IC package contains several number of logic gates, it is economical to use as many of the gates from an already used package, even if the total number of gates is increased by doing so. Moreover, some of the interconnections among the gates in many ICs are internal to the chip and it is more economical to use such types of ICs to minimize the external interconnections or wirings among the IC pins as much as possible. A typical example of this is if the circuit diagrams of Figures 5.23 and 5.24 are considered. Both circuit diagrams perform the function of Excess-3-to-BCD code conversion and consist of 13 logic gates. However, the circuit of Figure 5.23 needs six ICs (one 3-input OR, one 3-input AND, two 2-input AND, one 2-input OR, and one INVERTER, since one 3-input OR IC package contains three gates, one 3-input AND IC contains three gates, one 2-input AND IC contains four gates, one 2-input OR IC contains four gates, and one INVERTER IC contains six gates), but the circuit diagram of Figure 5.24 requires four ICs (two 2-input AND IC, one 2-input OR IC, and one INVERTER). So obviously, logic implementation of Figure 5.24 is economical because of its fewer number of IC packages. So for design with integrated circuits, it is not the count of logic gates that reduces the cost, but the number and type of IC packages used and the number of interconnections required to implement certain functions.

Though the classical method constitutes a general procedure, is very easy to understand, and certain to produce a result, on numerous occasions it does not achieve the best possible combinational circuit for a given function. Moreover, the truth table and simplification procedure in this method become too cumbersome if the number of input variables is excessively large and the final circuit obtained may require a relatively large number of ICs and interconnecting wires. In many cases the alternative design approach can lead to a far better combinational circuit for a given function with comparison to the classical method. The alternate design approach depends on the particular application and the ingenuity as well as experience of the designer. To handle a practical design problem, it should always be investigated which method is more suitable and efficient.

Design approach of a combinational circuit is first to analysis and to find out whether the function is already available as an IC package. Numerous ICs are commercially available, some of which perform specific functions and are commonly employed in the design of digital computer system. If the required function is not exactly matched with any of the commercially available devices, a good designer will formulate a method to incorporate the ICs that are nearly suitable to the function.

A large number of integrated circuit packages are commercially available nowadays. They can be widely categorized into three groups—SSI or small scale integration where the number of logic gates is limited to ten in one IC package, MSI or medium scale integration where the number of logic gates is eleven to one hundred in one IC package, and LSI or large-scale integration containing more than one hundred gates in one package. Some of them are fabricated for specific functions. VLSI or very large scale integration IC packages are also introduced, which perform dedicated functions achieving high circuit space reduction and interconnection reduction.

5.9 FOUR-BIT BINARY PARALLEL ADDER

In the preceding section, we discussed how two binary bits can be added and the addition of two binary bits with a carry. In practical situations it is required to add two data each containing more than one bit. Two binary numbers each of n bits can be added by means of a full adder circuit. Consider the example that two 4-bit binary numbers $B_4B_3B_2B_1$ and $A_4A_3A_2A_1$ are to be added with a carry input C_1. This can be done by cascading four full adder circuits as shown in Figure 5.48. The least significant bits A_1, B_1, and C_1 are added to the produce sum output S_1 and carry output C_2. Carry output C_2 is then added to the next significant bits A_2 and B_2 producing sum output S_2 and carry output C_3. C_3 is then added to A_3 and B_3 and so on. Thus finally producing the four-bit sum output $S_4S_3S_2S_1$ and final carry output Cout. Such type of four-bit binary adder is commercially available in an IC package.

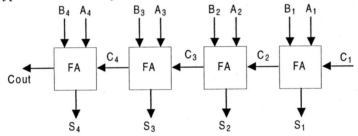

Figure 5.48

For the addition of two n bits of data, n numbers of full adders can be cascaded as demonstrated in Figure 5.48. It can be constructed with 4-bit, 2-bit, and 1-bit full adder IC packages. The carry output of one package must be connected to the carry input of the next higher order bit IC package of higher order bits.

The addition technique adopted here is a parallel type as all the bit addition operations are performed in parallel. Therefore, this type of adder is called a *parallel adder*. Serial types of adders are also available where a single full adder circuit can perform any n number of bit addition operations in association with shift registers and sequential logic network. This will be discussed in the later chapters.

The 4-bit parallel binary adder IC package is useful to develop combinational circuits. Some examples are demonstrated here.

Example 5.6. *Design a BCD-to-Excess-3 code converter.*

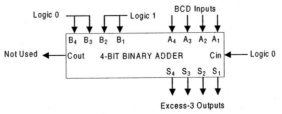

Figure 5.49

If we analyze the BCD code and Excess-3 code critically, you will see that Excess-3 code can be achieved by adding 0011 (decimal equivalent is 3) with BCD numbers. So a 4-bit binary adder IC can solve this very easily as shown in Figure 5.49.

It may be noticed that a BCD-to-Excess-3 converter has been implemented by classical method in Section 5.5.3, where four OR gates, four AND gates, and three INVERTER gates are employed. In terms of IC packages, three SSI packages (one AND gates IC, one OR gate IC, and one INVERTER IC) are used and a good amount of interconnections are present. In comparison to that the circuit developed in Figure 5.49 requires only one MSI IC of 4-bit binary adder and interconnections have reduced drastically. So the combinational circuit of Figure 5.49 is of low cost, trouble-free, less board, space consuming and less power dissipation.

5.9.1 Four-bit Binary Parallel Subtractor

It is interesting to note that a 4-bit binary adder can be employed to obtain the 4-bit binary subtraction. In Chapter 1, we saw how binary subtraction can be achieved using 1's complement or 2's complement. By 1's complement method, the bits of subtrahend are complemented and added to the minuend. If any carry is generated it is added to the sum output. Figure 5.50 demonstrates the subtraction of $B_4B_3B_2B_1$ from $A_4A_3A_2A_1$. Each bit of $B_4B_3B_2B_1$ is first complemented by using INVERTER gates and added to $A_4A_3A_2A_1$ by a 4-bit binary adder. End round carry is again added using the C in pin of the IC.

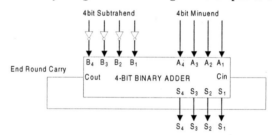

Figure 5.50

5.9.2 Four-bit Binary Parallel Adder/Subtractor

Due to the property of the 4-bit binary adder that it can perform the subtraction operation with external inverter gates, a single combinational circuit may be developed that can perform addition as well as the subtraction introducing a control bit. A little modification helps to obtain this dual operation. Figure 5.51 demonstrates this dual-purpose combinational logic circuit.

XOR gates are used at addend or subtrahend bits when one of the inputs of the XOR gate is connected to the ADD/SUBTRACT terminal, which is acting as control terminal. The same terminal is connected to Cin. When this terminal is connected to logic 0 the combinational circuit behaves like a 4-bit full adder, as at this instant Cin is logic low and XOR gates are acting as buffers whose outputs are an uncomplemented form of inputs. If logic 1 is applied to the ADD/SUBTRACT terminal, the XOR gates behave like INVERTER gates and data bits are complemented. The 4-bit adder now performs the addition operation of data $A_3A_2A_1A_0$ with complemented form of data $B_3B_2B_1B_0$ as well as with a single bit 1, as Cin is now logic 1. This operation is identical to a subtraction operation using 2's complment.

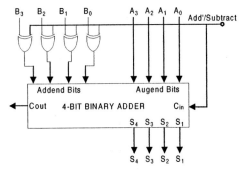

B_3 B_2 B_1 B_0 A_3 A_2 A_1 A_0 Add'/Subtract

Addend Bits Augend Bits

Cout 4-BIT BINARY ADDER C_{in}

S_4 S_3 S_2 S_1

S_4 S_3 S_2 S_1

Figure 5.51

5.9.3 Fast Adder

The addition of two binary numbers in parallel implies that all the bits of both augend and addend are available at the same time for computation. In any combinational network, the correct output is available only after the signal propagates through all the gates of its concern. Every logic gate offers some delay when the signal passes from its input to output, which is called the *propagation delay* of the logic gate. So every combinational circuit takes some time to produce its correct output after the arrival of all the input, which is called total propagation time and is equal to the propagation delay of individual gates times the number of gate levels in the circuit. In a 4-bit binary parallel adder, carry generated from the first full adder is added to the next full adder, carry generated form here is added to the next full adder and so on (refer to Figure 5.48). Therefore, the steady state of final carry is available after the signal propagating through four full adder stages and suffers the longest propagation delay with comparison to the sum outputs, as the sum outputs are produced after the signal propagation of only one full adder stage.

The number of gate levels for the carry propagation can be found from the circuit of full adder. The circuit shown in Figure 5.7 is redrawn in Figure 5.52 for convenience. The input and output variables use the subscript i to denote a typical stage in the parallel adder. In Figure 5.52, P_i and G_i represent the intermediate signals settling to their steady sate values after the propagation through the respective gates and common to all full adders and depends only on the input augend and addend bits. The signal from input carry C_i to output carry C_{i+1} propagates through two gate levels—an AND gate and an OR gate. Therefore, for a four-bit parallel adder, the final carry will be available after propagating through $2 \times 4 = 8$ gate levels. For an n-bit parallel adder there will be $2n$ number of gate levels to obtain the final carry to its steady state.

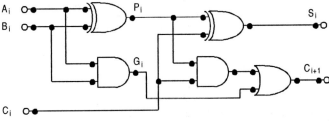

A_i P_i S_i

B_i

G_i C_{i+1}

C_i

Figure 5.52

Although any combinational network will always have some value at the output terminals, the outputs should not be considered correct unless the signals are given enough time to propagate through all the gates required for computation from input stage to output. For a 4-bit parallel binary adder, carry propagation plays an important role as it takes the longest propagation time. Since all other arithmetic operations are implemented by successive addition process, the time consumed during the addition process is very critical. One obvious method to reduce the propagation delay time is to use faster gates. But this is not always the practical solution because the physical circuits have a limit to their capability. Another technique is to employ a little more complex combinational circuit, which can reduce the carry propagation delay time. There are several techniques for the reduction of carry propagation delay time. However, the most widely used method employs the principle of *look ahead carry generation*, which is illustrated below.

5.9.4 Look-ahead Carry Generator

Consider the full adder circuit in Figure 5.52. Two intermediate variables are defined as Pi and Ci such that

$$P_i = A_i \oplus B_i \qquad \text{and} \qquad G_i = A_i B_i.$$

The output sum and carry can be expressed in terms of P_i and G_i as

$$S_i = P_i \oplus C_i \qquad \text{and} \qquad C_{i+1} = G_i + P_i C_i.$$

G_i is called the *carry generate* and it generates an output carry if both the inputs A_i and B_i are logic 1, regardless of the input carry. P_i is called the *carry propagate* because it is the term associated with the propagation of the carry from C_i to C_{i+1}.

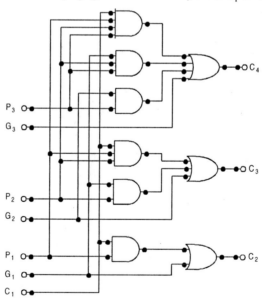

Figure 5.53

Now the Boolean expressions for the carry output of each stage can be written after substituting C_i and C_{i+1} as

$C_2 = G_1 + P_1C_1$

$C_3 = G_2 + P_2C_2 = G_2 + P_2(G_1 + P_1C_1) = G_2 + P_2G_1 + P_2P_1C_1$

$C_4 = G_3 + P_3C_3 = G_3 + P_3G_2 + P_3P_2G_1 + P_3P_2P_1C_1$

$C_5 = G_4 + P_4C_4 = G_4 + P_4G_3 + P_4P_3G_2 + P_4P_3P_2G_1 + P_4P_3P_2P_1C_1.$

Each of the above Boolean expressions are in sum of products form and each function can be implemented by one level of AND gates followed by one level of OR gates (or by two levels of NAND gates). So the final carry C_5 after 4-bit addition now has the propagation delay of only two level gates instead of eight levels as described earlier. In fact, all the intermediate carry as well as the final carry C_2, C_3, C_4, and C_5 can be implemented by only two levels of gates and available at the same time. The final carry C_5 need not have to wait for the intermediate carry to propagate. The three Boolean functions C_2, C_3, and C_4 are shown in Figure 5.53 which is called the *look ahead carry generator*.

The 4-bit parallel binary adder can be constructed with the association of a look-ahead carry generator as shown in Figure 5.54. P_i and G_i signals are generated with the help of XOR

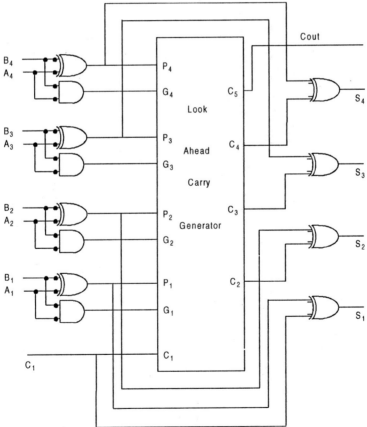

Figure 5.54

gates and AND gates, and sum outputs S_1 to S_4 are derived by using XOR gates. Thus, all sum outputs have equal propagation delay. Therefore, the 4-bit parallel binary adder realized with a look-ahead carry generator has reduced propagation delay and has a higher speed of operation.

5.9.5 Decimal Adder

Since computers and calculators perform arithmetic operations directly in the decimal number system, the arithmetic data employed in those devices must be in binary coded decimal form. The arithmetic circuit must accept data in coded decimal numbers and produce the outputs in the accepted code. For general binary addition, it is sufficient to consider two significant bits at a time and the previous carry.

But each decimal number of binary coded form consists of four bits. So the combinational network for addition of two decimal numbers involves at least nine input variables (two decimal numbers each of the four bits and a carry bit from the previous stage) and five output variables (four bits for the sum result and a carry bit).

There are a wide variety of combinational circuits for addition operations of decimal numbers depending on the code used. The design of nine-input five-output combinational circuits by classical method requires a truth table of $2^9 = 512$ entries. Many of the input conditions are don't-care conditions as binary code representing decimal numbers have nine valid combinations and six combinations are invalid. To obtain the simplified expression of each of the output is too lengthy and cumbersome by classical method. A computer-generated program for the tabulation method may be adopted, but that too will involve a lot of logic gates and interconnections. A 4-bit parallel binary adder may be employed for this purpose if illegal bit combinations are intelligently tackled.

5.9.5.1 BCD Adder

Consider the arithmetic addition of two decimal numbers in BCD (Binary Coded Decimal) form together with a possible carry bit from a previous stage. Since each input cannot exceed 9, the output sum must not exceed $9 + 9 + 1 = 19$ (1 in the sum is input carry from a previous stage). If a four-bit binary adder is used, the normal sum output will be of binary form and may exceed 9 or carry may be generated. So the sum output must be converted to BCD form. A truth table is shown in Figure 5.55 for the conversion of binary to BCD for numbers 0 to 19. Here, the sum outputs of a 4-bit binary adder are considered as $X_4 X_3 X_2 X_1$ with its carry output K and they are converted to BCD form $S_4 S_3 S_2 S_1$ with a final carry output C.

By examining the contents of the table, it may be observed that the output of the BCD form is identical to the binary sum when the binary sum is equal to or less than 1001 or 9, and therefore, no conversion is needed for these bit combinations. When the binary sum is greater than 1001, they are invalid data in respect to BCD form. The valid BCD form can be obtained with the addition of 0110 to the binary sum and also the required output carry is generated.

Decimal	Binary sum					BCD sum				
	K	X_4	X_3	X_2	X_1	C	S_4	S_3	S_2	S_1
0	0	0	0	0	0	0	0	0	0	0
1	0	0	0	0	1	0	0	0	0	1
2	0	0	0	1	0	0	0	0	1	0
3	0	0	0	1	1	0	0	0	1	1
4	0	0	1	0	0	0	0	1	0	0
5	0	0	1	0	1	0	0	1	0	1
6	0	0	1	1	0	0	0	1	1	0
7	0	0	1	1	1	0	0	1	1	1
8	0	1	0	0	0	0	1	0	0	0
9	0	1	0	0	1	0	1	0	0	1
10	0	1	0	1	0	1	0	0	0	0
11	0	1	0	1	1	1	0	0	0	1
12	0	1	1	0	0	1	0	0	1	0
13	0	1	1	0	1	1	0	0	1	1
14	0	1	1	1	0	1	0	1	0	0
15	0	1	1	1	1	1	0	1	0	1
16	1	0	0	0	0	1	0	1	1	0
17	1	0	0	0	1	1	0	1	1	1
18	1	0	0	1	0	1	1	0	0	0
19	1	0	0	1	1	1	1	0	0	1

Figure 5.55

A logic circuit is necessary to detect the illegal binary sum output and can be derived from the table entries. It is obvious that correction is needed when the binary sum produces an output carry K = 1, and for six illegal combinations from 1010 to 1111. Let us consider a logic function Y is generated when the illegal data is detected. A Karnaugh map is prepared for

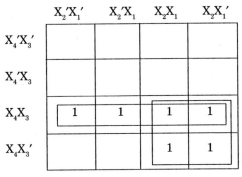

Figure 5.56

Y with the variables X_4, X_3, X_2, and X_1 in Figure 5.56. The output carry K is left aside as we know correction must be done when K = 1. The simplified Boolean expression for Y with variables X_4, X_3, X_2, and X_1 is

$$Y = X_4X_3 + X_4X_2.$$

As the detection logic is also 1 for K = 1, the final Boolean expression of Y taking the variable K into account will be

$$Y = K + X_4X_3 + X_4X_2.$$

The complete combinational circuit for a BCD adder network implemented with the help of a 4-bit binary adder is shown in Figure 5.57.

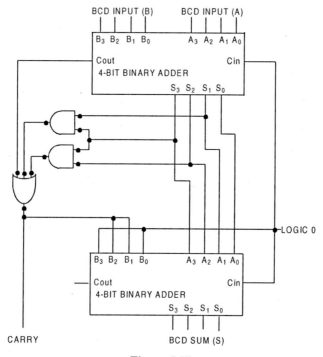

Figure 5.57

A BCD adder is a combinational circuit that adds two BCD numbers in parallel and produces a sum output also in BCD form. A BCD adder circuit must have the correction logic circuit in its internal construction. The correction logic is activated when the stage of binary sum is greater than 1001 and adds 0110 to the binary sum with the help of another binary adder. The output carry generated from the later stage of addition may be ignored as the final carry bit is already established.

The BCD adder circuit may be implemented by two 4-bit binary adder MSI ICs and one IC to generate the correction logic. However, a BCD adder is also available in an MSI package. To achieve shorter propagation delay, an MSI BCD adder includes the necessary look ahead carry generator circuit. The adder circuit for the correction logic does not need all four full adders and it is optimized within the IC package.

A decimal parallel adder of n decimal digits requires n numbers of BCD adder stages. The output carry from one stage must be connected to the input carry of the next higher order stage.

5.9.6 Parallel Multiplier

To understand the multiplication process, let us consider the multiplication of two 4-bit binary numbers, say 1101 and 1010.

```
        1 1 0 1  →  Multiplicand
      × 1 0 1 0  →  Multiplier
      ─────────────
        0 0 0 0
      1 1 0 1          Partial Products
    0 0 0 0
  1 1 0 1
      ─────────────
  1 0 0 0 0 0 1 0  →  Final Product
```

From the above multiplication process, one can easily understand that if the multiplier bit is 1, the multiplicand is simply copied as a partial product. If the multiplicand bit is 0, partial product is 0. Whenever a partial product is obtained, it is placed by shifting one bit left to the previous partial product. After obtaining all the partial products and placing them in the above manner, they are added to get the final product. The multiplication, as illustrated above, can be implemented by a 4-bit binary adder. Figure 5.58 demonstrates a 4-bit binary parallel multiplier using three 4-bit adders and sixteen 2-input AND gates. Here, each group of four AND gates is used to obtain partial products while 4-bit parallel adders are used to add the partial products.

The operation of the 4-bit parallel multiplier is explained in symbolic form of a binary multiplication process as follows.

			X_3	X_2	X_1	X_0		Multiplicand
			Y_3	Y_2	Y_1	Y_0		Multiplier
			$X_3 Y_0$	$X_2 Y_0$	$X_1 Y_0$	$X_0 Y_0$		Partial Product
		$X_3 Y_1$	$X_2 Y_1$	$X_1 Y_1$	$X_0 Y_1$			Partial Product
		C_2	C_1	C_0				
	C_3	S_3	S_2	S_1	S_0			Addition
	$X_3 Y_2$	$X_2 Y_2$	$X_1 Y_2$	$X_0 Y_2$				Partial Product
	C_6	C_5	C_4					
	C_7	S_7	S_6	S_5	S_4			Addition
	$X_3 Y_3$	$X_2 Y_3$	$X_1 Y_3$	$X_0 Y_3$				Partial Product
	C_{10}	C_9	C_8					
C_{11}	S_{11}	S_{10}	S_9	S_8				Addition
↓	↓	↓	↓	↓	↓	↓	↓	
M_7	M_6	M_5	M_4	M_3	M_2	M_1	M_0	Final Product

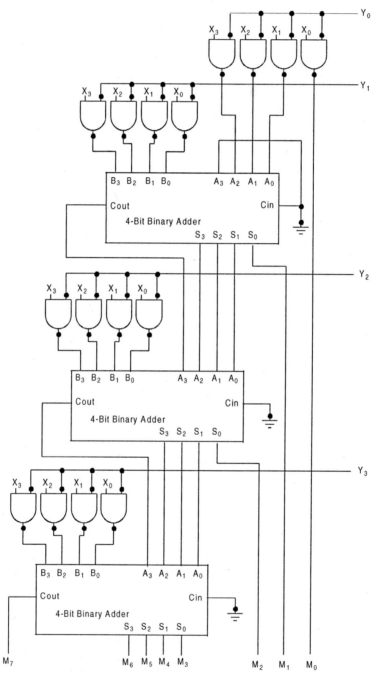

Figure 5.58

5.10 MAGNITUDE COMPARATOR

A *magnitude comparator* is one of the useful combinational logic networks and has wide applications. It compares two binary numbers and determines if one number is greater than, less than, or equal to the other number. It is a multiple output combinational logic circuit. If two binary numbers are considered as A and B, the magnitude comparator gives three outputs for A > B, A < B, and A = B.

For comparison of two *n*-bit numbers, the classical method to achieve the Boolean expressions requires a truth table of 2^{2n} entries and becomes too lengthy and cumbersome. It is also desired to have a digital circuit possessing with a certain amount of regularity, so that similar circuits can be applied for the comparison of any number of bits. Digital functions that follow an inherent well-defined regularity can usually be developed by means of algorithmic procedure if it exists. An *algorithm* is a process that follows a finite set of steps to arrive at the solution to a problem. A method is illustrated here by deriving an algorithm to design a 4-bit magnitude comparator.

The algorithm is the direct application of the procedure to compare the relative magnitudes of two binary numbers. Let us consider the two binary numbers A and B are expanded in terms of bits in descending order as

$$A = A_4 A_3 A_2 A_1$$
$$B = B_4 B_3 B_2 B_1,$$

where each subscripted letter represents one of the digits in the number. It is observed from the bit contents of the two numbers that A = B when $A_4 = B_4$, $A_3 = B_3$, $A_2 = B_2$, and $A_1 = B_1$. As the numbers are binary they possess the value of either 1 or 0, the equality relation of each pair can be expressed logically by the equivalence function as

$$X_i = A_i B_i + A_i' B_i' \qquad \text{for} \quad i = 1, 2, 3, 4.$$

Or, $X_i = (A \oplus B)'$.   Or, $X_i' = A \oplus B$.

Or, $X_i = (A_i B_i' + A_i' B_i)'$.

X_i is logic 1 when both A_i and B_i are equal *i.e.*, either 1 or 0 at the same instant. To satisfy the equality condition of two numbers A and B, it is necessary that all X_i must be equal to logic 1. This dictates the AND operation of all X_i variables. In other words, we can write the Boolean expression for two equal 4-bit numbers

$$F (A = B) = X_4 X_3 X_2 X_1.$$

To determine the relative magnitude of two numbers A and B, the relative magnitudes of pairs of significant bits are inspected from the most significant position. If the two digits of the most significant position are equal, the next significant pair of digits are compared. The comparison process is continued until a pair of unequal digits is found. It may be concluded that A>B, if the corresponding digit of A is 1 and B is 0. On the other hand, A<B if the corresponding digit of A is 0 and B is 1. Therefore, we can derive the logical expression of such sequential comparison by the following two Boolean functions,

$$F (A>B) = A_4 B_4' + X_4 A_3 B_3' + X_4 X_3 A_2 B_2' + X_4 X_3 X_2 A_1 B_1' \qquad \text{and}$$
$$F (A<B) = A_4' B_4 + X_4 A_3' B_3 + X_4 X_3 A_2' B_2 + X_4 X_3 X_2 A_1' B_1.$$

The logic gates implementation for the above expressions are not too complex as they contains many subexpressions of a repetitive nature and can be used at different places. The complete logic diagram of a 4-bit magnitude comparator is shown in Figure 5.59. This is a multilevel implementation and you may notice that the circuit maintains a regular pattern. Therefore, an expansion of binary magnitude comparator of higher bits can be easily obtained. This combinational circuit is also applicable to the comparison of BCD numbers.

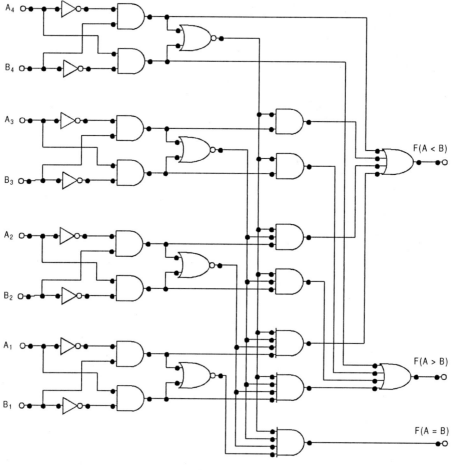

Figure 5.59

5.11 DECODERS

In a digital system, discrete quantities of information are represented with binary codes. A binary code of n bits can represent up to 2^n distinct elements of the coded information. A *decoder* is a combinational circuit that converts n bits of binary information of input lines to a maximum of 2^n unique output lines. Usually decoders are designated as an *n to m lines decoder*, where n is the number of input lines and m ($=2^n$) is the number of output lines. Decoders have a wide variety of applications in digital systems such as data demultiplexing, digital display, digital to analog converting, memory addressing, etc. A 3-to-8 line decoder is illustrated in Figure 5.60.

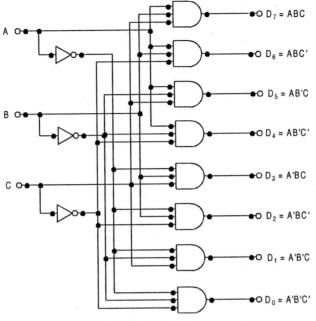

Figure 5.60

Input variables			Outputs							
A	B	C	D_0	D_1	D_2	D_3	D_4	D_5	D_6	D_7
0	0	0	1	0	0	0	0	0	0	0
0	0	1	0	1	0	0	0	0	0	0
0	1	0	0	0	1	0	0	0	0	0
0	1	1	0	0	0	1	0	0	0	0
1	0	0	0	0	0	0	1	0	0	0
1	0	1	0	0	0	0	0	1	0	0
1	1	0	0	0	0	0	0	0	1	0
1	1	1	0	0	0	0	0	0	0	1

Figure 5.61

The 3-to-8 line decoder consists of three input variables and eight output lines. Note that each of the output lines represents one of the minterms generated from three variables. The internal combinational circuit is realized with the help of INVERTER gates and AND gates.

The operation of the decoder circuit may be further illustrated from the input output relationship as given in the table in Figure 5.61. Note that the output variables are mutually exclusive to each other, as only one output is possible to be logic 1 at any one time.

In this section, the 3-to-8 line decoder is illustrated elaborately. However, higher order decoders like 4 to 16 lines, 5 to 32 lines, etc., are also available in MSI packages, where the internal circuits are similar to the 3-to-8 line decoder.

5.11.1 Some Applications of Decoders

As we have seen that decoders give multiple outputs equivalent to the minterms corresponding to the input variables, it is obvious that any Boolean expression in the sum of the products form can be very easily implemented with the help of decoders. It is not necessary to obtain the minimized expression through simplifying procedures like a Karnaugh map, or tabulation method, or any other procedure. It is sufficient to inspect the minterm contents of a function from the truth table, or the canonical form of sum of the products of a Boolean expression and selected minterms obtained from the output lines of a decoder may be simply OR-gated to derive the required function. The following examples will demonstrate this.

Example 5.7. *Implement the function F (A,B,C) = Σ (1,3,5,6).*

Solution. Since the above function has three input variables, a 3-to-8 line decoder may be employed. It is in the sum of the products of the minterms m_1, m_3, m_5, and m_6, and so decoder output D_1, D_3, D_5, and D_6 may be OR-gated to achieve the desired function. The combinational circuit of the above functions is shown in Figure 5.62.

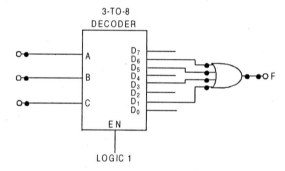

Figure 5.62

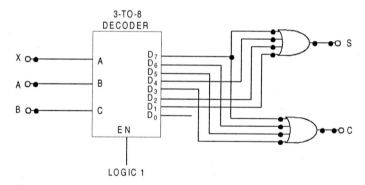

Figure 5.63

Example 5.8. *Design a full adder circuit with decoder IC.*

Solution. We have seen that full adder circuits are implemented with logic gates in Section 5.3.2. This can be very easily implemented with the help of a decoder IC. Observe the truth table of a full adder in Figure 5.4. In respect to minterms, the Boolean expression of sum output S and carry output C can be written as:

$$S = X'A'B + X'AB' + XA'B' + XAB \qquad \text{and}$$
$$C = X'AB + XA'B + XAB' + XAB.$$

The above expression can be realized in Figure 5.63.

Example 5.9. Similarly, a full-subtractor as described at Section 5.4.2 can be developed with the help of decoder. From the truth table in Figure 5.10 the Difference D and Borrow B outputs may be written as

$$D = X'Y'Z + X'YZ' + XY'Z' + XYZ \qquad \text{and}$$
$$B = X'Y'Z + X'YZ' + X'YZ + XYZ.$$

The combinational circuit with decoder is shown in Figure 5.64.

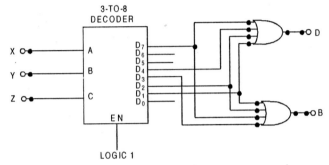

Figure 5.64

Example 5.10. *Design a BCD-to-decimal decoder with the use of a decoder.*

Solution. BCD code uses four bits to represent its different numbers from 0 to 9. So the decoder should have four input lines and ten output lines. By simple method a BCD-to-decimal coder may use a 4-to-16 line decoder. But at output, six lines are illegal and they are deactivated with the use of AND gates or any other means. However, a 3-to-8 line decoder may be employed for this purpose with its intelligent utilization. A partial truth table of a BCD-to-decimal decoder is shown in Figure 5.65.

Input variables				Output									
A	B	C	D	D_0	D_1	D_2	D_3	D_4	D_5	D_6	D_7	D_8	D_9
0	0	0	0	1	0	0	0	0	0	0	0	0	0
0	0	0	1	0	1	0	0	0	0	0	0	0	0
0	0	1	0	0	0	1	0	0	0	0	0	0	0
0	0	1	1	0	0	0	1	0	0	0	0	0	0
0	1	0	0	0	0	0	0	1	0	0	0	0	0
0	1	0	1	0	0	0	0	0	1	0	0	0	0
0	1	1	0	0	0	0	0	0	0	1	0	0	0
0	1	1	1	0	0	0	0	0	0	0	1	0	0
1	0	0	0	0	0	0	0	0	0	0	0	1	0
1	0	0	1	0	0	0	0	0	0	0	0	0	1

Figure 5.65

Since the circuit has ten outputs, ten Karnaugh maps are drawn to simplify each one of the outputs. However, it would be useful to construct a single map similar to a Karnaugh map indicating the outputs and don't-care conditions as in Figure 5.66. It can be seen that pairs and groups may be formed considering the don't-care conditions.

The Boolean expressions of the different outputs may be written as

$$D_0 = A'B'C'D', \quad D_1 = A'B'C'D, \quad\quad D_2 = B'CD',$$
$$D_3 = B'CD, \quad\quad D_4 = BC'D', \quad\quad\quad D_5 = BC'D,$$
$$D_6 = BCD', \quad\quad D_7 = BCD, \quad\quad\quad D_8 = AD'.$$

and $\quad\quad\quad\quad\quad\quad D_9 = AD.$

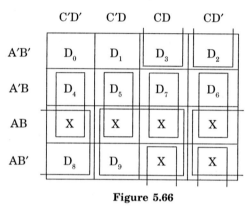

Figure 5.66

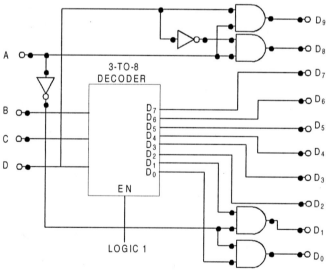

Figure 5.67

Figure 5.67 illustrates the complete circuit diagram of a BCD decoder implemented with a 3-to-8 decoder IC, with B, C, and D as input lines to the decoder.

Example 5.11. *Construct a 3-to-8 line decoder with the use of a 2-to-4 line decoder.*

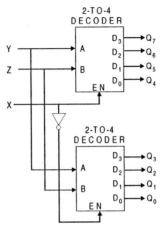

Figure 5.68

Solution. Lower order decoders can be cascaded to build higher order decoders. Normally every commercially available decoder ICs have a special input other than normal working input variables called ENABLE. The use of this ENABLE input is that when activated the complete IC comes to the working condition for its normal functioning. If ENABLE input is deactivated the IC goes to sleep mode, the normal functioning is suspended, and all the outputs become logic 0 irrespective of normal input variables conditions. This behavior of ENABLE input makes good use of a cascade connection as in Figure 5.69 where a 3-to-8 line decoder is demonstrated with a 2-to-4 line decoder. Here input variables are designated as X, Y, and Z, and outputs are denoted as Q_0 to Q_7. X input is connected to the ENABLE input of one decoder and X is used as an ENABLE input of another decoder. When X is logic 0, a lower decoder is activated and gives output Q_0 to Q_3 and an upper decoder is activated for X is logic 1, output Q_4 to Q_7 are available this time.

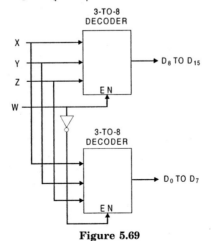

Figure 5.69

Example 5.12. *Construct a 4-to-16 line decoder using a 3-to-8 line decoder.*

Solution. A 4-to-16 line decoder has four input variables and sixteen outputs, whereas a 3-to-8 line decoder consists of three input variables and eight outputs. Therefore, one of the input variables is used as the ENABLE input as demonstrated in Example 5.11. Two 3-to-8 line decoders are employed to realize a 4-to-16 line decoder as shown in Figure 5.69. Input variables are designated as W, X, Y, and Z. W input is used as the ENABLE input of the upper 3-to-8 line decoder, which provides D_8 to D_{16} outputs depending on other input variables X, Y, and Z. W is also used as an ENABLE input at inverted mode to a lower decoder, which provides D_0 to D_7 outputs.

5.12 ENCODERS

An *encoder* is a combinational network that performs the reverse operation of the decoder. An encoder has 2^n or less numbers of inputs and n output lines. The output lines of an encoder generate the binary code for the 2^n input variables. Figure 5.70 illustrates an eight inputs/three outputs encoder. It may also be referred to as an octal-to-binary encoder where binary codes are generated at outputs according to the input conditions. The truth table is given in Figure 5.71.

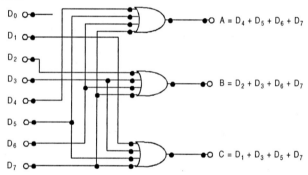

Figure 5.70

Inputs								Outputs		
D_0	D_1	D_2	D_3	D_4	D_5	D_6	D_7	A	B	C
1	0	0	0	0	0	0	0	0	0	0
0	1	0	0	0	0	0	0	0	0	1
0	0	1	0	0	0	0	0	0	1	0
0	0	0	1	0	0	0	0	0	1	1
0	0	0	0	1	0	0	0	1	0	0
0	0	0	0	0	1	0	0	1	0	1
0	0	0	0	0	0	1	0	1	1	0
0	0	0	0	0	0	0	1	1	1	1

Figure 5.71

The encoder in Figure 5.70 assumes that only one input line is activated to logic 1 at any particular time, otherwise the other circuit has no meaning. It may be noted that for eight inputs there are a possible $2^8 = 256$ combinations, but only eight input combinations are useful and the rest are don't-care conditions. It may also be noted that D_0 input is not connected to any of the gates. All the binary outputs A, B, and C must be all 0s in this case. All 0s output may also be obtained if all input variables D_0 to D_7 are logic 0. This is the main discrepancy of this circuit. This discrepancy can be eliminated by introducing another output indicating the fact that all the inputs are not logic 0.

However, this type of encoder is not available in an IC package because it is not easy to implement with OR gates and not much of the gates are used. The type of encoder available in IC package is called a *priority encoder*. These encoders establish an input priority to ensure that only highest priority input is encoded. As an example, if both D_2 and D_4 inputs are logic 1 simultaneously, then output will be according to D_4 only *i.e.*, output is 100.

5.13   MULTIPLEXERS OR DATA SELECTORS

A multiplexer is one of the important combinational circuits and has a wide range of applications. The term multiplex means "*many into one.*" Multiplexers transmit large numbers of information channels to a smaller number of channels. A *digital multiplexer* is a combinational circuit that selects binary information from one of the many input channels and transmits to a single output line. That is why the multiplxers are also called *data selectors*. The selection of the particular input channel is controlled by a set of select inputs. A digital multiplexer of 2^n input channels can be controlled by n numbers of select lines and an input line is selected according to the bit combinations of select lines.

Selection Inputs		Input Channels				Output
S_1	S_0	I_0	I_1	I_2	I_3	Y
0	0	0	X	X	X	0
0	0	1	X	X	X	1
0	1	X	0	X	X	0
0	1	X	1	X	X	1
1	0	X	X	0	X	0
1	0	X	X	1	X	1
1	1	X	X	X	0	0
1	1	X	X	X	1	1

Figure 5.72

A 4-to-1 line multiplexer is defined as the multiplexer consisting of four input channels and information of one of the channels can be selected and transmitted to an output line according to the select inputs combinations. Selection of one of the four input channels is possible by two selection inputs. Figure 5.72 illustrates the truth table. Input channels I_0, I_1, I_2, and I_3 are selected by the combinations of select inputs S_1 and S_0. The circuit diagram is shown in Figure 5.73. To demonstrate the operation, let us consider that select input combination S_1S_0 is 01. The AND gate associated with I_1 will have two of inputs equal to logic 1 and a third input is connected to I_1. Therefore, output of this AND gate is according to the information provided by channel I_1. The other three AND gates have logic 0 to at least one of their inputs which makes their outputs to logic 0. Hence, OR output (Y) is equal to the data provided by the channel I_1. Thus, information from I_1 is available at Y. Normally a multiplexer has an ENABLE input to also control its operation. The ENABLE input (also called STROBE) is useful to expand two or more multiplexer ICs to a digital multiplexer with a larger number of inputs, which will be demonstrated in a later part of this section. A multiplexer is often abbreviated as MUX. Its block diagram is shown in Figure 5.74.

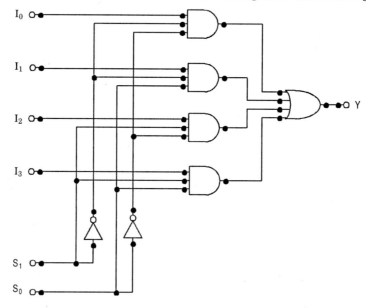

Figure 5.73

If the multiplexer circuit is inspected critically, it may be observed that the multiplexer circuit resembles the decoder circuit and indeed the n select lines are decoded to 2^n lines which are ANDed with the channel inputs. Figure 5.75 demonstrates how a decoder is employed to form a 4-to-1 multiplexer.

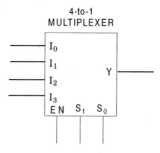

Figure 5.74

In some cases two or more multiplexers are accommodated within one IC package. The selection and ENABLE inputs in multiple-unit ICs may be common to all multiplexers.

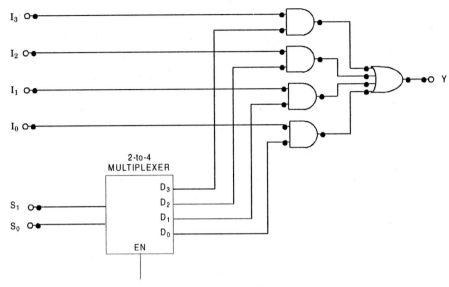

Figure 5.75

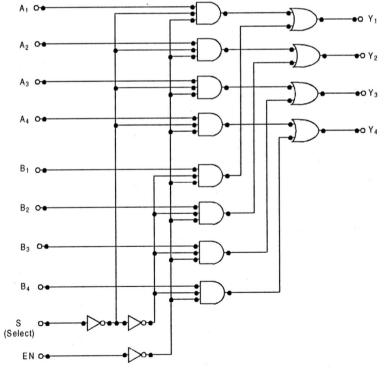

Figure 5.76

The internal circuit diagram of a quadruple 2-to-1 multiplexer IC is illustrated in Figure 5.76. It has four multiplexers, each capable of selecting one of two input lines. Either of the inputs A_1 or B_1 may be selected to provide output at Y_1. Similarly, Y_2 may have the value of A_2 or B_2 and so on. One input selection line S is sufficient to perform the selection operation of one of the two input lines in all four multiplexers. The control input EN enables the multiplexers for their normal function when it is at logic 0 state, and all the multiplexers suspend their functioning when EN is logic 1.

A function table is provided in Figure 5.77. When EN = 1, all the outputs are logic 0, irrespective of any data at inputs I_0, I_1, I_2, or I_4. When EN = 0, all the multiplexers become activated, outputs possess the A value if S = 0 and outputs are equal to data at B if S = 1.

E	S	Output Y
1	X	All 0's
0	0	Select A
0	1	Select B

Figure 5.77

5.13.1 Cascading of Multiplexers

As stated earlier, multiplexers of a larger number of inputs can be implemented by the multiplexers of a smaller number of input lines. Figure 5.78 illustrates that an 8-to-1 line multiplexer is realized by two 4-to-1 line multiplexers.

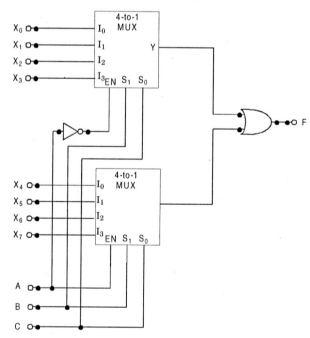

Figure 5.78

Here, variables B and C are applied to select inputs S_1 and S_0 of both multiplexers whereas the ENABLE input of the upper multiplexer is connected to A and the lower multiplexer is connected to A. So for A = 0, the upper multiplexer is selected and input lines X_0 to X_3 are selected according to the selected inputs and data is transmitted to an output through the OR gate. When A = 1, the lower multiplexer is activated and input lines X_4 to X_7 are selected according to the selected inputs.

Similarly, a 16-to-1 multiplexer may be developed by two 8-to-1 multiplexers as shown in Figure 5.79. Alternatively, a 16-to-1 multiplxer can be realized with five 4-to-1 multiplexers as shown in Figure 5.80.

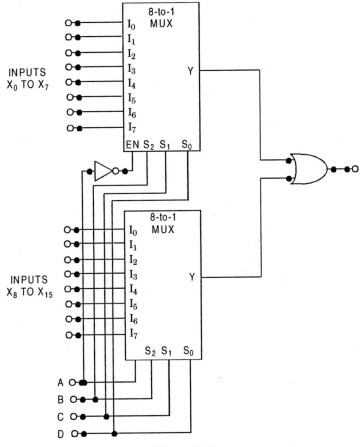

Figure 5.79

The multiplexer is a very useful MSI function and has various ranges of applications in data communication. Signal routing and data communication are the important applications of a multiplexer. It is used for connecting two or more sources to guide to a single destination among computer units and it is useful for constructing a common bus system. One of the general properties of a multiplexer is that Boolean functions can be implemented by this device, which will be demonstrated here.

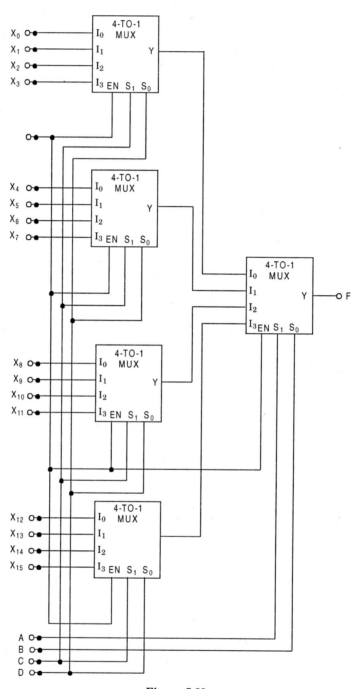

Figure 5.80

5.13.2 Boolean Function Implementation

In the previous section it was shown that decoders are employed to implement the Boolean functions by incorporating an external OR gate. It may be observed that multiplexers are constructed with decoders and OR gates. The selection of minterm outputs of the decoder can be controlled by the input lines. Hence, the minterms included in the Boolean function may be chosen by making their corresponding input lines to logic 1. The minterms not needed for the function are disabled by making their input lines equal to logic 0. By this method Boolean functions of n variables can be very easily implemented by using a 2^n-to-1 multiplexer. However, a better approach may be adopted with the judicious use of the function variables.

If a Boolean function consists of $n+1$ number of variables, n of these variables may be used as the select inputs of the multiplexer. The remaining single variable of the function is used as the input lines of the multiplexer. If X is the left-out variable, the input lines of the multiplexer may be chosen from four possible values, - X, X', logic 1, or logic 0. It is possible to implement any Boolean function with a multiplexer by intelligent assignment of the above values to input lines and other variables to selection lines. By this method a Boolean function of $n+1$ variables can be implemented by a 2^n-to-1 line multiplexer. Assignment of values to the input lines can be made through a typical procedure, which will be demonstrated by the following examples.

Example 5.13. *Implement the 3-variable function F(A,B,C) =(0,2,4,7) with a multiplexer.*

Solution. Here the function has three variables, A, B, and C and can be implemented by a 4-to-1 line multiplexer as shown in Figure 5.82. Figure 5.81 presents the truth table of the above Boolean function. Two of the variables, say B and C, are connected to the selection lines S_1 and S_0 respectively. When both B and C are 0, I_0 is selected. At this time, the output required is logic 1, as both the minterms m_0 (A'B'C') and m_4 (AB'C') produce output logic 1 regardless of the input variable A, so I_0 should be connected to logic 1. When select inputs BC=01, I_1 is selected and it should be connected to logic 0 as the corresponding minterms m_1 (A'B'C) and m_5 (AB'C) both produce output 0. For select inputs BC = 10, I_2 is selected and connected to variable A', as only one minterm m_2 (A'BC') associated with A' produce output logic 1, whereas the minterm m_6 (ABC') associated with A produces output 0. And finally, I_3 is selected and connected to variable A, when select inputs BC = 11, because only the minterm $m7$ (ABC) produce output 1, whereas output is 0 for the mintrem m_3 (A'BC). Multiplexer must be in ENABLE mode to be at its working condition. Hence EN input is connected to logic 1.

Minterms	A	B	C	F
0	0	0	0	1
1	0	0	1	0
2	0	1	0	1
3	0	1	1	0
4	1	0	0	1
5	1	0	1	0
6	1	1	0	0
7	1	1	1	1

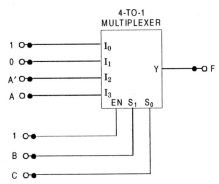

Figure 5.81   **Figure 5.82**

The above analysis describes how a Boolean function can be implemented with the help of multiplexers. However, there is a general procedure for the implementation of Boolean functions of n variables with a 2^{n-1}-to-1 multiplexer.

First, the function is expressed in its sum of the minterms form. Assume that the most significant variables will be used at input lines and the other $n-1$ variables will be connected to selection lines of the multiplexer in ordered sequence. This means the lowest significant variable is connected to S_0 input, the next higher significant variable is connected to S_1, the next higher variable to S_2, and so on. Now consider the single variable A. Since this variable represents the highest order position in the sequence of variables, it will be at complemented form in the minterms 0 to 2^{n-1}, which comprises the first half of the list of minterms. The variable A is at uncomplemented form in the second half of the list of the minterms. For a three-variable function like Example 5.13, among the possible eight minterms, A is complemented for the minterms 0 to 3 and at uncomplemented form for the minterms 4 to 7.

An implementation table is now formed, where the input designations of the multiplexer are listed in the first row. Under them the minterms where A is at complemented form are listed row-wise. At the next row other minterms of A at uncomplemented form are listed. Circle those minterms that produce output to logic 1.

If the two elements or minterms of a column are not circled, write 0 under that column.

If both the two elements or minterms of a column are circled, write 1 under that column.

If the upper element or minterm of a column is circled but not the bottom, write A' under that column.

If the lower element or minterm of a column is circled but not the upper one, write A under that column.

The lower most row now indicates input behavior of the corresponding input lines of the multiplexer as marked at the top of the column.

The above procedure can be more clearly understood if we consider Example 5.13 again. Since this function can be implemented by a multiplexer, the lower significant variables B and C are applied to S_1 and S_0 respectively. The inputs of multiplexer I_0 to I_3 are listed at the uppermost row. A' and its corresponding minterms 0 to 3 are placed at the next row. Variable A and the rest of the minterms 4 to 7 are placed next as in Figure 5.83. Now circle the minterms 0, 2, 4, and 7 as these minterms produce logic 1 output.

	I_0	I_1	I_2	I_3
A'	⓪	1	②	3
A	④	5	6	⑦
	1	0	A'	A

Figure 5.83

From Figure 5.83, it can seen that both the elements of the first column 0 and 4 are circled. Therefore, '1' is placed at the bottom of that column. At the second column no elements are circled and so '0' is placed at the bottom of the column. At the third column only '2' is circled. Its corresponding variable is A' and so A' is written at the bottom of this column. And finally, at the fourth column only '7' is circled and A is marked at the bottom of the column. The muxltiplexer inputs are now decided as $I_0 = 1$, $I_1 = 0$, $I_2 = A'$, and $I_3 = A$.

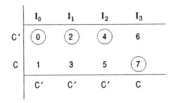

Figure 5.84

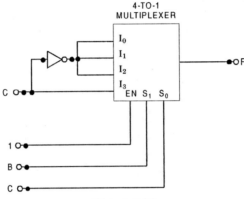

Figure 5.85

It may be noted that it is not necessary to reserve the most significant variable for use at multiplexer inputs. Example 5.13 may also be implemented if variable C is used at multiplexer inputs and, A and B are applied to selection inputs S_1 and S_0 respectively. In this case the function table is modified as in Figure 5.84 and circuit implementation is shown in Figure 5.85. Note that the places of minterms are changed in the implementation table in Figure 5.84 due to the change in assignment of selection inputs.

It should also be noted that it is not always necessary to assign the most significant variable or the least significant variable out of n variables to the multiplexer inputs and the rest to selection inputs. It is also not necessary that the selection inputs are connected in order. However, these types of connections will increase the complexities at preparation of an implementation table as well as circuit implementation.

Multiplexers are employed at numerous applications in digital systems. They are used immensely in the fields of data communication, data selection, data routing, operation sequencing, parallel-to-serial conversion, waveform generation, and logic function implementation.

Example 5.14. *Implement the following function using a multiplexer.*

$$F(A, B, C) = (1, 3, 5, 6)$$

Solution. The given function contains three variables. The function can be realized by one 4-to-1 multiplexer. The implementation table is shown in Figure 5.86 and the circuit diagram is given in Figure 5.87.

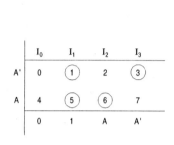

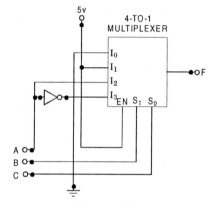

Figure 5.86   **Figure 5.87**

Example 5.15. *Implement the following function with a multiplexer.*

$$F\ (A,\ B,\ C,\ D)\ =\ (0,\ 1,\ 3,\ 4,\ 8,\ 9,\ 15)$$

Solution. The given function contains four variables. The function can be realized by one 8-to-1 multiplexer. The implementation table is shown in Figure 5.88 and the circuit diagram is given in Figure 5.89.

	I_0	I_1	I_2	I_3	I_4	I_5	I_6	I_7
A'	0	1	2	3	4	5	6	7
A	8	9	10	11	12	13	14	15
	1	1	0	A'	A'	0	0	A

Figure 5.88

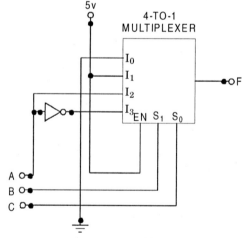

Figure 5.89

Example 5.16. *Implement a BCD-to-seven segment decoder with multiplexers.*

Solution. A BCD-to-seven segment decoder is already described by classical approach and realized with simple gates. The same circuit can be realized with the help of multiplexers. The truth table of a BCD-to-seven segment decoder (for common cathode type) is repeated here at Figure 5.90 for convenience. As there are four input variables, 4-to-1 multiplexers are employed to develop the combinational logic circuit. Implementation tables for each of the outputs a to g are shown in Figures 5.91(a)-(g). The logic diagram implementation of a BCD-to-seven segment decoder with 4-to-1 multiplexers is shown in Figure 5.92. Note that don't-care combinations (X) are judiciously considered as logic 1 or logic 0 in the implementation table.

Decimal	Input Variables				Output Variables as Seven Segment Display						
Numbers	A	B	C	D	a	b	c	d	e	f	g
0	0	0	0	0	1	1	1	1	1	1	0
1	0	0	0	1	0	1	1	0	0	0	0
2	0	0	1	0	1	1	0	1	1	0	1
3	0	0	1	1	1	1	1	1	0	0	1
4	0	1	0	0	0	1	1	0	0	1	1
5	0	1	0	1	1	0	1	1	0	1	1
6	0	1	1	0	0	0	1	1	1	1	1
7	0	1	1	1	1	1	1	0	0	0	0
8	1	0	0	0	1	1	1	1	1	1	1
9	1	0	0	1	1	1	1	0	0	1	1

Figure 5.90 (For a common cathode display.)

	I_0	I_1	I_2	I_3	I_4	I_5	I_6	I_7
A′	(0)	1	(2)	(3)	4	(5)	6	(7)
A	(8)	(9)	(10)	(11)	12	(13)	14	(15)
			X	X	X	X	X	X
	1	A	1	1	0	1	0	1

Figure 5.91(a) For *a*.

	I_0	I_1	I_2	I_3	I_4	I_5	I_6	I_7
A′	(0)	(1)	(2)	(3)	(4)	5	6	(7)
A	(8)	(9)	(10)	(11)	(12)	13	14	(15)
			X	X	X	X	X	X
	1	1	1	1	1	0	0	1

Figure 5.91(b) For *b*.

	I_0	I_1	I_2	I_3	I_4	I_5	I_6	I_7
A'	(0)	(1)	2	(3)	(4)	(5)	(6)	(7)
A	(8)	(9)	10	(11)	(12)	(13)	(14)	(15)
			X	X	X	X	X	X
	1	1	0	1	1	1	1	1

Figure 5.91(c) For c.

	I_0	I_1	I_2	I_3	I_4	I_5	I_6	I_7
A'	(0)	1	(2)	(3)	4	(5)	(6)	7
A	(8)	9	(10)	(11)	12	(13)	(14)	15
			X	X	X	X	X	X
	1	0	1	1	0	1	1	0

Figure 5.91(d) For d.

	I_0	I_1	I_2	I_3	I_4	I_5	I_6	I_7
A'	(0)	1	(2)	3	4	5	(6)	7
A	(8)	9	(10)	11	12	13	(14)	15
			X	X	X	X	X	X
	1	0	1	0	0	0	1	0

Figure 5.91(e) For e.

	I_0	I_1	I_2	I_3	I_4	I_5	I_6	I_7
A'	(0)	1	2	3	(4)	(5)	(6)	7
A	(8)	(9)	10	11	(12)	(13)	(14)	15
					X	X	X	X
	1	A	0	0	1	1	1	0

Figure 5.91(f) For f.

	I_0	I_1	I_2	I_3	I_4	I_5	I_6	I_7
A'	0	1	(2)	(3)	(4)	(5)	(6)	7
A	(8)	(9)	(10)	(11)	(12)	(13)	(14)	15
			X	X	X	X	X	X
	A	A	1	1	1	1	1	0

Figure 5.91(g) For g.

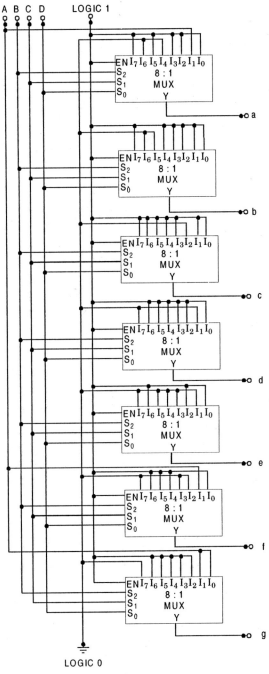

Figure 5.92

5.14 DEMULTIPLEXERS OR DATA DISTRIBUTORS

The term "demultiplex" means one into many. Demultiplexing is the process that receives information from one channel and distributes the data over several channels. It is the reverse operation of the multiplexer. A demultiplexer is the logic circuit that receives information through a single input line and transmits the same information over one of the possible 2^n output lines. The selection of a specific output line is controlled by the bit combinations of the selection lines.

Selection Inputs			Outputs							
A	B	C	Y_0	Y_1	Y_2	Y_3	Y_4	Y_5	Y_6	Y_7
0	0	0	$Y_0 = I$	0	0	0	0	0	0	0
0	0	1	0	$Y_1 = I$	0	0	0	0	0	0
0	1	0	0	0	$Y_2 = I$	0	0	0	0	0
0	1	1	0	0	0	$Y_3 = I$	0	0	0	0
1	0	0	0	0	0	0	$Y_4 = I$	0	0	0
1	0	1	0	0	0	0	0	$Y_5 = I$	0	0
1	1	0	0	0	0	0	0	0	$Y_6 = I$	0
1	1	1	0	0	0	0	0	0	0	$Y_7 = I$

Figure 5.93

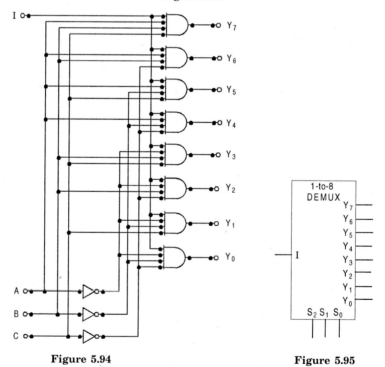

Figure 5.94

Figure 5.95

A 1-to-8 demultiplexer circuit is demonstrated in Figure 5.94. The selection input lines A, B, and C activate an AND gate according to its bit combination. The input line I is common to one of the inputs of all the AND gates. So information of I passed to the output line is activated by the particular AND gate. As an example, for the selection input combination 000, input I is transmitted to Y_0. A truth table is prepared in Figure 5.93 to illustrate the relation of selection inputs and output lines. The demultiplexer is symbolized in Figure 5.95 where S_2, S_1, and S_0 are the selection inputs.

It may be noticed that demultiplexer circuits may be derived from a decoder with the use of AND gates. As we have already seen, decoder outputs are equivalent to the minterms, these minterms can be used as the selection of output lines, and when they are ANDed with input line I, the data from input I is transmitted to output lines as activated according to the enabled minterms. Figure 5.96 demonstrates the construction of a 1-to-4 demultiplexer with a 2-to-4 decoder and four AND gates.

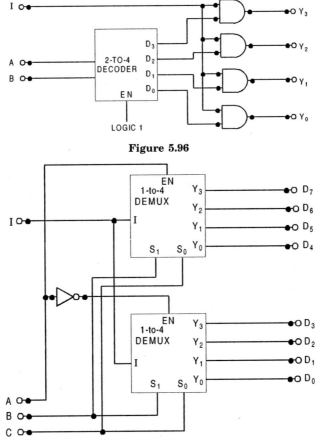

Figure 5.96

Figure 5.97

Like decoders and multiplexers, demultiplexers can also be cascaded to form higher order demultiplexers. Figure 5.97 demonstrates how a 1-to-8 demultiplexer can be formed with two 1-to-4 demultiplexers. Here, the highest significant bit A of the selection inputs is connected to the ENABLE inputs, one directly and the other one is complemented. When A is logic 0, one of the output lines D_0 to D_3 will be selected according to selection inputs B and C, and when A is logic 1, one of the output lines D_4 to D_7 will be selected.

5.15 CONCLUDING REMARKS

Various design methods of combinational circuits are described in this chapter. It is also illustrated and demonstrated that a number of SSI and MSI circuits can be used while designing more complicated digital systems. More complicated digital systems can be realized with LSI circuits, which will be discussed in Chapter 6.

The MSI functions discussed here are also described in the data books and catalog along with other commercially available ICs. IC data books contain exact descriptions of many MSI and other integrated circuits.

There are varieties of applications of combinational circuits in SSI or MSI or LSI form. A resourceful designer finds many applications to suit their particular needs. Manufactures of integrated circuits publish application notes to suggest the possible utilization of their products.

REVIEW QUESTIONS

5.1 What is a half-adder? Write its truth table.

5.2 Design a half-adder using NOR gates only.

5.3 What is a full-adder? Draw its logic diagram with basic gates.

5.4 Implement a full-adder circuit using NAND gates only.

5.5 Implement a full-adder circuit using NOR gates only.

5.6 What is the difference between a full-adder and full-subtractor?

5.7 Construct a half-subtractor using (a) basic gates, (b) NAND gates, and (c) NOR gates.

5.8 Construct a full-subtractor using (a) basic gates, (b) NAND gates, and (c) NOR gates.

5.9 Show a full-adder can be converted to a full-subtractor with the addition of an INVERTER.

5.10 Design a logic diagram for an addition/subtraction circuit, using a control variable P such that this operates as a full-adder when P = 0 and as a full-subtractor for P = 1.

5.11 What is a decoder? Explain a 3-to-8 decoder with logic diagram.

5.12 What is a priority encoder?

5.13 Can more than one output be activated for a decoder? Justify the answer.

5.14 Design a 4-bit binary subtractor using a 4-bit adder and INVERTERs.

5.15 What is a look ahead carry generator? What is its importance? Draw a circuit for a 3-bit binary adder using a look ahead carry generator and other gates.

5.16 What is a magnitude comparator?

5.17 What is a multiplexer? How is it different from a decoder?

5.18 How are multiplexers are useful in developing combinational circuits?

5.19 What is the function of enable input(s) for a decoder?

5.20 What are the major applications of multiplexers?

5.21 Design a combinational circuit for a BCD-to-gray code using (a) standard logic gates, (b) decoder, (c) 8-to-1 multiplexer, and (d) 4-to-1 multiplexer.

5.22 Design a combinational circuit for a gray-to-BCD code using (a) standard logic gates, (b) decoder, (c) 8-to-1 multiplexer, and (d) 4-to-1 multiplexer.

5.23 A certain multiplexer can switch one of 32 data inputs to output. How many different inputs does this MUX have?

5.24 An 8-to-1 MUX has inputs A, B, and C connected to selection lines S_2, S_1, and S_0 respectively. The data inputs I_0 to I_7 are connected as $I_1 = I_2 = I_7 = 0$, $I_3 = I_5 = 1$, $I_0 = I_4 = D$, and $I_6 = D'$. Determine the Boolean expression of the MUX output.

5.25 Design an 8-bit magnitude comparator using 4-bit comparators and other gates.

5.26 Implement the Boolean function F(A, B, C, D) = Σ (1, 3, 4, 11, 12, 13, 15) using (a) decoder and external gates, and (b) 8-to-1 MUX and external gates.

5.27 Is it possible to implement the Boolean function of problem 5.26 using one 4-to-1 MUX and external gates?

5.28 Design an Excess-3-to-8421 code converter using a 4-to-16 decoder with enable input E' and associated gates.

5.29 Repeat problem 5.28 using 8-to-1 multiplexers.

❏ ❏ ❏

Chapter 6

PROGRAMMABLE LOGIC DEVICES

6.1 INTRODUCTION

In Chapter 5, we discussed various combinational circuits that are commercially available in IC packages. We also saw how other combinational circuits and Boolean functions are realized with the help of these commercially available IC packages. With the advent of large-scale integration technology, it has become feasible to fabricate large circuits within a single chip. One such consequence of this technology is the *Programmable Logic Devices* or *PLD*s.

The advantages of using programmable logic devices are:

1. Reduced space requirements.

2. Reduced power requirements.

3. Design security.

4. Compact circuitary.

5. Short design cycle.

6. Low development cost.

7. Higher switching speed.

8. Low production cost for large-quantity production.

In earlier chapters, we have seen that any Boolean function or combinational circuit can be represented by sum of the products form or sum of the required minterms. It was also shown that a decoder generates 2^n minterms for n number of input variables and required minterm outputs of a decoder are fed to an OR gate to obtain a desired function. This fact leads to the development of IC packages with larger integration that contain decoders with a number of OR gates or one single chip containing a large number of basic gates—AND, OR, and NOT. These ICs are programmed according to desired functions by the manufacturers or the designers. Another advantage of employing these ICs is that one single IC can generate multiple outputs, thus reducing the board space, interconnections, and power consumption.

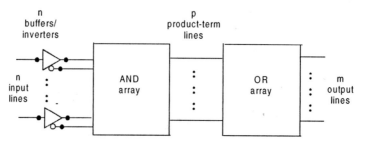

Figure 6.1

The general structure of programmable logic devices is illustrated in Figure 6.1. The inputs to the PLD are applied to a set of buffers/inverters. Buffers/inverters provide the true values of the inputs as well as the complemented values of the inputs. In addition, they also provide the necessary drive for the AND array, which consists of a large number of AND gates that follow next to buffers/inverters. The AND array produces p numbers of product terms from n numbers of input variables and their complements. These product terms are fed to the OR array, which follows next. The OR array also consists of several numbers of OR gates and realizes a set of m numbers of outputs at sum of the products form.

Programmable logic devices are broadly classified as three types of devices—*Read Only Memory* or ROM, *Programmable Logic Array* or PLA, and *Programmable Array Logic* or PAL. PLDs serve as the general circuits for realization of a set of Boolean functions. One or both of the arrays of PLDs are programmable in the sense that the logic designer can select the connections within the array. In ROM and PAL, one of the arrays are programmable whereas both the arrays are programmable for PLA. The following table summarizes which arrays are programmable for the various PLDs.

Device type	AND array	OR array
ROM	Fixed	Programmable
PLA	Programmable	Programmable
PAL	Programmable	Fixed

In a programmable array, the connections of gates can be selected. The simple approach for fabricating the programmable gate is to employ fuse links at each of the inputs of the gate as demonstrated in Figure 6.2(a). Some of the fuses are programmed to blow out to achieve the desired output from the gate. As an example, if the desired output of the gate is BC, then fuses at A and D are to be blown out as shown in Figure 6.2(b). Similarly, the same gate may be programmed for the function ACD, if only the fuse at input B is blown out. Therefore, with the blowing of fuses with proper programming, the same gate can generate several Boolean functions.

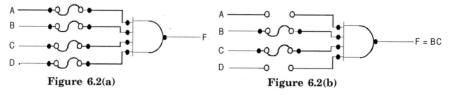

Figure 6.2(a)   **Figure 6.2(b)**

Although various schemes are used at fabrication of these types of gate arrays, this simple approach is assumed here to understand the function of PLDs. It should also be assumed that the open inputs of an AND gate array are connected to logic 1 and open inputs of an OR gate are connected to logic 0.

6.2 PLD NOTATION

To indicate the connections to an AND array and an OR array of a PLD, a simplified notation is frequently used. The notation is illustrated in Figures 6.3(a) and 6.3(b). Rather than drawing all the inputs to the AND gate or OR gate, a single line is drawn to the input to the gate. The inputs are indicated by the right-angled lines. The connected input variables are indicated by cross (×) at junctions and unconnected inputs are left blank. The cross-marked junctions represent the fusible joints while junctions with dots indicate permanent junctions that are not fusible.

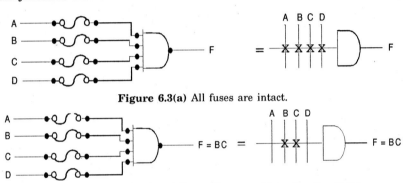

Figure 6.3(a) All fuses are intact.

Figure 6.3(b) Fuses A and D are blown to obtain function F=BC.

6.3 READ ONLY MEMORY (ROM)

A ROM is essentially a memory device for storage purpose in which a fixed set of binary information is stored. An user must first specify the binary information to be stored and then it is embedded in the unit to form the required interconnection pattern. ROM contains special internal links that can be fused or broken. Certain links are to be broken or blown out to realize the desired interconnections for a particular application and to form the required circuit path. Once a pattern is established for a ROM, it remained fixed even if the power supply to the circuit is switched off and then switched on again.

A block diagram of ROM is shown in Figure 6.4. It consists of n input lines and m output lines. Each bit combination of input variables is called an *address* and each bit combination that is formed at output lines is called a *word*. Thus, an address is essentially a binary number that denotes one of the minterms of n variables and the number of bits per word is equal to the number of output lines m. It is possible to generate $p = 2^n$ number of distinct addresses from n number of input variables. Since there are 2^n distinct addresses in a ROM, there are 2^n distinct words which are said to stored in the device and an output word can be selected by a unique address. The address value applied to the input lines specifies the word at output lines at any given time. A ROM is characterized by the number of words 2^n and number of bits per word m and denoted as $2^n \times m$ ROM.

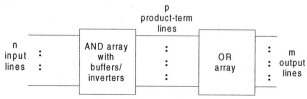

Figure 6.4

As an example, consider a 32 × 8 ROM. The device contains 32 words of 8 bits each. This means there are eight output lines and there are 32 numbers of distinct words stored in that unit, each of which is applied to the output lines. The particular word selected from the presently available output lines is determined by five input variables, as there are five input lines for a 32 × 8 ROM, because $2^5 = 32$. Five input variables can specify 32 addresses or minterms and for each address input there is a unique selected word. Thus, if the input address is 0000, word number 0 is selected. For address 0001, word number 1 is selected and so on.

A ROM is sometimes specified by the total number of bits it contains, which is $2^n \times m$. For example, a 4,096-bit ROM may be organized as 512 words of 8 bits each. That means the device has 9 input lines ($2^9 \times m = 512$) and 8 output lines.

In Figure 6.4, the block consisting of an AND array with buffers/inverters is equivalent to a decoder. The decoder basically is a combinational circuit that generates 2^n numbers of minterms from n number of input lines as already discussed in Chapter 5. 2^n or p numbers of minterms are realized from n number of input variables with the help of n numbers of buffers, n numbers of inverters, and 2^n numbers of AND gates. Each of the minterms is applied to the inputs of m number of OR gates through fusible links. Thus, m numbers of output functions can be produced after blowing of some selected fuses. The equivalent logic diagram of a $2^n \times m$ ROM is shown in Figure 6.5(a) and its logic diagram with PLD notation is redrawn in Figure 6.5(b).

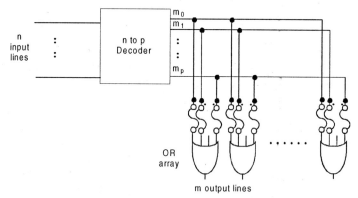

Figure 6.5(a)

The ROM is a two-level logic representation in the sum of products form. It may not be essentially an AND-OR realization, but it can be realized by other two-level minterm implementations. The second level is usually a wired-logic connection to facilitate the blowing of the links.

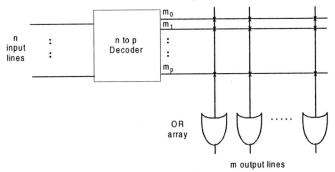

Figure 6.5(b)

ROM has many important applications in the design of digital computer systems. Realization of complex combinational circuits, code conversions, generating bit patterns, performing arithmetic functions like multipliers, forming look-up tables for arithmetic functions, and bit patterns for characters are some of its applications. They are particularly useful for the realization of multiple output combinational circuits with the same set of inputs. As such, they are used to store fixed bit patterns that represent the sequence of control variables needed to enable the various operations in the system. They are also used in association with microprocessors and microcontrollers.

6.3.1 Implementation of Combinational Logic Circuits

The implementation of Boolean functions using decoders was already discussed in Chapter 5. The same approach is applicable in using ROM, since ROM is the device that includes both a decoder and OR gates within the same chip. Given a set of Boolean expressions in minterms canonical form or a set of expressions in truth table form, first it is only necessary to select a ROM according to the input variables and number of output lines, and then to identify which links of the ROM are to be retained and which are to be blown. The blowing off of appropriate fuses or opening the links is referred to as *programming*. The designer needs only to specify a ROM program table that provides information for the required paths in the ROM. Some examples of ROM-based design are demonstrated here.

Example 6.1. Consider that the following Boolean functions are to be developed using ROM.

$$F_1 (A, B, C) = (0,1,2,5,7) \qquad \text{and}$$
$$F_2 (A, B, C) = (1,4,6).$$

When a combinational circuit is developed by means of a ROM, the functions must be expressed in the sum of minterms or by a truth table. The truth table of the above functions is shown in Figure 6.6. Since there are three input variables, a ROM containing a 3-to-8 line decoder is needed. In addition, since there are two output functions, the OR array must contain at least two OR gates. That means, a $2^3 \times 2$ ROM or 8×2 ROM is to be employed to realize the above functions. The logic diagram of the ROM after blowing off the appropriate fuses is illustrated in Figure 6.7. Obviously, this is too simple a combinational circuit to be implemented with a ROM. This example is merely for illustration purpose only. From the practical point of view, the real advantage of a ROM is in implementation of complex combinational networks having a large number of inputs and outputs.

Some ROM units are available with INVERTERs after each of the OR gates and they are specified as having initially all 0s at their outputs. The programming procedure in such ROMs require to blow off the link paths of the minterms (or addresses) that specify an output of 1 in the truth table. The outputs of the OR gates will then generate the complements of the functions, but the INVERTERs placed after OR gates complement the functions once more to provide the desired outputs. This is shown in Figure 6.8 for implementation of the logic functions as described in the previous example.

Decimal	Input Variables			Outputs	
Equivalent	A	B	C	F_1	F_2
0	0	0	0	1	0
1	0	0	1	1	1
2	0	1	0	1	0
3	0	1	1	0	0
4	1	0	0	0	1
5	1	0	1	1	0
6	1	1	0	0	1
7	1	1	1	1	0

Figure 6.6

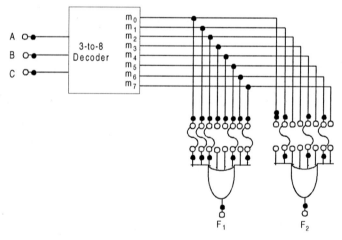

Figure 6.7

The previous example demonstrates the general procedure for implementing any combinational circuit with a ROM. From the number of inputs and outputs, the size of the ROM is determined first and then the programming for blowing off the appropriate fuse links is required with the help of the truth table or minterms. No further manipulation or simplification of Boolean functions is required. In practice, while designing with ROM, it is not essential to show the internal gate connections of links inside the unit. The designer simply has to specify the particular ROM and provide the ROM truth table as in Figure 6.6. The truth table provides all the information for programming of ROM. No internal logic diagram is necessary to accompany the truth table.

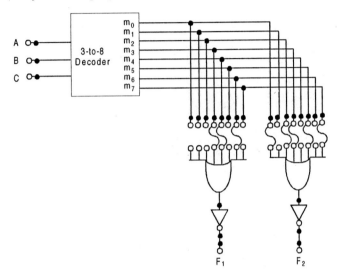

Figure 6.8

Example 6.2. *Find the squares of 3-bit numbers.*

Solution. This example has already been discussed and implemented with the classical method in Chapter 5. There are three input variables and six output functions. To implement with ROM, a $2^3 \times 6$ ROM or 8×6 ROM is required. The truth table is again shown in Figure 6.9 for convenience. Figure 6.10 shows the inputs and outputs with ROM and the internal fusible junctions are shown in Figure 6.11 after programming.

Input variables				Output variables						
Decimal	X	Y	Z	Decimal	A	B	C	D	E	F
0	0	0	0	0	0	0	0	0	0	0
1	0	0	1	1	0	0	0	0	0	1
2	0	1	0	4	0	0	0	1	0	0
3	0	1	1	9	0	0	1	0	0	1
4	1	0	0	16	0	1	0	0	0	0
5	1	0	1	25	0	1	1	0	0	1
6	1	1	0	36	1	0	0	1	0	0
7	1	1	1	49	1	1	0	0	0	1

Figure 6.9

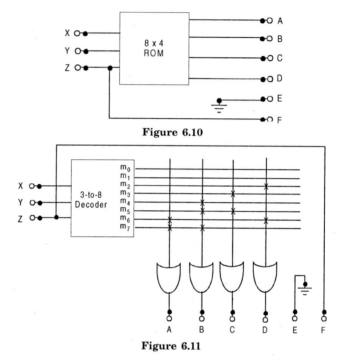

Figure 6.10

Figure 6.11

Example 6.3. *Design a code converter circuit for BCD-to-Excess-3 as well as BCD-to-2421 code using ROM.*

Solution.

Decimal	BCD code				Excess-3 code				2421 code			
Equivalent	A	B	C	D	W	X	Y	Z	P	Q	R	S
0	0	0	0	0	0	0	1	1	0	0	0	0
1	0	0	0	1	0	1	0	0	0	0	0	1
2	0	0	1	0	0	1	0	1	0	0	1	0
3	0	0	1	1	0	1	1	0	0	0	1	1
4	0	1	0	0	0	1	1	1	0	1	0	0
5	0	1	0	1	1	0	0	0	1	0	1	1
6	0	1	1	0	1	0	0	1	1	1	0	0
7	0	1	1	1	1	0	1	0	1	1	0	1
8	1	0	0	0	1	0	1	1	1	1	1	0
9	1	0	0	1	1	1	0	0	1	1	1	1

Figure 6.12

Here, two code converter circuits are housed in one single device. There are four input variables and eight output lines (four outputs for Excess-3 and four outputs for 2421).

Therefore, the ROM size required is $2^4 \times 8$ or 16×8. The combined truth table is presented in Figure 6.12. A logic diagram with PLD notation using ROM is given in Figure 6.13.

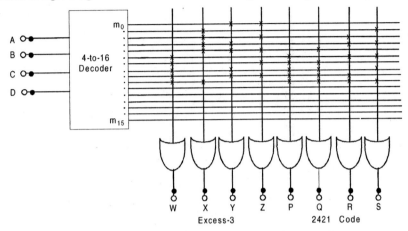

Figure 6.13

6.3.2 Types of ROM

The programming of ROM for selection of required paths may be done by two ways. The first is called *mask programming* and is done by the manufacturer during the last fabrication process of the device. The procedure for fabricating ROM is that the customer should   provide the truth table for the ROM to the manufacturer in a prescribed format. The manufacturer makes the corresponding mask for the links according to the truth table provided by the customer. This procedure is costly as the manufacturer demands a special charge from the customer for custom masking of a ROM. This procedure is economic only for large production of the same type of ROM. It is also less flexible because once it is programmed the functions cannot be modified by any means. With the advent of technology development various types of ROM are available nowadays.

1. *Programmable Read Only Memory* (PROM). It is more economic in cases requiring small quantities. In this method the manufacturer provides the PROM with all 0s (or all 1s) in every bit of the stored words. The required links are broken by application of current pulses. This allows the user to program the device in his own laboratory to obtain the desired relationship between input addresses and stored words. Special equipments called *PROM Programmers* are commercially available to facilitate this procedure. In any case, all procedures for programming ROMs are hardware procedures even though the word programming is used.

2. *Erasable PROM* (EPROM). The hardware procedure for programming of ROMs or PROMs as described above is irreversible, and once programmed, the configuration is fixed and cannot be altered. The device must be discarded if the bit pattern is required to be changed or modified. A third type of unit is available to overcome this disadvantage which is called *Erasable PROM* or EPROM. This device can reconstruct the initial bit patterns of all 0s or all 1s, though it is already programmed for some bit configuration. In other words, this device can be erased. This is achieved by placing

the *Erasable PROM* or EPROM under a special ultraviolet light for a given time. The short wave radiation discharges the internal gates that serve as links or contacts. After erasure, the device returns to its initial state and can be reprogrammed.

3. *Electrically Erasable PROM* (EEPROM). With the advancement of fabrication technology, further improvement of ROM has taken place, where ultraviolet light is not necessary to erase the programmed data. A new technique has been introduced to erase the bit pattern of ROM, where bit patterns are reset to their original state of all 0s or all 1s by applying a special electrical signal. Afterwards, the device can be reprogrammed with an alternate bit pattern. The equipment called *EPROM Programmer* serves the purpose of erasure as well as programming the bit patterns.

The function of a ROM may be interpreted two different ways. The first interpretation is of a device that realizes any combinational circuit. Each output terminal may be considered separately as the out of a Boolean function expressed in sum of the minterms. Secondly, it may be considered as a storage unit having a fixed pattern of bit strings called *words*. From this point of view, the inputs specify an address to a specific stored word which is then applied to the outputs. For example, the ROM in Figure 6.10 has three address lines specifying eight stored words, each of which is four bits long as given in the truth table. For this reason the device is called *read only memory*. Generally a storage device is called *memory* and the terminology *read* signifies the content in a specified location of a memory device, as addressed by the inputs available at the output. Thus, a ROM is a memory unit with a fixed word pattern that can be read out upon application of a given address. The bit pattern of ROM is permanent and cannot be altered during normal operation.

6.4 PROGRAMMABLE LOGIC ARRAY (PLA)

A combinational network may occasionally contain don't-care conditions. During the ROM implementation of this combinational circuit, this don't-care condition also forms an address input that will never occur. The words at the don't-care addresses need not be programmed and may be left in their original state of all 0s or all 1s. Since some of the bit patterns are not at all used, the address locations corresponding to don't-care conditions are considered a waste of memory.

Consider the simple case for Example 6.3, where code conversation from BCD to Excess-3 as well as 2421 code is demonstrated. It may be noted that for four input lines and eight output lines a 16 × 8 ROM has been used. This device has 16 addresses, though only 10 addresses are used because six addresses are attributed to don't-care conditions. That means, six words or 6 × 8 bit locations are wasted.

For the cases where don't-care conditions are excessive, it is more economical to use a second type of LSI device called a *Programmable Logic Array* or PLA. A PLA is similar to a ROM in concept. However, a PLA does not contain all AND gates to form the decoder or does not generate all the minterms like ROM. In the PLA, the decoder is replaced by a group of AND gates with buffers/inverters, each of which can be programmed to generate some product terms of input variable combinations that are essential to realize the output functions. The AND and OR gates inside the PLA are initially fabricated with the fusible links among them. The required Boolean functions are implemented in sum of the products form by opening the appropriate links and retaining the desired connections.

A block diagram of the PLA is shown in Figure 6.14. It consists of n inputs, m outputs, p product terms, and m sum terms. The product terms are obtained from an AND array containing p number of AND gates and the sum terms are developed by an OR array consisting of m number of OR gates. Fusible links are provided to each of the inputs of each of the AND gates as well as the OR gates. Additionally, outputs are provided with an INVERTER array with fusible links, so that the outputs are available at uncomplemented form as well as at complemented form. Therefore, the function is implemented in either AND-OR form when the output link across INVERTER is in place, or in AND-OR-INVERT form when the link is blown off. The general structure of a PLA with internal connections is shown Figure 6.15.

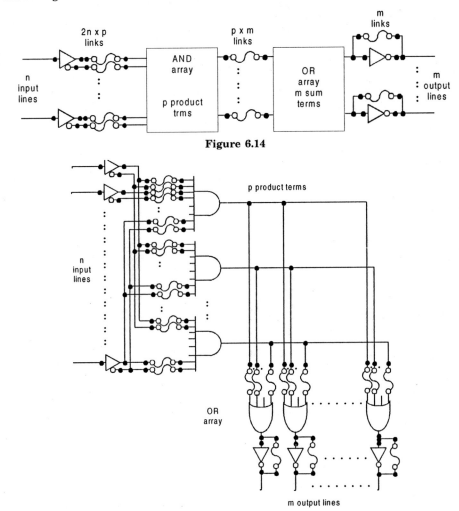

Figure 6.14

Figure 6.15

The size of a PLA is specified by the number of inputs, the number of product terms, and the number of outputs. The number of sum terms is equal to the number of outputs. The PLA described in Figure 6.14 or Figure 6.15 is specified as $n \times p \times m$ PLA. The number of programmable links is $2n \times p + p \times m + m$, whereas that of ROM is $2^n \times m$. A typical PLA of $16 \times 48 \times 8$ has 16 input variables, 48 product terms, and 8 output lines.

A comparison between ROM and PLA can be made to show how reduction in the number of gates is possible in PLA. Consider a typical example of implementation of a combinational circuit of 16 inputs, 8 outputs, and no more than 48 product terms. A $16 \times 48 \times 8$ PLA can serve the purpose, which consists of 48 product terms. To implement the same combinational circuit, a $2^{16} \times 8$ ROM is needed, which consists of $2^{16} = 65536$ minterms or product terms. So there is a drastic reduction in number of AND gates within the chip, thus reducing the fabrication time and cost. It should be noted that both complemented and uncomplemented inputs, i.e., 2^n number of inputs appear at each AND gate providing maximum flexibility in product term generation.

Like a ROM, the PLA may also be mask-programmable or field programmable. For a mask-programmable PLA, the user must submit a PLA program table to the manufacturer to produce a custom made PLA that has the required internal paths between inputs and outputs. The second type of PLA available is called a *field programmable logic array* or FPLA. The FPLA can be programmed by the users by means of certain recommended procedures. Programmer equipment is available commercially for use in conjunction with certain FPLAs.

6.4.1 Design Procedure with PLA

In the case of ROM-based design, we have seen that, since all the minterms are generated in a ROM, the realization of a set of Boolean functions is based on minterms canonical expressions. It is never necessary to minimize the expressions prior to obtaining the realization with a ROM. On the other hand, in the case of PLA, the product terms generated are not necessarily the minterms, as these product terms depend upon how the fuses are programmed. As a consequence, the realization using PLA is based on the sum of the products expressions. Also, it is significant that the number of product terms is limited for a PLA and the logic designer must utilize them most intelligently. This implies that it is necessary to obtain a set of expressions in such a way that the number of product terms does not exceed the number of AND gates in the PLA. Therefore, some degree of simplification of Boolean functions is needed. Several techniques of minimization of Boolean expressions have already been discussed in earlier chapters.

Example 6.4. To demonstrate the use of PLA to implement combinational logic circuits, consider the following expression

$$F_1 \ (A, B, C) = \ (0, 1, 3, 4) \qquad \text{and}$$
$$F_2 \ (A, B, C) = \ (1, 2, 3, 4, 5).$$

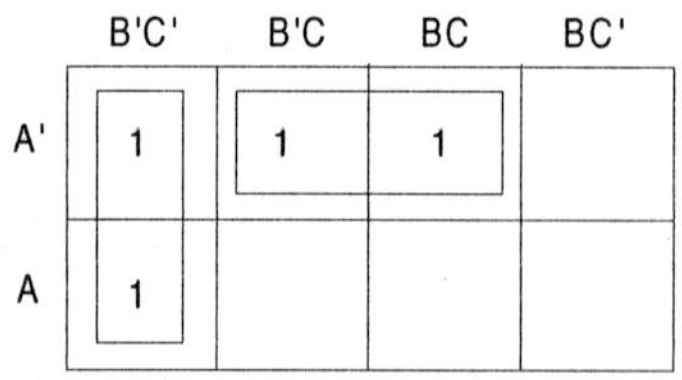

Figure 6.16(a) Map for function F_1.

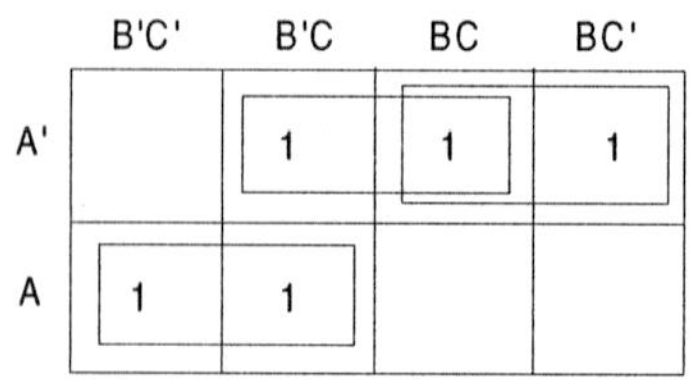

Figure 6.16(b) Map for function F_2.

Assume that a $3 \times 4 \times 2$ PLA is available for the realization of the above functions. It should be noted that according to the number of inputs and output, the specified PLA is sufficient to realize the functions. However, total distinct minterms in the functions are six, whereas available product terms or the number of AND gates in the specified PLA is four. So some simplification or minimization is required for the functions. Karnaugh maps are drawn in Figures 6.16(a) and 6.16(b) for this purpose.

The simplified Boolean expressions for the functions are

$$F_1 = B'C' + A'C \qquad \text{and}$$
$$F_2 = A'B + A'C + AB'.$$

In these expressions, there are four distinct product terms—B'C', A'C, A'B, and AB'. So these function can be realized by the specified 3 × 4 × 2 PLA. The internal connection diagram for the functions using PLA after fuse-links programming is demonstrated in Figure 6.17.

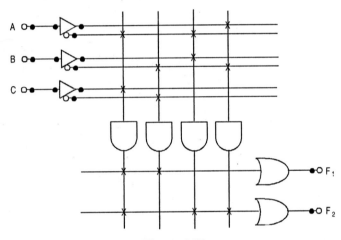

Figure 6.17

Programming the PLA means to specify the paths in its AND-OR-INVERT pattern. A PLA program table is a useful tool to specify the input-output relationship indicating the number of product terms and their expressions. It also specifies whether the output is complemented or not. The program table for the above example is shown in Figure 6.18.

Product		Inputs			Outputs		
Terms		A	B	C	F_1	F_2	
A'B	1	0	1	-	-	1	
A'C	2	0	-	1	1	1	
AB'	3	1	0	-	-	1	
B'C'	4	-	0	0	1	-	
					T	T	T/C

Figure 6.18

The first column lists the product terms numerically. The second column specifies the required paths between inputs and AND gates. The third column indicates the paths between the AND gates and OR gates. Under each output variable, T is written if output INVERTER is bypassed *i.e.*, the output at true form, and C is written if output is complemented with INVERTER. The Boolean terms listed at the leftmost are for reference only, they are not part of the table.

For each product term, the inputs are marked with 1, 0, or – (dash). If the input variable is present in the product term at its uncomplemented form, the corresponding input variable is marked with a 1. If the input variable appears in the product term at its complemented form, it is marked with a 0. If the variable does not at all appear in the product term, it is marked with a – (dash). Thus the paths between the inputs and the AND gates are specified under the column heading *inputs* and accordingly the links at the inputs of AND gates are to be retained or blown off. The AND gates produce the required product term. The open terminals of AND gates behave like logic 1.

The paths between the AND gates and OR gates are specified under the column heading *outputs*. Similar to the above, the output variables are also marked with 1, 0, or – (dash) depending upon the presence of product terms in the output expressions. Finally, a T (true) output dictates that links across the INVERTER are retained and for C (complemented) at output indicates that the link across the INVERTER is to be broken. The open terminals of OR gates are assumed to be logic 0.

While designing a digital system with PLA, there is no need to show the internal connections of the unit. The PLA program table is sufficient to specify the appropriate paths. For a custom made PLA chip this program table is needed to provide to the manufacturer.

Since for a given PLA, the number of AND gates is limited, careful investigation must be carried out, while implementing a combinational circuit with PLA, in order to reduce the total number of distinct product terms. This can be done by simplifying each function to a minimum number of terms. Note that the number of literals in a term is not important as all the inputs are available. It is required to obtain the simplified expressions both of true form and its complement form for each of the functions to observe which one can be expressed with fewer product terms and which one provides product terms that are common to other functions. The following example will clarify this.

Example 6.5. *Implement the following Boolean functions using a 3 × 4 × 2 PLA.*

$$F_1 \ (\ A, \ B, \ C) \ = \ (3, \ 5, \ 6, \ 7) \qquad and$$
$$F_2 \ (\ A, \ B, \ C) \ = \ (0, \ 2, \ 4, \ 7).$$

Solution. A total of seven minterms are present in the two functions above, whereas the number of AND gates is four in the specified PLA. So simplification of the above functions is necessary. Simplification is carried out for both the true form as well as the complement form for each of the functions. Karnaugh maps are drawn in Figure 6.19(a)-(d).

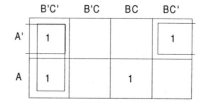

Figure 6.19(a) Map for function F_1.   Figure 6.19(b) Map for function F_2.

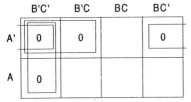

Figure 6.19(c) Map for function F_1'.

	B'C'	B'C	BC	BC'
A'	0	0		0
A	0			

Figure 6.19(d) Map for function F_2'.

The Boolean expressions are

$$F_1 = AC + AB + BC \qquad \text{and} \qquad F_2 = B'C' + A'C' + ABC$$
$$F_1' = B'C' + A'B' + A'C' \qquad \text{and} \qquad F_2' = A'C + B'C + ABC'.$$

From the Boolean expressions it can be observed that if both the true forms of F_1 and F_2 are selected for implementation, the total number of distinct product terms needed to be realized is six, which is not possible by the specified $3 \times 4 \times 2$ PLA. However, if F_1' and F_2 are selected, then the total number of distinct product terms reduces to four, which is now possible to be implemented by the specified PLA. F_1' can be complemented by the output INVERTER to obtain its true form of F_1. The PLA program table for these expressions is prepared in Figure 6.20. Note that the C (complement) is marked under the output F_1 indicating that output INVERTER exists at the output path of F_1. The logic diagram for the above combinational circuit is shown in Figure 6.21.

	Product Terms	Inputs			Outputs	
		A	B	C	F_1	F_2
B'C'	1	-	0	0	1	1
A'B'	2	0	0	-	1	-
A'C'	3	0	-	0	1	1
ABC	4	1	1	1	-	1
					C	T	T/C

Figure 6.20

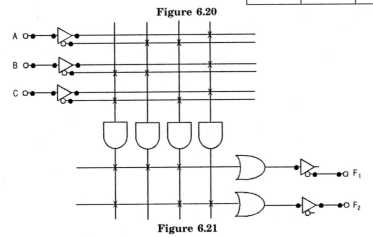

Figure 6.21

It should be noted that the combinational circuits for the examples presented here are too small and simple for practical implementation with PLA. But they do serve the purpose of demonstration and show the concept of PLA combinational logic design. A typical commercial PLA would have over 10 inputs and about 50 product terms. The simplification of so many variables are carried out by means of tabular method or other computer-based simplification methods. Thus, the computer program assists in designing the complex digital systems. The computer program simplifies each of the functions of the combinational circuit and its complements to a minimum number of terms. Then it optimizes and selects a minimum number of distinct product terms that cover all the functions in their true form or complement form.

6.5   PROGRAMMABLE ARRAY LOGIC (PAL) DEVICES

The final programmable logic device to be discussed is the *Programmable Array Logic* or PAL device. The general structure of this device is similar to PLA, but in a PAL device only AND gates are programmable. The OR array in this device is fixed by the manufacturer. This makes PAL devices easier to program and less expensive than PLA. On the other hand, since the OR array is fixed, it is less flexible than a PLA device.

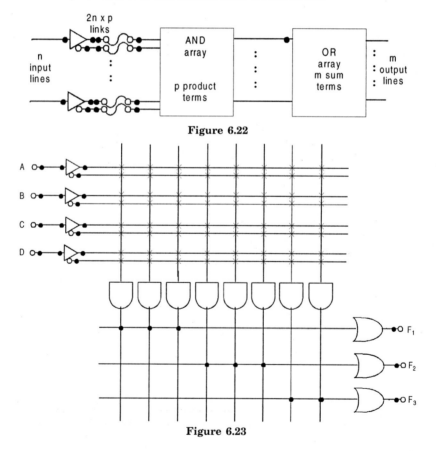

Figure 6.22

Figure 6.23

Figure 6.22 represents the general structure of a PAL device. It has n input lines which are fed to buffers/inverters. Buffers/inverters are connected to inputs of AND gates through programmable links. Outputs of AND gates are then fed to the OR array with fixed connections. It should be noted that, all the outputs of an AND array are not connected to an OR array. In contrast to that, only some of the AND outputs are connected to an OR array which is at the manufacturer's discretion. This can be clarified by Figure 6.23, which illustrates the internal connection of a four-input, eight AND-gates and three-output PAL device before programming. Note that while every buffer/inverter is connected to AND gates through links, F_1-related OR gates are connected to only three AND outputs, F_2 with three AND gates, and F_3 with two AND gates. So this particular device can generate only eight product terms, out of which two of the three OR gates may have three product terms each and the rest of the OR gates will have only two product terms. Therefore, while designing with PAL, particular attention is to be given to the fixed OR array.

6.5.1 Designing with Programmable Array Logic

Let us consider that the following functions are to be realized using a PAL device.

$$F_1 (A,B,C) = (1,2,4,5,7)$$
$$F_2 (A,B,C) = (0,1,3,5,7)$$

Similar to designing with PLA, in the case of a PAL device some simplification must be carried out to reduce the total number of distinct product terms. Karnaugh maps for the above functions are drawn in Figures 6.24(a) and 6.24(b).

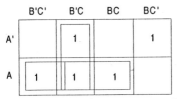

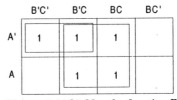

Figure 6.24(a) Map for function F_1.   **Figure 6.24(b)** Map for function F_2.

The Boolean expressions are

$$F_1 (A,B,C) = AB' + AC + B'C + A'BC' \qquad \text{and}$$
$$F_2 (A,B,C) = C + A'B'.$$

To use the PAL device as illustrated in Figure 6.23 for realization of the above expressions, it may be noted that a problem occurs that the specified PAL device has at the most three product terms associated with one OR gate, whereas one of the given functions F_1 has four product terms. However, realization of the functions are achievable with the specified PAL device by the following method.

Let the above expressions be rewritten as

$$F_1 (A,B,C) = F_3 + B'C + A'BC'$$
$$F_2 (A,B,C) = C + A'B'$$

where, $\qquad F_3 = AB' + AC.$

Now there are three functions each of which contains no more than three product terms and these can be realizable by the specified PAL. The connection diagram of PAL is illustrated in Figure 6.25.

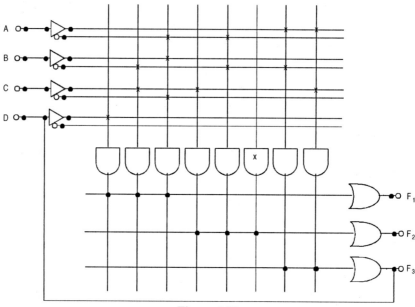

Figure 6.25

Here, one subfunction F_3 has been generated with two product terms, and this subfunction is connected to one of the inputs to realize the final function F_1. To realize F_2, only two terms need to be generated. Since a three-input OR gate is used, the input must be kept at logic 0, so as not to affect the F_2 output. This is achieved by keeping all the fuses intact to the AND gate that serves as the third input to the OR gate which is indicated by an 'x' mark on the AND gate in Figure 6.24. With a variable and its complement as inputs an AND gate always produces logic 0.

It should be noted that the PAL device as demonstrated here is too small for the practical point of view. Similar to a PLA device, a practical PAL device contains at least ten inputs and about fifty product terms. This small and simple PAL device has been illustrated here only to show its general internal architecture and how the combinational circuits are realized. Simplification of Boolean functions should be carried out and special attention must be given while selecting the minimal terms as the number of OR gates is limited as well as limited product terms are connected to them.

With the fast advancement of technology, various types of programmable logic devices are being developed to meet the users desire. Programmable logic devices are also available with flip-flops. Some of the useful programmable devices are mentioned here.

6.6 REGISTERED PAL DEVICES

Flip-flops are employed in sequential digital circuits in addition to the combinational circuits, and therefore for the design of sequential circuits, PALs have been developed with flip-flops in the outputs. These devices are referred to as *registered PALs*. The flip-flops are all controlled by a common clock and another dedicated input pin is provided for output ENABLE control of INVERTERs.

6.7 CONFIGURABLE PAL DEVICES

Development in the design of programmable array logic devices led to the introduction of configurable outputs enhancing the output capabilities of such devices. The configurable device architecture is achieved by providing some special circuitry at the output stage, known as *macrocells*. Each of the macrocells are provided with two fuses that can be programmed for four different configurations of outputs. The output configuration may be of true output without flip-flop or complemented output without flip-flop or true output with flip-flop or complemented output with flip-flop. So it may be observed that this type of device can function for sequential logic circuits as well as for combinational logic circuits and in each case outputs are available in inverted form or noninverted form.

6.8 GENERIC ARRAY LOGIC DEVICES

The *Generic Array Logic* or GAL device is another type of configurable PAL device. GAL devices are intended as pin-to-pin replacements for a wide variety of PAL devices. It is designed to be compatible, all the way to the fuse level, for any simpler PAL which can be directly implemented in the GAL device. In this device, the OR gate is considered to be a part of a macrocell to obtain various types of I/O configurations found in the PAL devices that it is designed to replace.

Another family of devices that are intended for PAL replacements are *programmable electrically erasable logic* or PEEL devices. Its output macrocell can be programmed for numerous types of I/O configurations.

6.9 FIELD-PROGRAMMABLE GATE ARRAY (FPGA)

These type of programmable devices are based on the basic structure equivalent to programmable logic array or PLA. Over the years, programmable arrays have increased in size and complexity. Highly configurable macrocells have been induced to enhance their flexibility and capability. *Field-programmable gate array* or FPGA has been developed with the concept of alternate architecture, to increase the effective size and to provide more functional flexibility in a single programmable device. The densities of FPGAs are much higher than any other PLDs. Each FPGA accommodates 1,200 to 20,000 equivalent gates whereas PLDs range in size from a few hundred to 2,000 equivalent gates.

An FPGA contains a number of relatively independent configurable logic modules, configurable I/Os and programmable interconnection paths or routing channels. All the resources of this device are uncommitted and these must be selected, configured, and interconnected by the user to form a logic system for his application. FPGAs are specified by their size, configuration of their logic modules, and interconnection requirements. FPGA with larger logic modules may not be sufficiently utilized to perform simple logic functions and thereby wasting the logic modules. Use of smaller logic modules leads to a larger number of interconnections with the device causing significant propagation delay as well as consuming a large percentage of FPGA area. The designer must optimize the logic module size and interconnection requirements according to the application of logic system design. For a given FPGA device, there are many possible ways to configure to meet the design requirements. Different types of FPGAs are available that differ in their architecture, technologies, and programming techniques.

6.10 CONCLUDING REMARKS

The basic concepts of programmable logic devices and programmable gate arrays have been discussed. With the development of these devices, complex digital systems have become possible to be designed. However, high-level design techniques and computer-aided tools are required to realize efficient PLD and FPGA implementations. The emergence of these devices has revolutionized the design of digital systems similar to the emergence of microprocessor or microcontrollers. The programmable logic concept has provided the power to design one's custom ICs which cannot be copied by others.

REVIEW QUESTIONS

6.1 Define PLD. What are the advantages PLD?

6.2 What are the types of PLD?

6.3 List the applications of PLD.

6.4 What is PLA? How does it differ from ROM? Draw the block diagram of PLA.

6.5 What is PAL? How does it differ from ROM? Draw the block diagram of PAL.

6.6 What are the advantages of FPGA over other types of PLD?

6.7 Draw the internal logic construction of 32 × 4 ROM.

6.8 Give the comparison among PROM, PLA, and PAL.

6.9 How many words can be stored in a ROM of capacity 16K × 32?

6.10 What is the bit storage capacity of a 512 × 4 ROM?

6.11 State the differences among ROM, PROM, EPROM, and EEPROM.

6.12 Explain the difference between ROM and RAM.

6.13 What do a dot and an × represent in a PLD diagram?

6.14 How many memory locations are there for address values?

(a) 0000 to 7FFF, (b) C000 to C3FF, or (c) A000 to BFFF.

6.15 Specify the size of a ROM for implementation of the following combinational circuit.

(a) a binary multiplier for multiplication of two 4-bit numbers, or (b) a 4-bit adder/subtractor.

6.16 Implement the following Boolean expressions using ROM.

F_1 (A, B, C) = Σ (0, 2, 4, 7), F_2 (A, B, C) = Σ (1, 3, 5, 7)

6.17 Implement the following Boolean expressions using PLA.

F_1 (A, B, C) = Σ (0, 1, 3, 5), F_2 (A, B, C) = Σ (0, 3, 5, 7)

6.18 Implement the following Boolean expressions using PAL.

F_1 (A, B, C, D) = Σ (1, 2, 5, 7, 8, 10, 12, 13)

F_2 (A, B, C, D) = Σ (0, 2, 6, 8, 9, 14)

F_3 (A, B, C, D) = Σ (0, 3, 7, 9, 11, 12, 14)

F_4 (A, B, C, D) = Σ (1, 2, 4, 5, 9, 10, 14)

6.19 Tabulate the PLA programmable table for the four Boolean functions listed below.

A (X, Y, Z) = Σ (0, 1, 2, 4, 6)

B (X, Y, Z) = Σ (0, 2, 6, 7)

C (X, Y, Z) = Σ (3, 6)

D (X, Y, Z) = Σ (1, 3, 5, 7)

6.20 Design a BCD-to-Excess-3 code converter using (a) PROM, (b) PLA, and (c) PAL.

6.21 Design an Excess-3-to-BCD code converter using (a) PROM, (b) PLA, and (c) PAL.

6.22 Design a BCD-to-seven segment display decoder using (a) PROM, (b) PLA, and (c) PAL.

6.23 Tabulate the PLA programmable table for the four Boolean functions listed below.

A (X, Y, Z) = Σ (1, 2, 4, 6)

B (X, Y, Z) = Σ (0, 1, 6, 7)

C (X, Y, Z) = Σ (2, 6)

D (X, Y, Z) = Σ (1, 2, 3, 5, 7)

6.24 Following is a truth table of a three-input, four-output, combinational circuit. Tabulate the PAL programming table for the circuit and mark the fuse map in the diagram.

Inputs			Outputs			
X	Y	Z	A	B	C	D
0	0	0	0	1	0	0
0	0	1	1	1	1	1
0	1	0	1	0	1	1
0	1	1	0	1	0	1
1	0	0	1	0	1	0
1	0	1	0	0	0	1
1	1	0	1	1	1	0
1	1	1	0	1	1	1

6.25 Design a code converter that converts 2421 code to BCD as well as to Excess-3 code using (a) PROM, (b) PLA, and (c) PAL.

❑ ❑ ❑

Chapter 7

SEQUENTIAL LOGIC CIRCUITS

7.1 INTRODUCTION

So far, all of the logic circuits we have studied were basically based on the analysis and design of combinational digital circuits. Though these type of circuits are very important, they constitute only a part of digital systems. The other major aspect of a digital system is the analysis and design of sequential digital circuits. However, sequential circuit design depends, greatly, on the combinational circuit design.

The logic circuits whose outputs at any instant of time depend only on the input signals present at that time are known as combinational circuits. The output in combinational circuits does not depend upon any past inputs or outputs. Moreover, in a combinational circuit, the output appears immediately for a change in input, except for the propagation delay through circuit gates.

On the other hand, the logic circuits whose outputs at any instant of time depend on the present inputs as well as on the past outputs are called sequential circuits. In sequential circuits, the output signals are fed back to the input side. A block diagram of a sequential circuit is shown in Figure 7.1

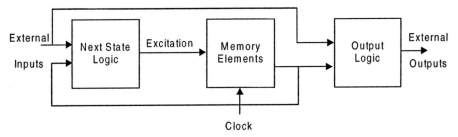

Figure 7.1 Block diagram of a sequential circuit.

From Figure 7.1, we find that it consists of combinational circuits, which accept digital signals from external inputs and from outputs of memory elements and generates signals for external outputs and for inputs to memory elements, referred to as excitation.

215

A memory element is a medium in which one bit of information (0 or 1) can be stored or retained until necessary, and thereafter its contents can be replaced by a new value. The contents of memory elements can be changed by the outputs of combinational circuits that are connected to its input.

Combinational circuits are often faster than sequential circuits since the combinational circuits do not require memory elements whereas the sequential circuit needs memory elements to perform its operations in sequence.

Sequential circuits are broadly classified into two main categories, known as synchronous or clocked and asynchronous or unclocked sequential circuits, depending on the timing of their signals.

A sequential circuit whose behavior can be defined from the knowledge of its signal at discrete instants of time is referred to as a synchronous sequential circuit. In these systems, the memory elements are affected only at discrete instants of time. The synchronization is achieved by a timing device known as a system clock, which generates a periodic train of clock pulses as shown in Figure 7.2. The outputs are affected only with the application of a clock pulse. The rate at which the master clock generates pulses must be slow enough to permit the slowest circuit to respond. This limits the speed of all circuits. Synchronous circuits have gained considerable domination and wide popularity.

A sequential circuit whose behavior depends upon the sequence in which the input signals change is referred to as an asynchronous sequential circuit. The output will be affected whenever the input changes. The commonly used memory elements in these circuits are time-delay devices. There is no need to wait for a clock pulse. Therefore, in general, asynchronous circuits are faster than synchronous sequential circuits. However, in an asynchronous circuit, events are allowed to occur without any synchronization. And in such a case, the system becomes unstable. Since the designs of asynchronous circuits are more tedious and difficult, their uses are rather limited. The memory elements used in sequential circuits are flip-flops which are capable of storing binary information.

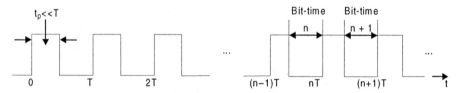

Figure 7.2 Train of pulses.

7.2 FLIP-FLOPS

The basic 1-bit digital memory circuit is known as a flip-flop. It can have only two states, either the 1 state or the 0 state. A flip-flop is also known as a bistable multivibrator. Flip-flops can be obtained by using NAND or NOR gates. The general block diagram representation of a flip-flop is shown in Figure 7.3. It has one or more inputs and two outputs. The two outputs are complementary to each other. If Q is 1 i.e., Set, then Q' is 0; if Q is 0 i.e., Reset, then Q' is 1. That means Q and Q' cannot be at the same state simultaneously. If it happens by any chance, it violates the definition of a flip-flop and hence is called an *undefined* condition. Normally, the state of Q is called the *state* of the flip-flop, whereas the state of Q' is called the *complementary state* of the flip-flop. When the output Q is either 1 or 0, it remains in that state unless one or more inputs are excited to effect a change in the output. Since the output of the flip-flop remains in the same state until the trigger pulse is applied to change the state, it can be regarded as a memory device to store one binary bit.

As mentioned earlier, a flip-flop is also known as a bistable multivibrator, whose circuit is shown in Figure 7.4, where the trigger inputs are named as Set and Reset.

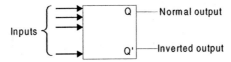

Figure 7.3 Block diagram of a flip-flop.

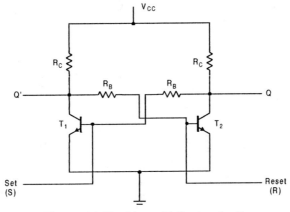

Figure 7.4 Bistable multivibrator circuit.

From the circuit shown in Figure 7.4, we find that the multivibrator is basically two cross-coupled inverting amplifiers, comparising of two transistors and four resistors. Obviously, if transistor T_1 is initially turned ON (saturated) by applying a positive signal through the Set input at its base, its collector will be at $V_{CE\ (sat)}$ (0.2 to 0.4 V). The collector of T_1 is connected to the base of T_2, which cannot turn T_2 On. Hence, T_2 remains OFF (cut off). Therefore, the voltage at the collector of T_2 tries to reach V_{CC}. This action only enhances the initial positive signal applied to the base of T_1. Now if the initial signal at the Set input is removed, the circuit will maintain T_1 in the ON state and T_2 in the OFF state indefinitely, $i.e.$, Q = 1 and Q' = 0. In this condition the bistable multivibrator is said to be in the Set state.

A positive signal applied to the Reset input at the base of T_2 turns it ON. As we have discussed earlier, in the same sequence T_2 turns ON and T_1 turns OFF, resulting in a second stable state, $i.e.$, Q = 0 and Q' = 1. In this condition the bistable multivibrator is said to be in the *Reset* state.

7.2.1 Latch

We consider the fundamental circuit shown in Figure 7.5. It consists of two inverters G_1 and G_2 (NAND gates are used as inverters). The output of G_1 is connected to the input of G_2 (A_2) and the output of G_2 is connected to the input of G_1 (A_1). Let us assume the output of G_1 to be Q = 0, which is also the input of G_2 ($A_2 = 0$). Therefore, the output of G_2 will be Q' = 1, which makes $A_1 = 1$ and consequently Q = 0 which is according to our assumption.

Similarly, we can demonstrate that if Q = 1, then Q' = 0 and this is also consistent with the circuit connections. Hence we see that Q and Q' are always complementary. And if the circuit is in 1 state, it continues to remain in this state and vice versa is also true. Since this information is locked or latched in this circuit, therefore, this circuit is also referred to as a *latch*. In this circuit there is no way to enter the desired digital information to be stored in it. To make that possible we have to modify the circuit by replacing the inverters by NAND gates and then it becomes a flip-flop.

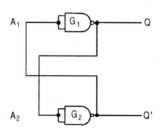

Figure 7.5 Cross-coupled inverters as a memory element.

7.3 TYPES OF FLIP-FLOPS

There are different types of flip-flops depending on how their inputs and clock pulses cause transition between two states. We will discuss four different types of flip-flops in this chapter, *viz.*, S-R, D, J-K, and T. Basically D, J-K, and T are three different modifications of the S-R flip-flop.

7.3.1 S-R (Set-Reset) Flip-flop

An S-R flip-flop has two inputs named Set (S) and Reset (R), and two outputs Q and Q'. The outputs are complement of each other, *i.e.*, if one of the outputs is 0 then the other should be 1. This can be implemented using NAND or NOR gates. The block diagram of an S-R flip-flop is shown in Figure 7.6.

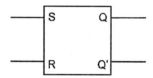

Figure 7.6 Block diagram of an S-R flip-flop.

S-R Flip-flop Based on NOR Gates

An S-R flip-flop can be constructed with NOR gates at ease by connecting the NOR gates back to back as shown in Figure 7.7. The cross-coupled connections from the output of gate 1 to the input of gate 2 constitute a feedback path. This circuit is not clocked and is classified as an asynchronous sequential circuit. The truth table for the S-R flip-flop based on a NOR gate is shown in the table in Figure 7.8.

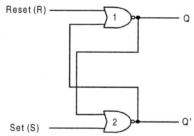

Figure 7.7 NOR-based S-R flip-flop.

To analyze the circuit shown in Figure 7.7, we have to consider the fact that the output of a NOR gate is 0 if any of the inputs are 1, irrespective of the other input. The output is 1 only if all of the inputs are 0. The outputs for all the possible conditions as shown in the table in Figure 7.8 are described as follows.

Inputs		Outputs		Action
S	R	Q_{n+1}	Q'_{n+1}	
0	0	Q_n	Q'_n	No change
0	1	0	1	Reset
1	0	1	0	Set
1	1	0	0	Forbidden (Undefined)
0	0	–	–	Indeterminate

Figure 7.8

Case 1. For $S = 0$ and $R = 0$, the flip-flop remains in its present state (Q_n). It means that the next state of the flip-flop does not change, i.e., $Q_{n+1} = 0$ if $Q_n = 0$ and vice versa. First let us assume that $Q_n = 1$ and $Q'_n = 0$. Thus the inputs of NOR gate 2 are 1 and 0, and therefore its output $Q'_{n+1} = 0$. This output $Q'_{n+1} = 0$ is fed back as the input of NOR gate 1, thereby producing a 1 at the output, as both of the inputs of NOR gate 1 are 0 and 0; so $Q_{n+1} = 1$ as originally assumed.

Now let us assume the opposite case, i.e., $Q_n = 0$ and $Q'_n = 1$. Thus the inputs of NOR gate 1 are 1 and 0, and therefore its output $Q_{n+1} = 0$. This output $Q_{n+1} = 0$ is fed back as the input of NOR gate 2, thereby producing a 1 at the output, as both of the inputs of NOR gate 2 are 0 and 0; so $Q'_{n+1} = 1$ as originally assumed. Thus we find that the condition $S = 0$ and $R = 0$ do not affect the outputs of the flip-flop, which means this is the memory condition of the S-R flip-flop.

Case 2. The second input condition is $S = 0$ and $R = 1$. The 1 at R input forces the output of NOR gate 1 to be 0 (i.e., $Q_{n+1} = 0$). Hence both the inputs of NOR gate 2 are 0 and 0 and so its output $Q'_{n+1} = 1$. Thus the condition $S = 0$ and $R = 1$ will always reset the flip-flop to 0. Now if the R returns to 0 with $S = 0$, the flip-flop will remain in the same state.

Case 3. The third input condition is $S = 1$ and $R = 0$. The 1 at S input forces the output of NOR gate 2 to be 0 (i.e., $Q'_{n+1} = 0$). Hence both the inputs of NOR gate 1 are 0 and 0 and so its output $Q_{n+1} = 1$. Thus the condition $S = 1$ and $R = 0$ will always set the flip-flop to 1. Now if the S returns to 0 with $R = 0$, the flip-flop will remain in the same state.

Case 4. The fourth input condition is $S = 1$ and $R = 1$. The 1 at R input and 1 at S input forces the output of both NOR gate 1 and NOR gate 2 to be 0. Hence both the outputs of NOR gate 1 and NOR gate 2 are 0 and 0; i.e., $Q_{n+1} = 0$ and $Q'_{n+1} = 0$. Hence this condition $S = 1$ and $R = 1$ violates the fact that the outputs of a flip-flop will always be the complement of each other. Since the condition violates the basic definition of flip-flop, it is called the *undefined* condition. Generally this condition must be avoided by making sure that 1s are not applied simultaneously to both of the inputs.

Case 5. If case 4 arises at all, then S and R both return to 0 and 0 simultaneously, and then any one of the NOR gates acts faster than the other and assumes the state. For example, if NOR gate 1 is faster than NOR gate 2, then Q_{n+1} will become 1 and this will make $Q'_{n+1} = 0$. Similarly, if NOR gate 2 is faster than NOR gate 1, then Q'_{n+1} will become 1 and this will make $Q_{n+1} = 0$. Hence, this condition is determined by the flip-flop itself. Since this condition cannot be controlled and predicted it is called the *indeterminate* condition.

S'-R' Flip-flop Based on NAND Gates

An S'-R' flip-flop can be constructed with NAND gates by connecting the NAND gates back to back as shown in Figure 7.9. The operation of the S'-R' flip-flop can be analyzed in a similar manner as that employed for the NOR-based S-R flip-flop. This circuit is also not clocked and is classified as an asynchronous sequential circuit. The truth table for the S'-R' flip-flop based on a NAND gate is shown in the table in Figure 7.10.

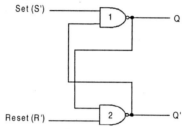

Figure 7.9 NAND-based S'-R' flip-flop.

To analyze the circuit shown in Figure 7.9, we have to remember that a LOW at any input of a NAND gate forces the output to be HIGH, irrespective of the other input. The output of a NAND gate is 0 only if all of the inputs of the NAND gate are 1. The outputs for all the possible conditions as shown in the table in Figure 7.10 are described below.

Inputs		Outputs		Action
S'	R'	Q_{n+1}	Q'_{n+1}	
1	1	Q_n	Q_n	No change
1	0	0	1	Reset
0	1	1	0	Set
0	0	1	1	Forbidden (Undefined)
1	1	–	–	Indeterminate

Figure 7.10

Case 1. For S' = 1 and R' = 1, the flip-flop remains in its present state (Q_n). It means that the next state of the flip-flop does not change, *i.e.*, $Q_{n+1} = 0$ if $Q_n = 0$ and vice versa. First let us assume that $Q_n = 1$ and $Q'_n = 0$. Thus the inputs of NAND gate 1 are 1 and 0, and therefore its output $Q_{n+1} = 1$. This output $Q_{n+1} = 1$ is fed back as the input of NAND gate 2, thereby producing a 0 at the output, as both of the inputs of NAND gate 2 are 1 and 1; so $Q'_{n+1} = 0$ as originally assumed.

Now let us assume the opposite case, *i.e.*, $Q_n = 0$ and $Q'_n = 1$. Thus the inputs of NAND gate 2 are 1 and 0, and therefore its output $Q'_{n+1} = 1$. This output $Q'_{n+1} = 1$ is fed back as the input of NAND gate 1, thereby producing a 0 at the output, as both of the inputs of NAND gate 1 are 1 and 1; so $Q_{n+1} = 0$ as originally assumed. Thus we find that the condition S' = 1 and R' = 1 do not affect the outputs of the flip-flop, which means this is the memory condition of the S'-R' flip-flop.

Case 2. The second input condition is S' = 1 and R' = 0. The 0 at R' input forces the output of NAND gate 2 to be 1 (*i.e.*, $Q'_{n+1} = 1$). Hence both the inputs of NAND gate 1 are 1 and 1 and so its output $Q_{n+1} = 0$. Thus the condition $S' = 1$ and $R' = 0$ will always reset the flip-flop to 0. Now if the R' returns to 1 with S' = 1, the flip-flop will remain in the same state.

Case 3. The third input condition is $S' = 0$ and $R' = 1$. The 0 at S' input forces the output of NAND gate 1 to be 1 (*i.e.*, $Q_{n+1} = 1$). Hence both the inputs of NAND gate 2 are 1 and 1 and so its output $Q'_{n+1} = 0$. Thus the condition $S' = 0$ and $R' = 1$ will always set the flip-flop to 1. Now if the S' returns to 1 with R' = 1, the flip-flop will remain in the same state.

Case 4. The fourth input condition is $S' = 0$ and $R' = 0$. The 0 at R' input and 0 at S' input forces the output of both NAND gate 1 and NAND gate 2 to be 1. Hence both the outputs of NAND gate 1 and NAND gate 2 are 1 and 1; *i.e.*, $Q_{n+1} = 1$ and $Q'_{n+1} = 1$. Hence this condition $S' = 0$ and $R' = 0$ violates the fact that the outputs of a flip-flop will always be the complement of each other. Since the condition violates the basic definition of a flip-flop, it is called the *undefined* condition. Generally, this condition must be avoided by making sure that 0s are not applied simultaneously to both of the inputs.

Case 5. If case 4 arises at all, then S' and R' both return to 1 and 1 simultaneously, and then any one of the NAND gates acts faster than the other and assumes the state. For example, if NAND gate 1 is faster than NAND gate 2, then Q_{n+1} will become 1 and this will make $Q'_{n+1} = 0$. Similarly, if NAND gate 2 is faster than NAND gate 1, then Q'_{n+1} will become 1 and this will make $Q_{n+1} = 0$. Hence, this condition is determined by the flip-flop itself. Since this condition cannot be controlled and predicted it is called the *indeterminate* condition.

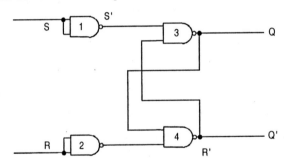

Figure 7.11 An S-R flip-flop using NAND gates.

Thus, comparing the NOR flip-flop and the NAND flip-flop, we find that they basically operate in just the complement fashion of each other. Hence, to convert a NAND-based S'-R' flip-flop into a NOR-based S-R flip-flop, we have to place an inverter at each input of the flip-flop. The resulting circuit is shown in Figure 7.11, which behaves in the same manner as an S-R flip-flop.

7.4 CLOCKED S-R FLIP-FLOP

Generally, synchronous circuits change their states only when clock pulses are present. The operation of the basic flip-flop can be modified by including an additional input to control the behaviour of the circuit. Such a circuit is shown in Figure 7.12.

The circuit shown in Figure 7.12 consists of two AND gates. The clock input is connected to both of the AND gates, resulting in LOW outputs when the clock input is LOW. In this situation the changes in S and R inputs will not affect the state (Q) of the flip-flop. On the other hand, if the clock input is HIGH, the changes in S and R will be passed over by the

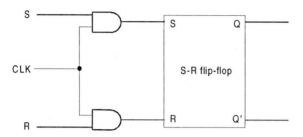

Figure 7.12 Block diagram of a clocked S-R flip-flop.

AND gates and they will cause changes in the output (Q) of the flip-flop. This way, any information, either 1 or 0, can be stored in the flip-flop by applying a HIGH clock input and be retained for any desired period of time by applying a LOW at the clock input. This type of flip-flop is called a *clocked S-R flip-flop*. Such a clocked S-R flip-flop made up of two AND gates and two NOR gates is shown in Figure 7.13.

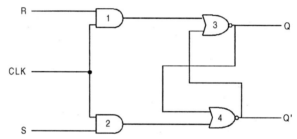

Figure 7.13 A clocked NOR-based S-R flip-flop.

Now the same S-R flip-flop can be constructed using the basic NAND latch and two other NAND gates as shown in Figure 7.14. The S and R inputs control the states of the flip-flop in the same way as described earlier for the unclocked S-R flip-flop. However, the flip-flop only responds when the clock signal occurs. The clock pulse input acts as an enable signal for the other two inputs. As long as the clock input remains 0 the outputs of NAND gates 1 and 2 stay at logic 1. This 1 level at the inputs of the basic NAND-based S-R flip-flop retains the present state.

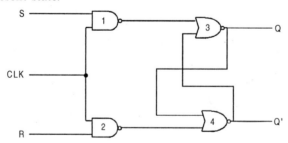

Figure 7.14 A clocked NAND-based S-R flip-flop.

The logic symbol of the S-R flip-flop is shown in Figure 7.15. It has three inputs: S, R, and CLK. The CLK input is marked with a small triangle. The triangle is a symbol that denotes the fact that the circuit responds to an edge or transition at CLK input.

Assuming that the inputs do not change during the presence of the clock pulse, we can express the working of the S-R flip-flop in the form of the truth table in Figure 7.16. Here, S_n and R_n denote the inputs and Q_n the output during the bit time n (Figure 7.2). Q_{n+1} denotes the output after the pulse passes, *i.e.*, in the bit time $n + 1$.

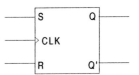

Figure 7.15 Logic symbol of a clocked S-R flip-flop.

Inputs		Output
S_n	R_n	Q_{n+1}
0	0	Q_n
0	1	0
1	0	1
1	1	–

Figure 7.16

Case 1. If $S_n = R_n = 0$, and the clock pulse is not applied, the output of the flip-flop remains in the present state. Even if $S_n = R_n = 0$, and the clock pulse is applied, the output at the end of the clock pulse is the same as the output before the clock pulse, *i.e.*, $Q_{n+1} = Q_n$. The first row of the table indicates that situation.

Case 2. For $S_n = 0$ and $R_n = 1$, if the clock pulse is applied (*i.e.*, CLK = 1), the output of NAND gate 1 becomes 1; whereas the output of NAND gate 2 will be 0. Now a 0 at the input of NAND gate 4 forces the output to be 1, *i.e.*, Q' = 1. This 1 goes to the input of NAND gate 3 to make both the inputs of NAND gate 3 as 1, which forces the output of NAND gate 3 to be 0, *i.e.*, Q = 0.

Case 3. For $S_n = 1$ and $R_n = 0$, if the clock pulse is applied (*i.e.*, CLK = 1), the output of NAND gate 2 becomes 1; whereas the output of NAND gate 1 will be 0. Now a 0 at the input of NAND gate 3 forces the output to be 1, *i.e.*, Q = 1. This 1 goes to the input of NAND gate 4 to make both the inputs of NAND gate 4 as 1, which forces the output of NAND gate 4 to be 0, *i.e.*, Q' = 0.

Case 4. For $S_n = 1$ and $R_n = 1$, if the clock pulse is applied (*i.e.*, CLK = 1), the outputs of both NAND gate 2 and NAND gate 1 becomes 0. Now a 0 at the input of both NAND gate 3 and NAND gate 4 forces the outputs of both the gates to be 1, *i.e.*, Q = 1 and Q' = 1. When the CLK input goes back to 0 (while S and R remain at 1), it is not possible to determine the next state, as it depends on whether the output of gate 1 or gate 2 goes to 1 first.

7.4.1 Preset and Clear

In the flip-flops shown in Figures 7.13 or figure 7.14, when the power is switched on, the state of the circuit is uncertain. It may come to reset (Q = 0) or set (Q = 1) state. But in many applications it is required to initially set or reset the flip-flop., *i.e.*, the initial state of the flip-flop is to be assigned. This is done by using the direct or asynchronous inputs. These inputs are referred to as *preset* (Pr) and *clear* (Cr) inputs. These inputs may be applied at any time between clock pulses and is not in synchronism with the clock. Such an S-R flip-flop containing preset and clear inputs is shown in Figure 7.17. From Figure 7.17, we see that if Pr = Cr = 1, the circuit operates according to the table in Figure 7.16.

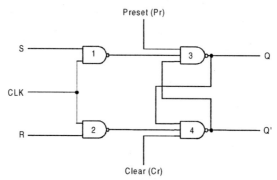

Figure 7.17 An S-R flip-flop with preset and clear.

If Pr = 1 and Cr = 0, the output of NAND gate 4 is forced to be 1, *i.e.*, Q' = 1 and the flip-flop is reset, overwriting the previous state of the flip-flop.

If Pr = 0 and Cr = 1, the output of NAND gate 3 is forced to be 1, *i.e.*, Q = 1 and the flip-flop is set, overwriting the previous state of the flip-flop. Once the state of the flip-flop is established asynchronously, the inputs Pr and Cr must be connected to logic 1 before the next clock is applied.

The condition Pr = Cr = 0 must not be applied, since this leads to an uncertain state.

The logic symbol of an S-R flip-flop with Pr and Cr inputs is shown in Figure 7.18. Here, bubbles are used for Pr and Cr inputs, which indicate these are active low inputs, which means that the intended function is performed if the signal applied to Pr and Cr is LOW. The operation of Figure 7.18 is shown in the table in Figure 7.19. The circuit can be designed such that the asynchronous inputs override the clock, *i.e.*, the circuit can be set or reset even in the presence of the clock pulse.

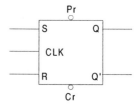

Figure 7.18 Logic symbol of an S-R flip-flop with preset and clear.

Inputs			Output	Operation
CLK	Cr	Pr	Q	performed
1	1	1	Q_{n+1} (Figure 7.3)	Normal flip-flop
0	1	0	1	Preset
0	0	1	0	Clear
0	0	0	–	Uncertain

Figure 7.19

7.4.2 Characteristic Table of an S-R Flip-flop

From the name itself it is very clear that the *characteristic table* of a flip-flop actually gives us an idea about the character, *i.e.*, the working of the flip-flop. Now, from all our above discussions, we know that the next state flip-flop output (Q_{n+1}) depends on the present inputs as well as the present output (Q_n). So in order to know the next state output of a flip-flop, we have to consider the present state output also. The characteristic table of an S-R flip-flop is given in the table in Figure 7.20. From the characteristic table we have to find out the characteristic equation of the S-R flip-flop.

Flip-flop inputs		Present output	Next output
S	R	Q_n	Q_{n+1}
0	0	0	0
0	0	1	1
0	1	0	0
0	1	1	0
1	0	0	1
1	0	1	1
1	1	0	X
1	1	1	X

Figure 7.20

Now we will find out the characteristic equation of the S-R flip-flop from the characteristic table with the help of the Karnaugh map in Figure 7.21.

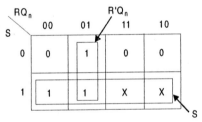

Figure 7.21

From the Karnaugh map above we find the expression for $Q_{n=1}$ as

$$Q_{n+1} = S + R'Q_n. \tag{7.1}$$

Along with the above equation we have to consider the fact that S and R cannot be simultaneously 0. In order to take that fact into account we have to incorporate another equation for the S-R flip-flop. The equation is given below.

$$SR = 0 \tag{7.2}$$

Hence the characteristic equations of an S-R flip-flop are

$$Q_{n+1} = S + R'Q_n$$
$$SR = 0.$$

7.5 CLOCKED D FLIP-FLOP

The D flip-flop has only one input referred to as the D input, or data input, and two outputs as usual Q and Q'. It transfers the data at the input after the delay of one clock pulse at the output Q. So in some cases the input is referred to as a delay input and the flip-flop gets the name *delay* (D) flip-flop. It can be easily constructed from an S-R flip-flop by simply incorporating an inverter between S and R such that the input of the inverter is at the S end and the output of the inverter is at the R end. We can get rid of the undefined condition, *i.e.*, S = R = 1 condition, of the S-R flip-flop in the D flip-flop. The D flip-flop is either used as a delay device or as a latch to store one bit of binary information. The truth table of D flip-flop is given in the table in Figure 7.23. The structure of the D flip-flop is shown in Figure 7.22, which is being constructed using NAND gates. The same structure can be constructed using only NOR gates.

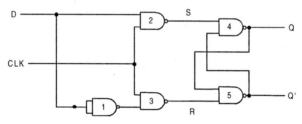

Figure 7.22 A D flip-flop using NAND gates.

Input	Output
D_n	Q_{n+1}
0	0
1	1

Figure 7.23

Case 1. If the CLK input is low, the value of the D input has no effect, since the S and R inputs of the basic NAND flip-flop are kept as 1.

Case 2. If the CLK = 1, and D = 1, the NAND gate 1 produces 0, which forces the output of NAND gate 3 as 1. On the other hand, both the inputs of NAND gate 2 are 1, which gives the output of gate 2 as 0. Hence, the output of NAND gate 4 is forced to be 1, *i.e.*, Q = 1, whereas both the inputs of gate 5 are 1 and the output is 0, *i.e.*, Q' = 0. Hence, we find that when D = 1, after one clock pulse passes Q = 1, which means the output follows D.

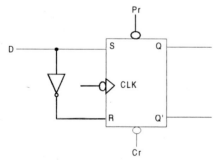

Figure 7.24(a) An S-R flip-flop converted into a D flip-flop.

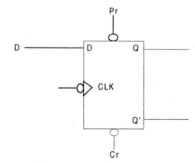

Figure 7.24 (b) the logic symbol of a D flip-flop.

Case 3. If the CLK = 1, and D = 0, the NAND gate 1 produces 1. Hence both the inputs of NAND gate 3 are 1, which gives the output of gate 3 as 0. On the other hand, D = 0 forces the output of NAND gate 2 to be 1. Hence the output of NAND gate 5 is forced to be 1, *i.e.*, Q' = 1, whereas both the inputs of gate 4 are 1 and the output is 0, *i.e.*, Q = 0. Hence, we find that when D = 0, after one clock pulse passes Q = 0, which means the output again follows D.

A simple way to construct a D flip-flop using an S-R flip-flop is shown in Figure 7.24(*a*). The logic symbol of a D flip-flop is shown in Figure 7.24(*b*). A D flip-flop is most often used in the construction of sequential circuits like registers.

7.5.1 Preset and Clear

In the flip-flops shown in Figure 7.22, we can incorporate two asynchronous inputs in order to initially set or reset the flip-flop, *i.e.*, in order to assign the initial state of the flip-flop. These inputs are referred to as *preset* (Pr) and *clear* (Cr) inputs as we did in the case of S-R flip-flops. These inputs may be applied at any time between clock pulses and is not in synchronism with the clock. Such a D flip-flop containing preset and clear inputs is shown in Figure 7.25. From Figure 7.25, we see that if Pr = Cr = 1, the circuit operates according to the table in Figure 7.23.

If Pr = 1 and Cr = 0, the output of NAND gate 5 is forced to be 1, *i.e.*, Q' = 1 and the flip-flop is reset, overwriting the previous state of the flip-flop.

If Pr = 0 and Cr = 1, the output of NAND gate 4 is forced to be 1, *i.e.*, Q = 1 and the flip-flop is set, overwriting the previous state of the flip-flop. Once the state of the flip-flop is established asynchronously, the inputs Pr and Cr must be connected to logic 1 before the next clock is applied.

The condition Pr = Cr = 0 must not be applied, since this leads to an uncertain state.

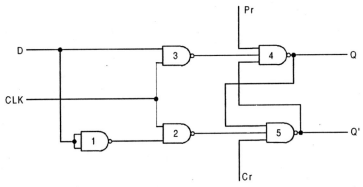

Figure 7.25 A D-type flip-flop with preset and clear.

The logic symbol of a D flip-flop with Pr and Cr inputs is shown in Figure 7.24. Here, bubbles are used for Pr and Cr inputs, which indicate these are active low inputs, which means that the intended function is performed if the signal applied to Pr and Cr is LOW. The operation of Figure 7.25 is shown in the table in Figure 7.26. The circuit can be designed such that the asynchronous inputs override the clock, *i.e.*, the circuit can be set or reset even in the presence of the clock pulse.

Inputs			Output	Operation
CLK	Cr	Pr	Q	performed
1	1	1	Q_{n+1}	Normal flip-flop
0	1	0	1	Preset
0	0	1	0	Clear
0	0	0	–	Uncertain

Figure 7.26

7.5.2 Characteristic Table of a D Flip-flop

As we have already discussed the characteristic equation of an S-R flip-flop, we can similarly find out the characteristic equation of a D flip-flop. The characteristic table of a D flip-flop is given in the table in Figure 7.27. From the characteristic table we have to find out the characteristic equation of the D flip-flop.

Flip-flop inputs	Present output	Next output
D	Q_n	Q_{n+1}
0	0	0
0	1	0
1	0	1
1	1	1

Figure 7.27

Now we will find out the characteristic equation of the D flip-flop from the characteristic table with the help of the Karnaugh map in Figure 7.28.

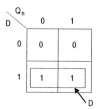

Figure 7.28

From the map, we obtain

$$Q_{n+1} = D. \tag{7.3}$$

Hence, the characteristic equation of a D flip-flop is

$$Q_{n+1} = D.$$

7.6 J-K FLIP-FLOP

A J-K flip-flop has very similar characteristics to an S-R flip-flop. The only difference is that the undefined condition for an S-R flip-flop, i.e., $S_n = R_n = 1$ condition, is also included in this case. Inputs J and K behave like inputs S and R to set and reset the flip-flop respectively. When J = K = 1, the flip-flop is said to be in a *toggle state*, which means the output switches to its complementary state every time a clock passes.

The data inputs are J and K, which are ANDed with Q' and Q respectively to obtain the inputs for S and R respectively. A J-K flip-flop thus obtained is shown in Figure 7.29. The truth table of such a flip-flop is given in Figure 7.32, which is reduced to Figure 7.33 for convenience.

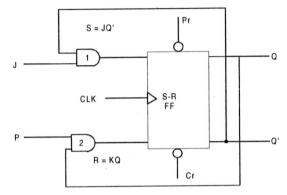

Figure 7.29 An S-R flip-flop converted into a J-K flip-flop.

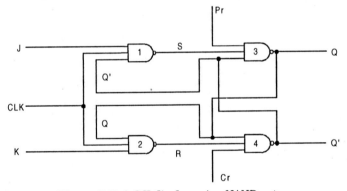

Figure 7.30 A J-K flip-flop using NAND gates.

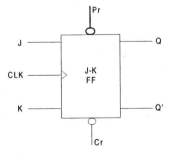

Figure 7.31 Logic symbol of a J-K flip-flop.

It is not necessary to use the AND gates of Figure 7.29, since the same function can be performed by adding an extra input terminal to each of the NAND gates 1 and 2. With this modification incorporated, we get the J-K flip-flop using NAND gates as shown in Figure 7.30. The logic symbol of a J-K flip-flop is shown in Figure 7.31.

Data inputs		Outputs		Inputs to S-R FF		Output
J_n	K_n	Q_n	Q'_n	S_n	R_n	Q_{n+1}
0	0	0	1	0	0	0
0	0	1	0	0	0	1
0	1	0	1	0	0	0
0	1	1	0	0	1	0
1	0	0	1	1	0	1
1	0	1	0	0	0	1
1	1	0	1	1	0	1
1	1	1	0	0	1	0

Figure 7.32

Case 1. When the clock is applied and J = 0, whatever the value of Q'_n (0 or 1), the output of NAND gate 1 is 1. Similarly, when K = 0, whatever the value of Q_n (0 or 1), the output of gate 2 is also 1. Therefore, when J = 0 and K = 0, the inputs to the basic flip-flop are S = 1 and R = 1. This condition forces the flip-flop to remain in the same state.

Inputs		Output
J_n	K_n	Q_{n+1}
0	0	Q_n
0	1	0
1	0	1
1	1	Q'_n

Figure 7.33

Case 2. When the clock is applied and J = 0 and K = 1 and the previous state of the flip-flop is reset (*i.e.*, Q_n = 0 and Q'_n = 1), then S = 1 and R = 1. Since S = 1 and R = 1, the basic flip-flop does not alter the state and remains in the reset state. But if the flip-flop is in set condition (*i.e.*, Q_n = 1 and Q'_n = 0), then S = 1 and R = 0. Since S = 1 and R = 0, the basic flip-flop changes its state and resets.

Case 3. When the clock is applied and J = 1 and K = 0 and the previous state of the flip-flop is reset (*i.e.*, Q_n = 0 and Q'_n = 1), then S = 0 and R = 1. Since S = 0 and R = 1, the basic flip-flop changes its state and goes to the set state. But if the flip-flop is already in set condition (*i.e.*, Q_n = 1 and Q'_n = 0), then S = 1 and R = 1. Since S = 1 and R = 1, the basic flip-flop does not alter its state and remains in the set state.

Case 4. When the clock is applied and J = 1 and K = 1 and the previous state of the flip-flop is reset (*i.e.*, Q_n = 0 and Q'_n = 1), then S = 0 and R = 1. Since S = 0 and R = 1, the basic flip-flop changes its state and goes to the set state. But if the flip-flop is already in set condition (*i.e.*, $Q_n = 1$ and $Q'_n = 0$), then $S = 1$ and $R = 0$. Since $S = 1$ and $R = 0$, the basic flip-flop changes its state and goes to the reset state. So we find that for $J = 1$ and $K = 1$, the flip-flop toggles its state from *set* to *reset* and vice versa. Toggle means to switch to the opposite state.

7.6.1 Characteristic Table of a J-K Flip-flop

As we have already discussed the characteristic equation of an S-R flip-flop, we can similarly find out the characteristic equation of a J-K flip-flop. The characteristic table of a J-K flip-flop is given in the table in Figure 7.34. From the characteristic table we have to find out the characteristic equation of the J-K flip-flop.

Flip-flop inputs		Present output	Next output
J	*K*	Q_n	Q_{n+1}
0	0	0	0
0	0	1	1
0	1	0	0
0	1	1	0
1	0	0	1
1	0	1	1
1	1	0	1
1	1	1	0

Figure 7.34

Now we will find out the characteristic equation of the J-K flip-flop from the characteristic table with the help of the Karnaugh map in Figure 7.35.

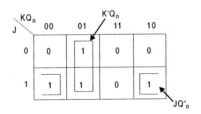

Figure 7.35

From the Karnaugh map, we obtain

$$Q_{n+1} = JQ'_n + K'Q_n. \tag{7.4}$$

Hence, the characteristic equation of a J-K flip-flop is

$$Q_{n+1} = JQ'_n + K'Q_n.$$

7.6.2 Race-around Condition of a J-K Flip-flop

The inherent difficulty of an S-R flip-flop (*i.e.*, $S = R = 1$) is eliminated by using the feedback connections from the outputs to the inputs of gate 1 and gate 2 as shown in Figure

7.30. Truth tables in Figure 7.32 and Figure 7.33 were formed with the assumption that the inputs do not change during the clock pulse (CLK = 1). But the consideration is not true because of the feedback connections.

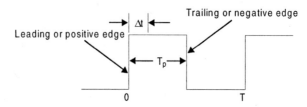

Figure 7.36 A clock pulse.

Consider, for example, that the inputs are J = K = 1 and Q = 1, and a pulse as shown in Figure 7.36 is applied at the clock input. After a time interval Δt equal to the propagation delay through two NAND gates in series, the outputs will change to Q = 0. So now we have J = K = 1 and Q = 0. After another time interval of Δt the output will change back to Q = 1. Hence, we conclude that for the time duration of t_p of the clock pulse, the output will oscillate between 0 and 1. Hence, at the end of the clock pulse, the value of the output is not certain. This situation is referred to as a *race-around condition.*

Generally, the propagation delay of TTL gates is of the order of nanoseconds. So if the clock pulse is of the order of microseconds, then the output will change thousands of times within the clock pulse. This race-around condition can be avoided if $t_p < \Delta t < T$. Due to the small propagation delay of the ICs it may be difficult to satisfy the above condition. A more practical way to avoid the problem is to use the master-slave (M-S) configuration as discussed below.

7.6.3 Master-Slave J-K Flip-flop

A master-slave (M-S) flip-flop is shown in Figure 7.37. Basically, a master-slave flip-flop is a system of two flip-flops—one being designated as *master* and the other is the *slave.* From the figure we see that a clock pulse is applied to the master and the inverted form of the same clock pulse is applied to the slave.

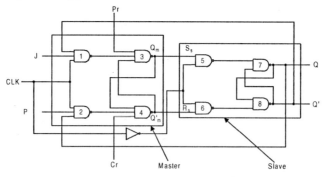

Figure 7.37 A master-slave J-K flip-flop.

When CLK = 1, the first flip-flop (*i.e.,* the master) is enabled and the outputs Q_m and Q'_m respond to the inputs J and K according to the table shown in Figure 7.13. At this time the second flip-flop (*i.e.,* the slave) is disabled because the CLK is LOW to the second flip-flop. Similarly, when CLK becomes LOW, the master becomes disabled and the slave becomes active, since now the CLK to it is HIGH. Therefore, the outputs Q and Q' follow the outputs Q_m and Q'_m respectively. Since the second flip-flop just follows the first one, it is referred to as a slave and the first one is called the master. Hence, the configuration is referred to as a master-slave (M-S) flip-flop.

In this type of circuit configuration the inputs to the gates 5 and 6 do not change at the time of application of the clock pulse. Hence the race-around condition does not exist. The state of the master-slave flip-flop, shown in Figure 7.37, changes at the negative transition (trailing edge) of the clock pulse. Hence, it becomes negative triggering a master-slave flip-flop. This can be changed to a positive edge triggering flip-flop by adding two inverters to the system—one before the clock pulse is applied to the master and an additional one in between the master and the slave. The logic symbol of a negative edge master-slave is shown in Figure 7.38.

The system of master-slave flip-flops is not restricted to J-K master-slave only. There may be an S-R master-slave or a D master-slave, etc., in all of them the slave is an S-R flip-flop, whereas the master changes to J-K or S-R or D flip-flops.

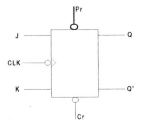

Figure 7.38 A negative edge-transition master-slave J-K flip-flop.

7.7 T FLIP-FLOP

With a slight modification of a J-K flip-flop, we can construct a new flip-flop called a T flip-flop. If the two inputs J and K of a J-K flip-flop are tied together it is referred to as a T flip-flop. Hence, a T flip-flop has only one input T and two outputs Q and Q'. The name T flip-flop actually indicates the fact that the flip-flop has the ability to toggle. It has actually only two states—*toggle state* and *memory state*. Since there are only two states, a T flip-flop is a very good option to use in counter design and in sequential circuits design where switching an operation is required. The truth table of a T flip-flop is given in Figure 7.39.

T	Q_n	Q_{n+1}
0	0	0
0	1	1
1	0	1
1	1	0

Figure 7.39

If the T input is in 0 state (*i.e.*, J = K = 0) prior to a clock pulse, the Q output will not change with the clock pulse. On the other hand, if the T input is in 1 state (*i.e.*, J = K = 1) prior to a clock pulse, the Q output will change to Q' with the clock pulse. In other words, we may say that, if T = 1 and the device is clocked, then the output toggles its state.

The truth table shows that when T = 0, then $Q_{n+1} = Q_n$, i.e., the next state is the same as the present state and no change occurs. When T = 1, then $Q_{n+1} = Q'_n$, i.e., the state of the flip-flop is complemented. The circuit diagram of a T flip-flop is shown in Figure 7.40 and the block diagram of the flip-flop is shown in Figure 7.41.

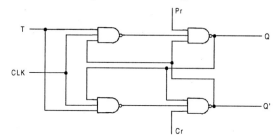

Figure 7.40 A T flip-flop.

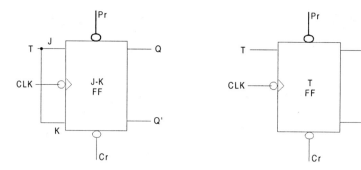

Figure 7.41(a) A J-K flip-flop converted into a
T flip-flop.

Figure 7.41(b) the logic symbol of a
T flip-flop.

7.7.1 Characteristic Table of a T Flip-flop

As we have already discussed the characteristic equation of a J-K flip-flop, we can similarly find out the characteristic equation of a T flip-flop. The characteristic table of a T flip-flop is given in Figure 7.42. From the characteristic table we have to find out the characteristic equation of the T flip-flop.

Flip-flop inputs	Present output	Next output
T	Q_n	Q_{n+1}
0	0	0
0	1	1
1	0	1
1	1	0

Figure 7.42

Now we will find out the characteristic equation of the T flip-flop from the characteristic table with the help of the Karnaugh map in Figure 7.43.

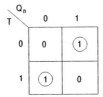

Figure 7.43

From the Karnaugh map, the Boolean expression of Q_{n+1} is derived as

$$Q_{n+1} = TQ'_n + T'Q_n. \qquad (7.5)$$

Hence, the characteristic equation of a T flip-flop is

$$Q_{n+1} = TQ'_n + T'Q_n.$$

7.8 TOGGLING MODE OF S-R AND D FLIP-FLOPS

Though an S-R flip-flop cannot be converted into a T flip-flop since S = R = 1 is not allowed, but an S-R flip-flop can be made to work in toggle mode, where the output Q changes with every clock pulse. The circuit is shown in Figure 7.44. The toggle mode of operation for a D flip-flop is also shown in Figure 7.45.

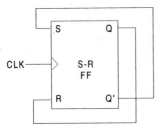   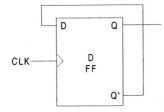

Figure 7.44 An S-R flip-flop in toggle mode.   **Fig 7.45** A D flip-flop in toggle mode.

If at any instant, Q = 1 and Q' = 0, then S = 0 and R = 1. Hence, with the clock pulse the S-R flip-flop gets reset, *i.e.*, Q = 0 and Q' = 1. Again, if at any instant, Q=0 and Q' = 1, then S = 1 and R = 0. Hence, with the clock pulse the S-R flip-flop gets set, *i.e.*, Q = 1 and Q' = 0. Hence, the flip-flop acts like a toggle flip-flop where the output is changing with each clock pulse.

Similarly, for a D flip-flop, if at any instant, Q = 1 and Q' = 0, then D = 0. Hence, with the clock pulse the D flip-flop gets reset, *i.e.*, Q = 0 and Q' = 1. Again, if at any instant, Q = 0 and Q' = 1, then D = 1. Hence, with the clock pulse the D flip-flop gets set, *i.e.*, Q = 1 and Q' = 0. Hence, the flip-flop acts like a toggle flip-flop where the output is changing with each clock pulse.

7.9 TRIGGERING OF FLIP-FLOPS

Flip-flops are synchronous sequential circuits. This type of circuit works with the application of a synchronization mechanism, which is termed as a *clock*. Based on the specific interval or point in the clock during or at which triggering of the flip-flop takes place, it can be classified into two different types—*level triggering* and *edge triggering*.

A clock pulse starts from an initial value of 0, goes momentarily to 1, and after a short interval, returns to the initial value.

7.9.1 Level Triggering of Flip-flops

If a flip-flop gets enabled when a clock pulse goes HIGH and remains enabled throughout the duration of the clock pulse remaining HIGH, the flip-flop is said to be a *level triggered flip-flop*. If the flip-flop changes its state when the clock pulse is positive, it is termed as a *positive level triggered flip-flop*. On the other hand, if a NOT gate is introduced in the clock input terminal of the flip-flop, then the flip-flop changes its state when the clock pulse is negative, it is termed as a *negative level triggered flip-flop*.

The main drawback of level triggering is that, as long as the clock pulse is active, the flip-flop changes its state more than once or many times for the change in inputs. If the inputs do not change during one clock pulse, then the output remains stable. On the other hand, if the frequency of the input change is higher than the input clock frequency, the output of the flip-flop undergoes multiple changes as long as the clock remains active. This can be overcome by using either master-slave flip-flops or the edge-triggered flip-flop.

7.9.2 Edge-triggering of Flip-flops

A clock pulse goes from 0 to 1 and then returns from 1 to 0. Figure 7.46 shows the two transitions and they are defined as the *positive edge* (0 to 1 transition) and the *negative edge* (1 to 0 transition). The term *edge-triggered* means that the flip-flop changes its state only at either the positive or negative edge of the clock pulse.

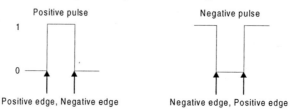

Figure 7.46 Clock pulse transition.

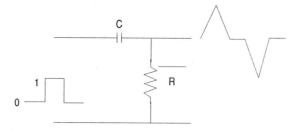

Figure 7.47 RC differentiator circuit for edge triggering.

One way to make the flip-flop respond to only the edge of the clock pulse is to use capacitive coupling. An RC circuit is shown in Figure 7.47, which is inserted in the clock input of the flip-flop. By deliberate design, the RC time constant is made much smaller than the clock pulse width. The capacitor can charge fully when the clock goes HIGH. This exponential charging produces a narrow positive spike across the resistor. Later, the trailing edge of the pulse results in a narrow negative spike. The circuit is so designed that one of the spikes (either the positive or negative) is neglected and the edge triggering occurs due to the other spike.

7.10 EXCITATION TABLE OF A FLIP-FLOP

The truth table of a flip-flop is also referred to as the characteristic table of a flip-flop, since this table refers to the operational characteristics of the flip-flop. But in designing sequential circuits, we often face situations where the present state and the next state of the flip-flop is specified, and we have to find out the input conditions that must prevail for the desired output condition. By present and next states we mean to say the conditions before and after the clock pulse respectively. For example, the output of an S-R flip-flop before the clock pulse is $Q_n = 1$ and it is desired that the output does not change when the clock pulse is applied.

Now from the characteristic table of an S-R flip-flop (Figure 7.20), we obtain the following conditions:

1. S = R = 0 (second row)
2. S = 1, R = 0 (sixth row).

We come to the conclusion from the above conditions that the R input must be 0, whereas the S input may be 0 or 1 (*i.e.*, don't-care). Similarly, for all possible situations, the input conditions can be found out. A tabulation of these conditions is known as an *excitation table*. The table in Figure 7.48 gives the excitation table for S-R, D, J-K, and T flip-flops. These conditions are derived from the corresponding characteristic tables of the flip-flops.

Present State (Q_n)	Next State (Q_{n+1})	S-R FF		D-FF	J-K FF		T-FF
		S_n	R_n	D_n	J_n	K_n	T_n
0	0	0	X	0	0	X	0
0	1	1	0	1	1	X	1
1	0	0	1	0	X	1	1
1	1	X	0	1	X	0	0

Figure 7.48 Excitation table of different flip-flops.

7.11 INTERCONVERSION OF FLIP-FLOPS

In many applications, we are being given a type of flip-flop, whereas we may require some other type. In such cases we may have to convert the given flip-flop to our required flip-flop. Now we may follow a general model for such conversions of flip-flops. The model is shown in Figure 7.49.

From the model we see that it is required to design the conversion logic for converting new input definitions into input codes that will cause the given flip-flop to work like the desired flip-flop. To design the conversion logic we need to combine the excitation table for both flip-flops and make a truth table with data input(s) and Q as the inputs and the input(s) of the given flip-flop as the output(s).

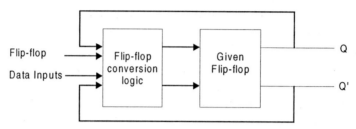

Figure 7.49 General model for conversion of one type of flip-flop to another type

7.11.1 Conversion of an S-R Flip-flop to a D Flip-flop

The excitation tables of S-R and D flip-flops are given in the table in Figure 7.48 from which we make the truth table given in Figure 7.50.

FF data inputs	Output	S-R FF inputs	
D	Q	S	R
0	0	0	X
1	0	1	0
0	1	0	1
1	1	X	0

Figure 7.50

From the table in Figure 7.50, we make the Karnaugh maps for inputs S and R as shown in Figure 7.51(a) and Figure 7.51(b).

Figure 7.51(a)   **Figure 7.51(b)**

Simplifying with the help of the Karnaugh maps, we obtain $S = D$ and $R = D'$.

Hence the circuit may be designed as in Figure 7.52.

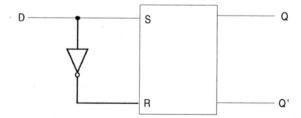

Figure 7.52 A D flip-flop using an S-R flip-flop.

7.11.2 Conversion of an S-R Flip-flop to a J-K Flip-flop

The excitation tables of S-R and J-K flip-flops are given in the table in Figure 7.48 from which we make the truth table given in Figure 7.53.

FF data inputs		Output	S-R FF inputs	
J	K	Q	S	R
0	0	0	0	X
0	1	0	0	X
1	0	0	1	0
1	1	0	1	0
0	1	1	0	1
1	1	1	0	1
0	0	1	X	0
1	0	1	X	0

Figure 7.53

From the truth table in Figure 7.53, the Karnaugh map is prepared as shown in Figure 7.54(a) and Figure 7.54(b).

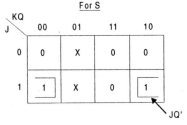

Figure 7.54(a)

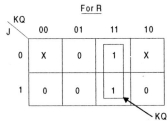

Figure 7.54(b)

Hence we get the Boolean expression for S and R as

$$S = JQ'$$

and
$$R = KQ.$$

Hence the circuit may be realized as in Figure 7.55.

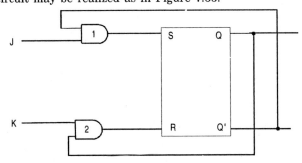

Figure 7.55 A J-K flip-flop using an S-R flip-flop.

7.11.3 Conversion of an S-R Flip-flop to a T Flip-flop

The excitation tables of S-R and T flip-flops are given in Figure 7.48 from which we make the truth table given in Figure 7.56.

FF data inputs	Output	S-R FF inputs	
T	Q	S	R
0	0	0	X
1	0	1	0
1	1	0	1
0	1	X	0

Figure 7.56

From the table in Figure 7.56, we prepare the Karnaugh maps as per Figure 7.57(a) and Figure 7.57(b).

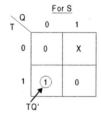

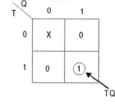

Figure 7.57(a)   **Figure 7.57(b)**

Hence we get,

$$S = TQ'$$

and $$R = TQ.$$

Hence the circuit may be designed as in Figure 7.58.

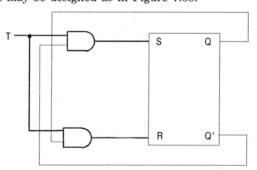

Figure 7.58 A T flip-flop using an S-R flip-flop.

7.11.4 Conversion of a D Flip-flop to an S-R Flip-flop

The excitation tables of S-R and D flip-flops are given in the table in Figure 7.48 from which we derive the truth table in Figure 7.59.

FF data inputs		Output	D FF inputs
S	R	Q	D
0	0	0	0
0	1	0	0
1	0	0	1
0	1	1	0
0	0	1	1
1	0	1	1

Figure 7.59

The Karnaugh map is shown in Figure 7.60 according to the truth table in Figure 7.59.

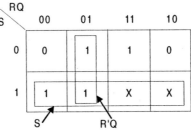

Figure 7.60

Hence we get

$$D = S + R'Q.$$

Hence the circuit may be realized as in Figure 7.61.

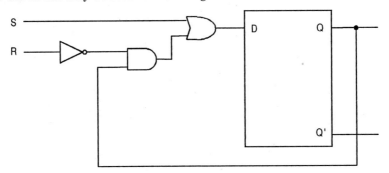

Figure 7.61 An S-R flip-flop using a D flip-flop.

7.11.5 Conversion of a D Flip-flop to a J-K Flip-flop

The excitation tables of J-K and D flip-flops are given in the table in Figure 7.48 from which we make the truth table given in Figure 7.62.

From the truth table in Figure 7.62, the Karnaugh map is drawn as in Figure 7.63.

FF data inputs		Output	D FF inputs
J	K	Q	D
0	0	0	0
0	1	0	0
1	0	0	1
1	1	0	1
0	1	1	0
1	1	1	0
0	0	1	1
1	0	1	1

Figure 7.62

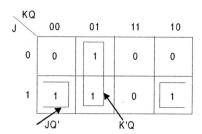

Figure 7.63

From the Karnaugh map above, the Boolean expression for D is derived as

D = JQ' + K'Q.

Hence the circuit may be designed as in Figure 7.64.

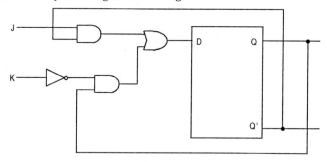

Figure 7.64 A J-K flip-flop using a D flip-flop.

7.11.6 Conversion of a D Flip-flop to a T Flip-flop

The excitation tables of D and T flip-flops are given in the table in Figure 7.48 from which we derive the required truth table as given in Figure 7.65.

From the truth table in Figure 7.65, we make the Karnaugh map in Figure 7.66.

FF data inputs	Output	D FF inputs
T	Q	D
0	0	0
1	0	1
1	1	0
0	1	1

Figure 7.65

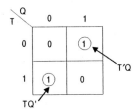

Figure 7.66

The Boolean expression we get is

$$D = TQ' + T'Q.$$

Hence the circuit may be designed as in Figure 7.67.

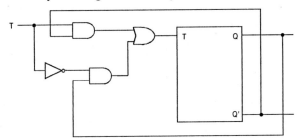

Figure 7.67 A T flip-flop using a D flip-flop.

7.11.7 Conversion of a J-K Flip-flop to a D Flip-flop

The excitation tables of J-K and D flip-flops are given in Figure 7.48 from which we make the truth table in Figure 7.68.

FF data inputs	Output	J-K FF inputs	
D	Q	J	K
0	0	0	X
1	0	1	X
0	1	X	1
1	1	X	0

Figure 7.68

The Karnaugh maps are prepared for J and K as in Figure 7.69(a) and Figure 7.69(b).

Figure 7.69(a)   Figure 7.69(b)

The Boolean expression for J and K are obtained as,

$$J = D$$

and $$K = D'.$$

Hence the circuit may be designed as in Figure 7.70.

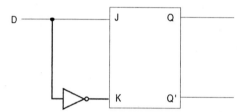

Figure 7.70 A D flip-flop using a J-K flip-flop.

7.11.8 Conversion of a J-K Flip-flop to a T Flip-flop

The excitation tables of J-K and T flip-flops are given in Figure 7.48 from which the required truth table is derived in Figure 7.71.

FF data inputs	Output	J-K FF inputs	
T	Q	J	K
0	0	0	X
1	0	1	X
1	1	X	1
0	1	X	0

Figure 7.71

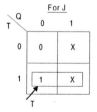

Figure 7.72(a)

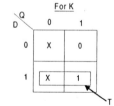

Figure 7.72(b)

The Karnaugh maps for J and K are prepared as in Figure 7.72(a) and Figure 7.72(b), according to the truth table described above in Figure 7.71.

Hence we get,

$$J = T$$

and
$$K = T.$$

The circuit may be realized as in Figure 7.73.

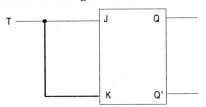

Figure 7.73 A T flip-flop using a J-K flip-flop.

7.11.9 Conversion of a T Flip-flop to an S-R Flip-flop

The truth table for the relation between S-R and T flip-flops is derived as in Figure 7.74, from the excitation table mentioned in Figure 7.48.

FF data inputs		Output	D FF inputs
S	R	Q	T
0	0	0	0
0	1	0	0
1	0	0	1
0	1	1	1
0	0	1	0
1	0	1	0

Figure 7.74

From the truth table in Figure 7.74, we make the Karnaugh map as shown in Figure 7.75.

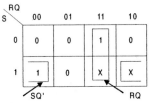

Figure 7.75

We obtain the Boolean expression,

$$T = SQ' + RQ.$$

Hence the circuit may be designed as shown in Figure 7.76.

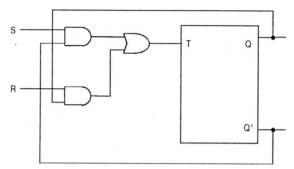

Figure 7.76 An S-R flip-flop using a T flip-flop.

7.11.10 Conversion of a T Flip-flop to a J-K Flip-flop

The excitation tables of J-K and T flip-flops are given in Figure 7.48, from which we make the truth table given in Figure 7.77.

FF data inputs		Output	T FF inputs
J	K	Q	T
0	0	0	0
0	1	0	0
1	0	0	1
1	1	0	1
0	1	1	1
1	1	1	1
0	0	1	0
1	0	1	0

Figure 7.77

From the truth table in Figure 7.77, we make the Karnaugh map in Figure 7.78.

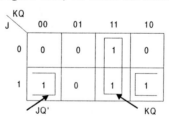

Figure 7.78

We get,

$$T = JQ' + KQ.$$

Hence the circuit may be designed as in Figure 7.79.

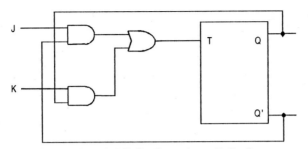

Figure 7.79 A J-K flip-flop using a T flip-flop.

7.11.11 Conversion of a T Flip-flop to a D Flip-flop

The excitation tables of D and T flip-flops are given in Figure 7.48 from which we make the truth table given in Figure 7.80.

FF data inputs	Output	T FF inputs
D	Q	T
0	0	0
1	0	1
0	1	1
1	1	0

Figure 7.80

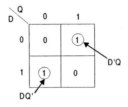

Figure 7.81

The Karnaugh map is shown in Figure 7.81 and the Boolean expression is derived as

$$T = D'Q + DQ'.$$

Hence the circuit may be designed as shown in Figure 7.82.

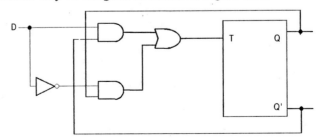

Figure 7.82 A D flip-flop using a T flip-flop.

7.12 SEQUENTIAL CIRCUIT MODEL

The model for a general sequential circuit is shown in Figure 7.83. The present state of the circuit is stored in the memory element. The memory can be any device capable of storing enough information to specify the state of the circuit.

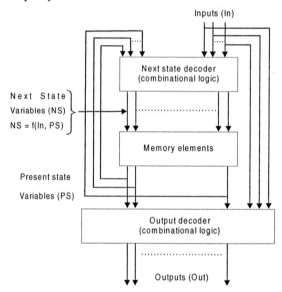

Figure 7.83 General sequential circuit model.

The next state (NS) of the circuit is determined by the present state (PS) of the circuit and by the inputs (In). The function of the Next state decoder logic is to decode the external inputs and the present state of the circuit and to generate an output called the Next state variable. These next state variables will become the present state variables when the memory loads them. This process is called a state change. Thus, sequential circuit is a feedback system where the present state of the circuit is fed back to the next state decoder and used along with the input to determine the next state.

The output (Out) of the circuit is determined by the present state of the machine and possibly by the input of the circuit. The output of a synchronous machine may be clocked, just as the state transition is clocked.

7.13 CLASSIFICATION OF SEQUENTIAL CIRCUITS

From the general sequential circuit model discussed in the preceding section, shown in Figure 7.83, sequential circuits are generally classified into five different classes:

1. Class A circuits
2. Class B circuits
3. Class C circuits
4. Class D circuits
5. Class E circuits.

The Class A circuit is defined as a MEALY machine, named after G. H. Mealy. The basic property of a Mealy machine is that the output is a function of the present input conditions and the present state of the circuit. The model of a Mealy machine is the same as shown in Figure 7.83.

The Class B and Class C circuits are generally defined as a MOORE machine, named after E. F. Moore. In these types of circuits the output is strictly a function of the present state (PS) of the circuit inputs. The block diagram of Class B and Class C circuits are shown in Figure 7.84 and Figure 7.85 respectively.

Both Mealy and Moore circuits are widely used. Even in some circuits a combination of both types are used. Class A, B, and C circuits with a single input form the general model for a counter circuit in which the events to be counted are entered directly into the memory element or through the next state decoder. Also, these circuits are equally applicable to both synchronous and asynchronous circuits.

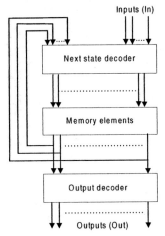

Figure 7.84 Class B circuit (MOORE machine with an output decoder).

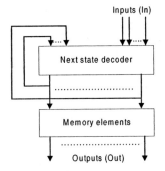

Figure 7.85 Class C circuit (MOORE machine without an output decoder).

The minimum number of inputs to any of these circuits is one. For synchronous circuits, the single input is clock input. The block diagram connections for Class D and Class E sequential circuits are shown in Figure 7.86 and Figure 7.87 respectively.

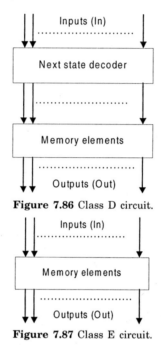

Figure 7.86 Class D circuit.

Figure 7.87 Class E circuit.

7.14 ANALYSIS OF SEQUENTIAL CIRCUITS

The behavior of a sequential circuit is determined from the inputs, the outputs, and the states of the flip-flops. Both the outputs and the next state are a function of the inputs and the present state. The analysis of sequential circuits consists of obtaining a table or a diagram for the time sequence of inputs, outputs, and internal states. Boolean expressions can be written that describe the behavior of the sequential circuits. We first introduce a specific example of a clocked sequential circuit to understand its behavior.

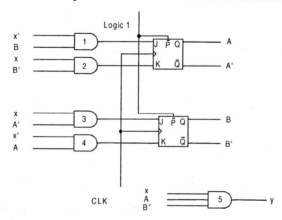

Figure 7.88 Example of a clocked sequential circuit.

7.14.1 State Table

The time sequence of inputs, outputs, and flip-flop states may be enumerated in a state table. The state table for the circuit in Figure 7.88 is shown in the table in Figure 7.89. Here in the table there are three sections designated as present state, next state, and output. The present state designates the states of the flip-flops before the occurrence of the clock pulse. The next state designates the states of the flip-flops after the application of the clock pulse. The output section shows the values of the output variables during the present state. Again, both the output and the next state sections have two columns, one for $x = 0$ and the other for $x = 1$.

The analysis of the circuit can start from any arbitrary state. In our example, we start the analysis from the initial state 00. When the present state is 00, A = 0 and B = 0. From the logic diagram, with $x = 0$, we find both AND gates 1 and 2 produce logic 0 signal and hence the next state remains unchanged. Also, B flip-flop for both AND gates 3 and 4 produce logic 0 signal and hence the next state of B also remains unchanged. Hence, with the clock pulse, flip-flop A and B are both in the memory state, making the next state 00. Similarly, with A = 0 and B = 0, with $x = 1$, we find that gate 1 produces logic 0, whereas gate 2 produces logic 1. Again, with the same condition, gate 3 produces logic 1 whereas gate 4 produces logic 0. Hence, with the clock pulse, flip-flop A is cleared and B is set, making the next state 01. This information is listed in the first row of the state table.

Present	Next state		Output	
state	$x = 0$	$x = 1$	$x = 0$	$x = 1$
AB	AB	AB	y	y
00	00	01	0	0
01	11	01	0	0
10	10	00	0	1
11	10	11	0	0

Figure 7.89 State table.

In a similar manner, we can derive the other conditions of the state table also. When the present state is 01, A = 0 and B = 1. From the logic diagram, with $x = 0$, we find gate 1 produces logic 1 signal and gate 2 produces logic 0. For B flip-flop both gates 3 and 4 produce logic 0 signal and hence the next state of B remains unchanged. Hence, with the clock pulse, flip-flop A is set and B remains in the memory state, making the next state 11. Similarly, with A = 0 and B = 1, with $x = 1$, we find that both gates 1 and 2 produce logic 0. Again, with the same condition, both gates 3 and 4 produce logic 0. Hence, with the clock pulse, both flip-flops A and B remain in the memory state, making the next state 01. This information is listed in the second row of the state table.

When the present state is 10, A = 1 and B = 0. From the logic diagram, with $x = 0$, we find both gates 1 and 2 produce logic 0. For B flip-flop gate 3 produces logic 0 signal but gate 4 produces logic 1. Hence, with the clock pulse, flip-flop A remains in the memory state and B is reset, making the next state 10. Similarly, with A = 1 and B = 0, with $x = 1$, we find that gate 1 produces logic 0, whereas gate 2 produces logic 1. Again, with the same condition, both gates 3 and 4 produce logic 0. Hence, with the clock pulse, A is reset and B remains in the memory state, making the next state 00. This information is listed in the third row of the state table.

Finally when the present state is 11, A = 1 and B = 1. From the logic diagram, with $x = 0$, we find gate 1 produces logic 1 and gate 2 produces logic 0. For B flip-flop gate 3 produces logic 0 signal but gate 4 produces logic 1. Hence, with the clock pulse, flip-flop A remains in the memory state and B is reset, making the next state 10. Similarly, with A = 1 and B = 1, with $x = 1$, we find that both gates 1 and 2 produce logic 0. Again, with the same condition, both gates 3 and 4 produce logic 0. Hence, with the clock pulse, both A and B remain in the memory state, making the next state 11. This information is listed in the last row of the state table.

The entries in the output section are easier to derive. In our example, output y is equal to 1 only when $x = 1$, A = 1, and B = 0. Hence the output columns are marked with 0s except when the present state is 10 and input $x = 1$, for which y is marked as 1.

The state table of any sequential circuit is obtained by the same procedure used in the example. In general, a sequential circuit with m flip-flops and n input variables will have $2m$ rows, one for each state. The next state and output sections will have $2n$ columns, one for each input combination.

The external output of a sequential circuit may come from memory elements or logic gates. The output section is only included in the state table if there are outputs from logic gates. Any external output taken directly from a flip-flop is already listed in the present state of the state table.

7.14.2 State Diagram

All the information available in the state table may be represented graphically in the state diagram.

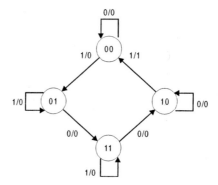

Figure 7.90 State diagram for the circuit in Figure 7.88.

In the diagram, a state is represented by a circle and the transitions between states is indicated by direct arrows connecting the circles. The binary number inside each circle identifies the state the circle represents. The direct arrows are labeled with two binary numbers separated by a /. The number before the / represents the value of the external input, which causes the state transition, and the number after the / represents the value of the output during the present state. For example, the directed arrow from the state 11 to 10 while $x = 0$ and $y = 0$, and that on the termination of the next clock pulse, the circuit goes to the next state 10. A directed arrow connecting a circle with itself indicates that no change of the state occurs.

There is no difference between a state table and a state diagram except in the manner of representation. The state table is easier to derive from a given logic diagram and the state diagram directly follows the state table. The state diagram gives a pictorial form of the state transitions and hence is easier to interpret.

7.14.3 State Equation

A state equation is an algebraic expression that specifies the conditions for a flip-flop state transition. The left side of the equation denotes the next state of the flip-flop and the right side a Boolean function that specifies the present state conditions that make the next state equal to 1. The state equation is derived directly from a state table. For example, the state equation for flip-flop A can be derived from the table in Figure 7.89. From the next state columns we find that flip-flop A goes to the 1 state four times: when $x = 0$ and AB = 01 or 10 or 11, or when $x = 1$ and AB = 11. This can be expressed algebraically in a state equation as follows:

$$A \, (t + 1) \; = \; (A'B + AB' + AB)x' + ABx.$$

Similarly, from the next state columns we find that flip-flop B goes to the 1 state four times: when $x = 0$ and AB = 01, or when $x = 1$ and AB = 00 or 01 or 11. This can be expressed algebraically in a state equation as follows:

$$B \, (t + 1) \; = \; A'Bx' + (A'B' + A'B + AB)x.$$

The right-hand side of the state equation is a Boolean function for the present state. When this function is equal to 1, the occurrence of a clock pulse causes flip-flop A or flip-flop B to have a next state of 1. When this function is equal to 0, the occurrence of a clock pulse causes flip-flop A or flip-flop B to have a next state of 0. The left side of the equation identifies the flip-flop by its letter symbol, followed by the time function designation $(t + 1)$, to emphasize that this value is to be reached by the flip-flop one pulse sequence later. The state equation for flip-flop A and B are simplified algebraically below.

Hence, we get

$$
\begin{aligned}
A \, (t + 1) \; &= \; (A'B + AB' + AB)x' + ABx \\
&= \; (Bx')A' + AB'x' + AB \\
&= \; (Bx')A' + (B + B'x')A \\
&= \; (Bx')A' + (B + x')A \\
&= \; (Bx')A' + (B'x)A.
\end{aligned}
$$

If we let $Bx' = J$ and $B'x = K$, we obtain the relationship:

$$A \, (t + 1) \; = \; JA' + KA.$$

which is the characteristic equation of the J-K flip-flop. This relationship between the state equation and the characteristic equation can be justified from inspection of the logic diagram in Figure 7.88. In it we find that the J input of flip-flop A is equal to the Boolean function Bx' and the K input is equal to $B'x$.

Similarly, for flip-flop B we get

$$
\begin{aligned}
B \, (t + 1) \; &= \; A'Bx' + (A'B' + A'B + AB)x \\
&= \; (A'x)B' + A'Bx' + Bx \\
&= \; (A'x)B' + (x + A'x')B \\
&= \; (A'x)B' + (x + A')B \\
&= \; (A'x)B' + (Ax')B.
\end{aligned}
$$

If we let A'x = J and Ax' = K, we obtain the relationship:

$$B(t + 1) = JB' + KB$$

which is the characteristic equation of the J-K flip-flop. In the diagram in Figure 7.88, we find that the J input of flip-flop B is equal to the Boolean function A'x and the K input is equal to Ax'.

7.15 DESIGN PROCEDURE OF SEQUENTIAL CIRCUITS

The design of a sequential circuit follows certain steps. The steps may be listed as follows:

1. The word description of a circuit may be given accompanied with a state diagram, or timing diagram, or other pertinent information.

2. Then from the given state diagram the state table has to be prepared.

3. If the state reduction mechanism is possible, then the number of states may be reduced.

4. After state reduction, assign binary values to the states if the states contain letter symbols.

5. Then the number of flip-flops required is to be determined. Each flip-flop is assigned a letter symbol.

6. Then the choice has to be made regarding the type of flip-flop to be used.

7. With the help of a state table and the flip-flop excitation table the circuit excitation and the output tables have to be determined.

8. Then using some simplification technique *e.g.*, a Karnaugh map or some other method, the circuit output functions and the flip-flop input functions have to be determined.

9. Then the logic diagram has to be drawn.

Although certain steps have been specified for designing the sequential circuit, the procedure can be shortened with experience. A sequential circuit is made up of flip-flops and combinational gates. One of the most important parts is the choice of flip-flop. From the excitation table of different flip-flops we see that the J-K flip-flop excitation table contains the maximum number of don't-care conditions. Hence, for designing any sequential circuit, it will be most simplified if the circuit is designed with, J-K flip-flop.

The number of flip-flops is determined by the number of states. A circuit may have unused binary states if the total number of states is less than $2m$. The unused states are taken as don't-care conditions during the design of the combinational part of the circuit.

Any design process must consider the problem of minimizing the cost of the final circuit. The most obvious cost reductions are reductions in the number of flip-flops and the number of gates. The reduction of the number of flip-flops in a sequential circuit is referred to as the state reduction. Since m flip-flops produce $2m$ states, a reduction in the number of states may (or may not) result in a reduction of the number of flip-flops. State reduction algorithms are concerned with procedures for reducing the number of states in a state table while keeping the external input-output requirements unchanged.

An algorithm for the state reduction is given here. If two states in a state table are equivalent, one of them can be removed without altering the input-output relationships.

Now two states are said to be equivalent if, for each member of the set of inputs, they give exactly the same output and send the circuit to the same state or to an equivalent state. We will discuss the state reduction problem with an example in this section later on.

In certain cases the states are specified in letter symbols. In such cases there comes another factor, called state assignment. State assignment procedures are concerned with methods for assigning binary values to states in such a way as to reduce the cost of the combinational circuit that drives the flip-flop. For any problem there may be a number of different state assignments leading to different combinational parts of the sequential circuit. The most common criterion is that the chosen assignment should result in a simple combinational circuit for the flip-flop inputs. However, to date, there are no state assignment procedures that guarantee a minimal-cost combinational circuit. In fact, state assignment is one of the most challenging problems of sequential circuit design.

We now wish to design the clocked sequential circuit whose state diagram is given below.

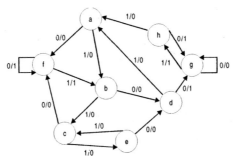

Figure 7.91 State diagram.

The state table for the state diagram shown in Figure 7.91 is shown in the table in Figure 7.92.

Present	*Next state*		*Output*	
state	*x = 0*	*x = 1*	*x = 0*	*x = 1*
a	f	b	0	0
b	d	c	0	0
c	f	e	0	0
d	g	a	1	0
e	d	c	0	0
f	f	b	1	1
g	g	h	0	1
h	g	a	1	0

Figure 7.92 State table.

We now look for two equivalent states, and find that *d* and *h* are two such states; they both go to *g* and *a* and have outputs of 1 and 0 for $x = 0$ and $x = 1$, respectively. Therefore, states *d* and *h* are equivalent; one can be removed. Similarly, we find that *b* and *e* are again two such states;

they both go to d and c and have outputs of 0 and 0 for $x = 0$ and $x = 1$, respectively. Therefore, states b and e are also equivalent; and one can be removed. The procedure of removing a state and replacing it by its equivalent is demonstrated in the table in Figure 7.93.

From the table in Figure 7.93 we find that present state c now has next states f and b and outputs 0 and 0 for $x = 0$ and $x = 1$, respectively. The same next states and outputs appear in the row with present state a. Therefore, states a and c are equivalent; state c can be removed and replaced by a. The final reduced state table is shown in Figure 7.94. The state diagram for the reduced state table consists of only five states and is shown in Figure 7.95.

Present	Next state		Output	
state	$x = 0$	$x = 1$	$x = 0$	$x = 1$
a	f	b	0	0
b	d	~~e~~ a	0	0
~~c~~	f	~~e~~ b	0	0
d	g	a	1	0
~~e~~	d	c	0	0
f	f	b	1	1
g	g	~~h~~ d	0	1
~~h~~	g	a	1	0

Figure 7.93 Reducing the state table.

Present	Next state		Output	
state	$x = 0$	$x = 1$	$x = 0$	$x = 1$
a	f	b	0	0
b	d	a	0	0
d	g	a	1	0
f	f	b	1	1
g	g	d	0	1

Figure 7.94 Reduced state table.

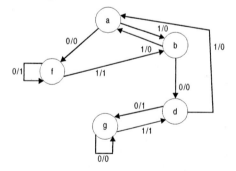

Figure 7.95 Reduced state diagram.

We now assign the different states the binary values. As we have already discussed, there may be a variety of state assignments. Some of them are shown in the table in Figure 7.96. Among them we may choose any of them and accordingly design the circuit.

State	Assignment 1	Assignment 2	Assignment 3	Assignment 4
a	000	001	111	011
b	001	010	001	101
d	010	011	110	111
f	011	100	101	001
g	100	101	010	000

Figure 7.96 Four possible binary assignments.

Present	Next state		Output	
state	$x = 0$	$x = 1$	$x = 0$	$x = 1$
000	011	001	0	0
001	010	000	0	0
010	100	000	1	0
011	011	001	1	1
100	100	010	0	1

Figure 7.97 Reduced state table with binary assignment 1.

In the table in Figure 7.97, we have used binary assignment 1 to substitute the letter symbols of the five states. It is obvious that a different binary assignment will result in a state table, with completely new binary values for the states while the input-output relationships will remain the same. We will now show the procedure for obtaining the excitation table and the combinational gate structure.

Present state			Input	Next state			Flip-flop inputs						Output
A	B	C	x	A	B	C	JA	KA	JB	KB	JC	KC	y
0	0	0	0	0	1	1	0	X	1	X	1	X	0
0	0	0	1	0	0	1	0	X	0	X	1	X	0
0	0	1	0	0	1	0	0	X	1	X	X	1	0
0	0	1	1	0	0	0	0	X	0	X	X	1	0
0	1	0	0	1	0	0	1	X	X	1	0	X	1
0	1	0	1	0	0	0	0	X	X	1	0	X	0
0	1	1	0	0	1	1	0	X	X	0	X	0	1
0	1	1	1	0	0	1	0	X	X	1	X	0	1
1	0	0	0	1	0	0	X	0	0	X	0	X	0
1	0	0	1	0	1	0	X	1	1	X	0	X	1

Figure 7.98

The derivation of the excitation table is facilitated if we arrange the state table in a different form. This form is shown in the table in Figure 7.98, where the present state and the input variables are arranged in the form of a truth table. As we have previously said, we may use any flip-flop, but the simplest form of the circuit is possible with J-K flip-flops. So we now design the circuit using J-K flip-flops.

There are three unused states in this circuit: binary states 101, 110, and 111. When an input of 0 or 1 is included with these unused states, we obtain six don't-care terms. These six binary combinations are not listed in the table under the present state or input and are treated as don't-care terms.

Karnaugh maps are prepared for JA, KA, JB, KB, JC, and KC in Figures 7.99(a), 7.99(b), 7.99(c), 7.99(d), 7.99(e), and 7.99(f).

For JA

AB \ Cx	00	01	11	10
00	0	0	0	0
01	1	0	0	0
11	X	X	X	X
10	X	X	X	X

Figure 7.99(a)

For KA

AB \ Cx	00	01	11	10
00	X	X	X	X
01	X	X	X	X
11	X	X	X	X
10	0	1	X	X

Figure 7.99(b)

From the Karnaugh maps for JA and KA, we obtain

$$JA = BC'x' \quad \text{and}$$

$$KA = x.$$

For JB

AB \ Cx	00	01	11	10
00	1	0	0	1
01	X	X	X	X
11	X	X	X	X
10	0	1	X	X

Figure 7.99(c)

For KB

AB \ Cx	00	01	11	10
00	X	X	X	X
01	1	1	1	0
11	X	X	X	X
10	0	X	X	X

Figure 7.99(d)

The Boolean expressions are derived for JB and KB from the Karnaugh maps as

$$JB = Ax + A'x' \text{ and}$$

$$KB = C' + x.$$

For JC

Cx AB	00	01	11	10
00	1	1	X	X
01	0	0	X	X
11	X	X	X	X
10	0	0	X	X

Figure 7.99(e)

For KC

Cx AB	00	01	11	10
00	X	X	1	1
01	X	X	0	0
11	X	X	X	X
10	X	X	X	X

Figure 7.99(f)

Similarly, the expressions for JC and KC we obtain as

$$JC = A'B' \qquad \text{and}$$
$$KC = B'.$$

A Karnaugh map has been also prepared for output y in Figure 7.99(g) and the Boolean expression for y is obtained as

$$Y = Bx' + BC + Ax.$$

For Y

Cx AB	00	01	11	10
00	0	0	0	0
01	1	0	1	1
11	X	X	X	X
10	0	1	X	X

Figure 7.99(g)

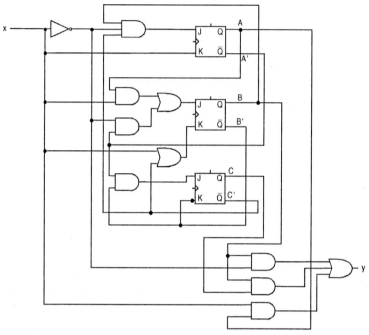

Figure 7.100 Logic diagram for the circuit.

The circuit diagram of the desired sequential logic network is shown in Figure 7.100.

REVIEW QUESTIONS

7.1 Show the logic diagram of a clocked D flip-flop with four NAND gates.

7.2 What is the difference between a level-triggered clock and an edge-triggered clock?

7.3 What is the difference between a latch and a flip-flop?

7.4 What is the race-around condition of a J-K flip-flop? How can it be avoided?

7.5 Draw the logic diagram of master-slave D flip-flop. Use NAND gates.

7.6 Derive the state table and the state diagram of the sequential circuit shown below.

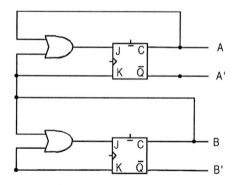

7.7 Obtain the excitation table of the J'K flip-flop.

7.8 A sequential circuit has one input and one output. The state diagram is shown below. Design the circuit with (*a*) JK flip-flops, (*b*) D flip-flops, (*c*) SR flip-flops, and (*d*) T flip-flops.

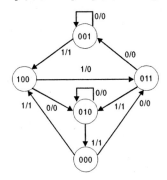

□ □ □

Chapter 8 REGISTERS

8.1 INTRODUCTION

A *register* is a group of binary storage cells capable of holding binary information. A group of flip-flops constitutes a register, since each flip-flop can work as a binary cell. An n-bit register, has n flip-flops and is capable of holding n-bits of information. In addition to flip-flops a register can have a combinational part that performs data-processing tasks.

Various types of registers are available in MSI circuits. The simplest possible register is one that contains no external gates, and is constructed of only flip-flops. Figure 8.1 shows such a type of register constructed of four S-R flip-flops, with a common clock pulse input. The clock pulse enables all the flip-flops at the same instant so that the information available at the four inputs can be transferred into the 4-bit register. All the flip-flops in a register should respond to the clock pulse transition. Hence they should be either of the edge-triggered type or the master-slave type. A group of flip-flops sensitive to the pulse duration is commonly called a *gated latch*. Latches are suitable to temporarily store binary information that is to be transferred to an external destination. They should not be used in the design of sequential circuits that have feedback connections.

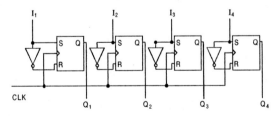

Figure 8.1 4-bit register

8.2 SHIFT REGISTER

A register capable of shifting its binary contents either to the left or to the right is called a *shift register*. The shift register permits the stored data to move from a particular location

263

to some other location within the register. Registers can be designed using discrete flip-flops (S-R, J-K, and D-type).

The data in a shift register can be shifted in two possible ways: (a) serial shifting and (b) parallel shifting. The serial shifting method shifts one bit at a time for each clock pulse in a serial manner, beginning with either LSB or MSB. On the other hand, in parallel shifting operation, all the data (input or output) gets shifted simultaneously during a single clock pulse. Hence, we may say that parallel shifting operation is much faster than serial shifting operation.

There are two ways to shift data into a register (serial or parallel) and similarly two ways to shift the data out of the register. This leads to the construction of four basic types of registers as shown in Figures 8.2(a) to 8.2(d). All of the four configurations are commercially available as TTL MSI/LSI circuits. They are:

1. Serial in/Serial out (SISO) – 54/74L91, 8 bits
2. Serial in/Parallel out (SIPO) – 54/74164, 8 bits
3. Parallel in/Serial out (PISO) – 54/74265, 8 bits
4. Parallel in/Parallel out (PIPO) – 54/74198, 8 bits.

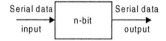

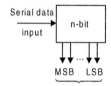

(a) Serial in/Serial out.

(b) Serial in/Parallel out.

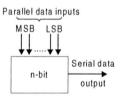

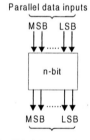

(c) Parallel in/Serial out.

(d) Parallel in/Parallel out.

Figure 8.2 Four types of shift registers.

8.3   SERIAL-IN–SERIAL-OUT SHIFT REGISTER

From the name itself it is obvious that this type of register accepts data serially, *i.e.*, one bit at a time at the single input line. The output is also obtained on a single output line in a serial fashion. The data within the register may be shifted from left to right using *shift-left* register, or may be shifted from right to left using *shift-right* register.

8.3.1 Shift-right Register

A shift-right register can be constructed with either J-K or D flip-flops as shown in Figure 8.3. A J-K flip-flop–based shift register requires connection of both J and K inputs. Input data are connected to the J and K inputs of the left most (lowest order) flip-flop. To input a 0, one should apply a 0 at the J input, *i.e.*, J = 0 and K = 1 and vice versa. With the application of a clock pulse the data will be shifted by one bit to the right.

In the shift register using D flip-flop, D input of the left most flip-flop is used as a serial input line. To input 0, one should apply 0 at the D input and vice versa.

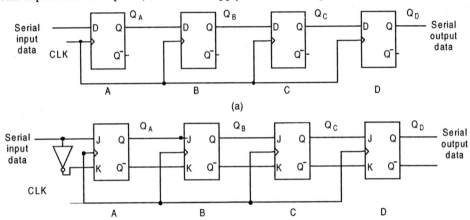

Figure 8.3 Shift-right register (a) using D flip-flops, (b) using J-K flip-flops.

The clock pulse is applied to all the flip-flops simultaneously. When the clock pulse is applied, each flip-flop is either set or reset according to the data available at that point of time at the respective inputs of the individual flip-flops. Hence the input data bit at the serial input line is entered into flip-flop A by the first clock pulse. At the same time, the data of stage A is shifted into stage B and so on to the following stages. For each clock pulse, data stored in the register is shifted to the right by one stage. New data is entered into stage A, whereas the data present in stage D are shifted out (to the right).

Table 8.1 Operation of the Shift-right Register

Timing pulse	Q_A	Q_B	Q_C	Q_D	Serial output at Q_D
Initial value	0	0	0	0	0
After 1st clock pulse	1	0	0	0	0
After 2nd clock pulse	1	1	0	0	0
After 3rd clock pulse	0	1	1	0	0
After 4th clock pulse	1	0	1	1	1

For example, consider that all the stages are reset and a logical input 1011 is applied at the serial input line connected to stage A. The data after four clock pulses is shown in Table 8.1.

Let us now illustrate the entry of the 4-bit number 1011 into the register, beginning with the right-most bit. A 1 is applied at the serial input line, making D = 1. As the first clock pulse is applied, flip-flop A is SET, thus storing the 1. Next, a 1 is applied to the serial input, making D = 1 for flip-flop A and D = 1 for flip-flop B also, because the input of flip-flop B is connected to the Q_A output.

When the second clock pulse occurs, the 1 on the data input is "shifted" to the flip-flop A and the 1 in the flip-flop A is "shifted" to flip-flop B. The 0 in the binary number is now applied at the serial input line, and the third clock pulse is now applied. This 0 is entered in flip-flop A and the 1 stored in flip-flop A is now "shifted" to flip-flop B and the 1 stored in flip-flop B is now "shifted" to flip-flop C. The last bit in the binary number that is the 1 is now applied at the serial input line and the fourth clock pulse is now applied. This 1 now enters the flip-flop A and the 0 stored in flip-lop A is now "shifted" to flip-flop B and the 1 stored in flip-flop B is now "shifted" to flip-flop C and the 1 stored in flip-flop C is now "shifted" to flip-flop D. Thus the entry of the 4-bit binary number in the shift-right register is now completed.

From the third column of Table 8.1 we can get the serial output of the data that is being entered in the register. We find that after the first, second, and the third clock pulses the output at the serial output line *i.e.*, Q_D is 0. After the fourth clock pulse the output at the serial output line is 1. If we want to get the total data that we have entered in the register in a serial manner from Q_D, then we have to apply another three clock pulses. After the fifth clock pulse we will gate another 1 at Q_D. After the sixth clock pulse the output at Q_D will be 0 and after the seventh clock pulse the output at Q_D will be 1. In this process of the fifth, sixth, and the seventh clock pulses if no data is being supplied at the serial input line then the A, B, and C flip-flops will again be RESET with output 0.

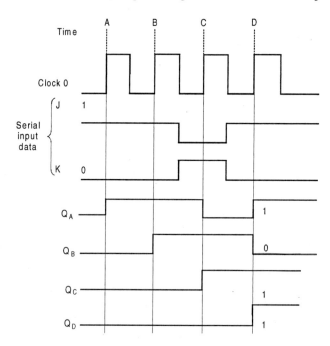

Figure 8.4 Waveforms of 4-bit serial input shift-right register.

The waveforms shown in Figure 8.4 illustrate the entry of a 4-bit number 1011. For a J-K flip-flop, the data bit to be shifted into the flip-flop must be present at the J and K inputs when the clock transitions from low to high occur. Since the data bit is either 1 or 0, there can be two different cases:

1. To shift a 1 into the flip-flop, J = 1 and K = 0,
2. To shift a 0 into the flip-flop, J = 0 and K = 1.

At time A: All the flip-flops are reset. At the serial data input line a 1 is given and with the first clock pulse this 1 is shifted at Q_A making Q_A = 1. At the same time the 0 in Q_A is shifted to Q_B, and the 0 in Q_B is shifted to Q_C and the 0 in Q_C is shifted to Q_D. Hence the flip-flop outputs just after time A are $Q_A Q_B Q_C Q_D$ = 1000.

At time B: The flip-flop A contains 1, and all other flip-flop contains 0. Now, again, 1 is given at the serial data input line. With the second clock pulse this 1 is shifted to Q_A. The 1 in Q_A is shifted to Q_B and the 0 in Q_B is shifted to Q_C and the 0 in Q_C is shifted to Q_D. Hence the flip-flop outputs just after time B are $Q_A Q_B Q_C Q_D$ = 1100.

At time C: The flip-flop A and flip-flop B contain 1, and all other flip-flops contain 0. Now a 0 is given at the serial data input line. With the third clock pulse this 0 is shifted to Q_A. The 1 in Q_A is shifted to Q_B and the 1 in Q_B is shifted to Q_C and the 0 in Q_C is shifted to Q_D. Hence the flip-flop outputs just after time C are $Q_A Q_B Q_C Q_D$ = 0110.

At time D: The flip-flop B and flip-flop C contain 1, and all other flip-flops contain 0. Now another 1 is given at the serial data input line. With the fourth clock pulse this 1 is shifted to Q_A. The 0 in Q_A is shifted to Q_B and the 1 in Q_B is shifted to Q_C and the 1 in Q_C is shifted to Q_D. Hence the flip-flop outputs just after time C are $Q_A Q_B Q_C Q_D$ = 1011.

To summarize, we have shifted 4 data bits in a serial manner into four flip-flops. These 4 data bits could represent a 4-bit binary number 1011, assuming that we began shifting with the LSB first. Notice that the LSB is in D and the MSB is in A. These four flip-flops could be defined as a 4-bit shift register.

8.3.2 Shift-left Register

A shift-left register can also be constructed with either J-K or D flip-flops as shown in Figure 8.5. Let us now illustrate the entry of the 4-bit number 1110 into the register, beginning with the right-most bit. A 0 is applied at the serial input line, making D = 0. As the first clock pulse is applied, flip-flop A is RESET, thus storing the 0. Next a 1 is applied to the serial input, making D = 1 for flip-flop A and D = 0 for flip-flop B, because the input of flip-flop B is connected to the Q_A output.

When the second clock pulse occurs, the 1 on the data input is "shifted" to the flip-flop A and the 0 in the flip-flop A is "shifted" to flip-flop B. The 1 in the binary number is now applied at the serial input line, and the third clock pulse is now applied. This 1 is entered in flip-flop A and the 1 stored in flip-flop A is now "shifted" to flip-flop B and the 0 stored in flip-flop B is now "shifted" to flip-flop C. The last bit in the binary number that is the 1 is now applied at the serial input line and the fourth clock pulse is now applied. This 1 now enters the flip-flop A and the 1 stored in flip-flop A is now "shifted" to flip-flop B and the 1 stored in flip-flop B is now "shifted" to flip-flop C and the 0 stored in flip-flop C is now "shifted" to flip-flop D. Thus the entry of the 4-bit binary number in the shift-right register is now completed.

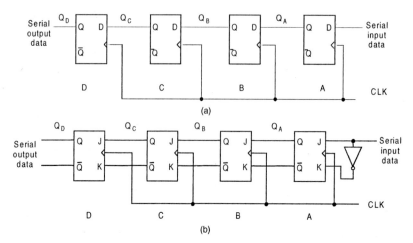

Figure 8.5 Shift-left register (*a*) using D flip-flops, (*b*) using J-K flip-flops.

Table 8.2 Operation of the Shift-left Register

Timing pulse	Q_D	Q_C	Q_B	Q_A	Serial output at Q_D
Initial value	0	0	0	0	0
After 1st clock pulse	0	0	0	0	0
After 2nd clock pulse	0	0	0	1	0
After 3rd clock pulse	0	0	1	1	0
After 4th clock pulse	0	1	1	1	0

8.3.3 8-bit Serial-in–Serial-out Shift Register

The pinout and logic diagram of IC 74L91 is shown in Figure 8.6. IC 74L91 is actually an example of an 8-bit serial-in–serial-out shift register. This is an 8-bit TTL MSI chip. There are eight S-R flip-flops connected to provide a serial input as well as a serial output. The clock input at each flip-flop is negative edge-triggered. However, the applied clock signal is passed through an inverter. Hence the data will be shifted on the positive edges of the input clock pulses.

An inverter is connected in between R and S on the first flip-flop. This means that this circuit functions as a D-type flip-flop. So the input to the register is a single liner on which the data can be shifted into the register appears serially. The data input is applied at either A (pin 12) or B (pin 11). The data level at A (or B) is complemented by the NAND gate and then applied to the R input of the first flip-flop. The same data level is complemented by the NAND gate and then again complemented by the inverter before it appears at the S input. So, a 0 at input A will *reset* the first flip-flop (in other words this 0 is shifted into the first flip-flop) on a positive clock transition.

The NAND gate with A and B inputs provide a gating function for the input data stream if required, if gating is not required, simply connect pins 11 and 12 together and apply the input data stream to this connection.

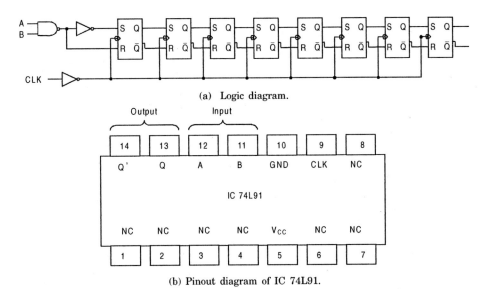

(a) Logic diagram.

(b) Pinout diagram of IC 74L91.

Figure 8.6 8-bit shift register—IC 74L91.

8.4 SERIAL-IN–PARALLEL-OUT REGISTER

In this type of register, the data is shifted in serially, but shifted out in parallel. To obtain the output data in parallel, it is required that all the output bits are available at the same time. This can be accomplished by connecting the output of each flip-flop to an output pin. Once the data is stored in the flip-flop the bits are available simultaneously. The basic configuration of a serial-in–parallel-out shift register is shown in Figure 8.2 (*b*).

8.4.1 8-bit Serial-in–Parallel-out Shift Register

The pinout and logic diagram of IC 74164 is shown in Figure 8.7. IC 74164 is an example of an 8-bit serial-in–parallel-out shift register. There are eight S-R flip-flops, which are all sensitive to negative clock transitions. The logic diagram in Figure 8.7 is almost the same as shown in Figure 8.6 with only two exceptions: (1) each flip-flop has an asynchronous CLEAR input; and (2) the true side of each flip-flop is available as an output—thus all 8 bits of any number stored in the register are available simultaneously as an output (this is a parallel data output).

Hence, a low level at the CLR input to the chip (pin 9) is applied through an amplifier and will reset every flip-flop. As long as the CLR input to the chip is LOW, the flip-flop outputs will all remain low. It means that, in effect, the register will contain all zeros.

Shifting of data into the register in a serial fashion is exactly the same as the IC 74L91. Data at the serial input may be changed while the clock is either low or high, but the usual hold and setup times must be observed. The data sheet for this device gives hold time as 0.0 ns and setup time as 30 ns.

Now we try to analyze the gated serial inputs A and B. Suppose that the serial data is connected to B; then A can be used as a control line. Here's how it works:

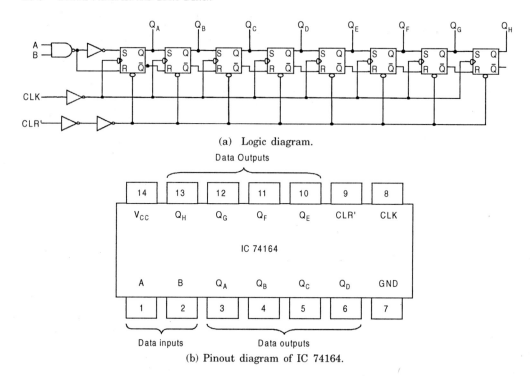

(a) Logic diagram.

(b) Pinout diagram of IC 74164.

Figure 8.7 8-bit shift register—IC 74164.

A is held high: The NAND gate is enabled and the serial input data passes through the NAND gate inverted. The input data is shifted serially into the register.

A is held low: The NAND gate output is forced high, the input data steam is inhibited, and the next clock pulse will shift a 0 into the first flip-flop. Each succeeding positive clock pulse will shift another 0 into the register. After eight clock pulses, the register will be full of zeros.

Example 8.1. *How long will it take to shift an 8-bit number into a 74164 shift register if the clock is set at 1 MHz?*

Solution. A minimum of eight clock Pulses will be required since the data is entered serially. One clock pulse period is 1000 ns, so it will require 8000 ns minimum.

8.5 PARALLEL-IN–SERIAL-OUT REGISTER

In the preceding two cases the data was shifted into the registers in a serial manner. We now can develop an idea for the parallel entry of data into the register. Here the data bits are entered into the flip-flops simultaneously, rather than a bit-by-bit basis.

A 4-bit parallel-in–serial-out register is illustrated in Figure 8.8. A, B, C, and D are the four parallel data input lines and *SHIFT / $\overline{LOAD}$* (*SH / $\overline{LD}$*) is a control input that allows the four bits of data at A, B, C, and D inputs to enter into the register in parallel or shift the data in serial. When *SHIFT / $\overline{LOAD}$* is HIGH, AND gates G_1, G_3, and G_5 are enabled, allowing the data bits to shift right from one stage to the next. When *SHIFT / $\overline{LOAD}$* is LOW, AND gates G_2, G_4, and G_6 are enabled, allowing the data bits at the parallel inputs. When a clock pulse is applied, the flip-flops with D = 1 will be set and the flip-flops with D = 0 will be reset, thereby storing all the four bits simultaneously. The OR gates allow either the normal shifting operation or the parallel data-entry operation, depending on which of the AND gates are enabled by the level on the *SHIFT / $\overline{LOAD}$* input.

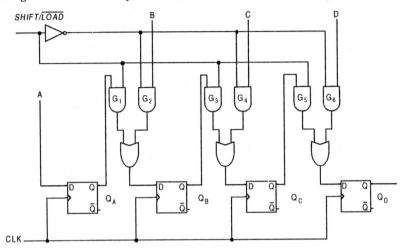

Figure 8.8 A 4-bit parallel-in–serial-out shift register.

8.5.1 8-bit Parallel-in–Serial-out Shift Register

The pinout and logic diagram of IC 74165 is shown in Figure 8.9. IC 74165 is an example of an 8-bit serial/parallel-in and serial-out shift register. The data can be loaded into

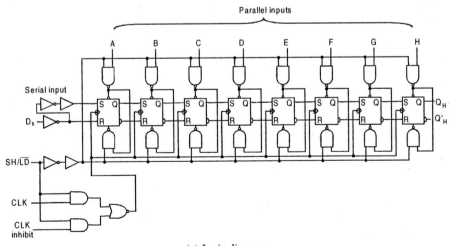

(a) Logic diagram.

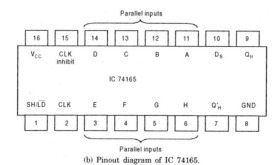

(b) Pinout diagram of IC 74165.

Figure 8.9 8-bit serial/parallel-in and serial-out shift register—IC 74165.

the register in parallel and shifted out serially at Q_H using either of two clocks (CLK or CLK inhibit). It also contains a serial input, D_S through which the data can be serially shifted in.

When the input *SHIFT / $\overline{LOAD}$* (*SH / $\overline{LD}$*) is LOW, it enables all the NAND gates for parallel loading. When an input data bit is a 0, the flip-flop is asynchronously RESET by a LOW output of the lower NAND gate. Similarly, when the input data bit is a 1, the flip-flop is asynchronously SET by a LOW output of the upper NAND gate. The clock is inhibited during parallel loading operation. A HIGH on the *SHIFT / $\overline{LOAD}$* input enables the clock causing the data in the register to shift right. With the low to high transitions of either clock, the serial input data (D_S) are shifted into the 8-bit register.

8.6 PARALLEL-IN–PARALLEL-OUT REGISTER

There is a fourth type of register already mentioned in Section 8.2, which is designed such that data can be shifted into or out of the register in parallel. The parallel input of data has already been discussed in the preceding section of parallel-in–serial-out shift register. Also, in this type of register there is no interconnection between the flip-flops since no serial shifting is required. Hence, the moment the parallel entry of the data is accomplished the data will be available at the parallel outputs of the register. A simple parallel-in–parallel-out shift register is shown in Figure 8.10.

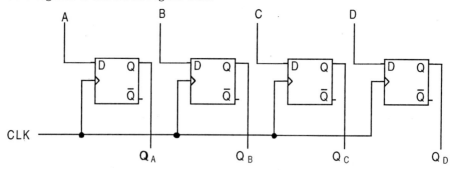

Figure 8.10 A 4-bit parallel-in–parallel-out shift register.

Here the parallel inputs to be applied at A, B, C, and D inputs are directly connected to the D inputs of the respective flip-flops. On applying the clock transitions, these inputs are entered into the register and are immediately available at the outputs Q_1, Q_2, Q_3, and Q_4.

8.6.1 8-bit Serial/Parallel-in and Serial/Parallel-out Shift Register

The pinout diagram of IC 74198 is shown in Figure 8.11. IC 74198 is an example of an 8-bit parallel-in and parallel-out shift register. IC 74198 is a 24-pin package where 16 pins are needed just for the input and output data lines. This chip can even shift data through the register in either direction, *i.e.*, shift-right and shift-left.

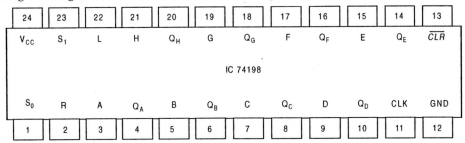

Figure 8.11 Pinout diagram of IC 74198, 8-bit parallel-in–parallel-out shift register.

L *i.e.*, pin 22 in Figure 8.11, represents the shift-left serial input and R (pin 2) represents the shift-right serial input. An 8-bit register can be created by either connecting two 4-bit registers in series or by manufacturing the two 4-bit registers on a single chip and placing the chip in a 24-pin package such as IC 74198.

There are a number of 4-bit parallel-input–parallel-output shift registers available since they can be conveniently packaged in a 16-pin dual-inline package. IC 74195 is a 4-bit TTL MSI having both serial/parallel input and serial/parallel output capability. The pinout diagram of IC 74195 is shown in Figure 8.12. Since this IC has a serial input, it can also be used for serial-in–serial-out and serial-in–parallel-out operation. This IC can be used for parallel-in–serial-out operation by using Q_D as the output.

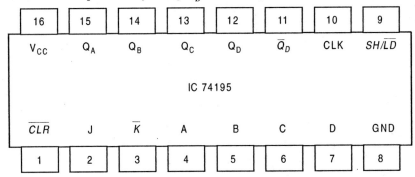

Figure 8.12 Pinout diagram of IC 74195.

When the $SH/\overline{LD}$ input is LOW, the data on the parallel inputs, *i.e.*, A, B, C, and D are entered synchronously on the positive transition of the clock. When $SH/\overline{LD}$ is HIGH, the stored data will shift right (Q_A to Q_D) synchronously with the clock. J and $\overline{K}$ are the serial inputs to the first stage of the register (Q_A); Q_D can be used for getting a serial output data. The active low clear is asynchronous.

8.7 UNIVERSAL REGISTER

A register that is capable of transfering data in only one direction is called a 'unidirectional shift register,' whereas the register that is capable of transfering data in both left and right direction is called a 'bidirectional shift register.' Now if the register has both the shift-right and shift-left capabilities, along with the necessary input and output terminals for parallel transfer, then it is called a *shift register* with *parallel load* or 'universal shift register.'

The most general shift register has all the capabilities listed below. Others may have only some of these functions, with at least one shift operation.

1. A shift-right control to enable the shift-right operation and the serial input and output lines associated with the shift-right.

2. A shift-left control to enable the shift-left operation and the serial input and output lines associated with the shift-left.

3. A parallel-load control to enable a parallel transfer and the n input lines associated with the parallel transfer.

4. n parallel output lines.

5. A *clear* control to clear the register to 0.

6. A *CLK* input for clock pulses to synchronize all operations.

7. A control state that leaves the information in the register unchanged even though clock pulses are continuously applied.

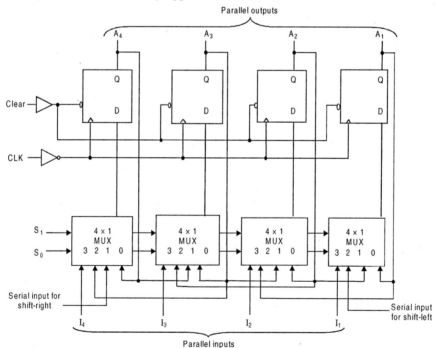

Figure 8.13 4-bit universal shift register.

The diagram of a shift-register with all the capabilities listed above is shown in Figure 8.13. This is similar to IC type 74194. Though it consists of four D flip-flops, S-R flip-flops can also be used with an inverter inserted between the S and R terminals. The four multiplexers drawn are also part of the register. The four multiplexers have two common selection lines S_1 and S_0. When $S_1S_0 = 00$, the input 0 is selected for each of the multiplexers. Similarly, when $S_1S_0 = 01$, the input 1, when $S_1S_0 = 10$, the input 2 and for $S_1 S_0 = 11$, the input 3, is selected for each of the multiplexers.

The S_1 and S_0 inputs control the mode of operation of the register as specified in the entries of functions in Table 8.3. When $S_1S_0 = 00$, the present value of the register is applied to the D inputs of the flip-flops. Hence this condition forms a path from the output of each flip-flop into the input of the same flip-flop. The next clock pulse transition transfers into each flip-flop the binary value held previously, and no change of state occurs. When $S_1S_0 = 01$, terminals 1 of each of the multiplexer inputs have a path to the D inputs of each of the flip-flops. This causes a shift-right operation, with the serial input transferred into flip-flop A_4. Similarly, with $S_1S_0 = 10$, a shift-left operation results, with the other serial input going into flip-flop A_1. Finally, when $S_1S_0 = 11$, the binary information on the parallel input lines is transferred into the register simultaneously during the next clock pulse.

Table 8.3 Function table for the universal register

Mode control		Register operation
S_1	S_0	
0	0	No change
0	1	Shift-right
1	0	Shift-left
1	1	Parallel load

A universal register is a general-purpose register capable of performing three operations: shift-right, shift-left, and parallel load. Not all shift registers available in MSI circuits have all these capabilities. The particular application dictates the choice of one MSI circuit over another. As we have already mentioned IC 74194 is a 4-bit bidirectional shift register with parallel load. The pinout diagram of IC 74194 is shown in Figure 8.14.

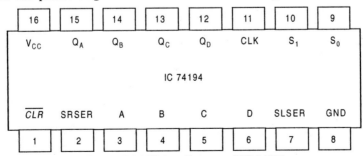

Figure 8.14 Pinout diagram of IC 74194.

The parallel loading of data is accomplished with a positive transition of the clock and by applying the four bits of data to the parallel inputs and a HIGH to the S_1 and S_0 inputs. Similarly, shift-right is accomplished synchronously with the positive edge of the clock when S_0 is HIGH and S_1 is LOW. In this mode the serial data is entered at the shift-right serial input. In the same manner, when S_0 is LOW and S_1 is HIGH, data bits shift left synchronously with the clock pulse and new data is entered at the shift-left serial input.

8.8 SHIFT REGISTER COUNTERS

Shift registers may be arranged to form different types of counters. These shift registers use *feedback*, where the output of the last flip-flop in the shift register is fed back to the first flip-flop. Based on the type of this feedback connection, the shift register counters are classified as (*i*) ring counter and (*ii*) twisted ring or Johnson or Shift counter.

8.8.1 Ring Counter

It is possible to devise a counter-like circuit in which each flip-flop reaches the state $Q = 1$ for exactly one count, while for all other counts $Q = 0$. Then Q indicates directly an occurrence of the corresponding count. Actually, since this does not represent binary numbers, it is better to say that the outputs of the flip-flops represent a code. Such a circuit is shown in Figure 8.15, which is known as a *ring counter*. The Q output of the last stage in the shift register is fed back as the input to the first stage, which creates a ring-like structure.

Hence a ring *counter* is a circular shift register with only one flip-flop being set at any particular time and all others being cleared. The single bit is shifted from one flip-flop to the other to produce the sequence of timing signals. Such encoding where there is a single 1 and the rest of the code variables are 0, is called a *one-hot code*.

Table 8.4 Truth table for a 4-bit ring counter

INIT	*CLK*	Q_A	Q_B	Q_C	Q_E
L	X	0	0	0	1
H	↑	1	0	0	0
H	↑	0	1	0	0
H	↑	0	0	1	0
H	↑	0	0	0	1

The circuit shown in Figure 8.15 consists of four flip-flops and their outputs are Q_A, Q_B, Q_C, and Q_E respectively. The PRESET input of the last flip-flop and the CLEAR inputs of the other three flip-flops are connected together. Now, by applying a LOW pulse at this line, the last flip-flop is SET and all the others are RESET, i.e., $Q_A Q_B Q_C Q_E = 0001$. Hence, from the circuit it is clear that $D_A = 1$, $D_B = 0$, $D_C = 0$, and $D_E = 0$. Therefore, when a clock pulse is applied, the first flip-flop is set to 1, while the other three flip-flops are reset to 0 *i.e.*, the output of the ring counter is $Q_A Q_B Q_C Q_E = 1000$. Similarly, when the second clock pulse is applied, the 1 in the first flip-flop is shifted to the second flip-flop and the output of the ring counter becomes $Q_A Q_B Q_C Q_E = 0100$; on occurrence of the third clock pulse, the output will be $Q_A Q_B Q_C Q_E = 0010$; on occurrence of the fourth clock pulse the output becomes $Q_A Q_B Q_C Q_E = 0001$, *i.e.*, the initial state. Thus, the 1 is shifted around the register as long as the clock pulses are applied. The truth table that describes the operation of the above 4-bit ring counter is shown in Table 8.4.

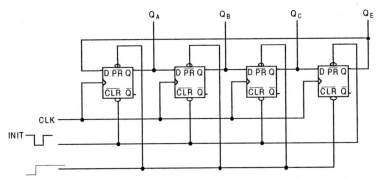

Figure 8.15 A 4-bit ring counter using D flip-flops.

8.8.2 Johnson Counter

A k-bit ring counter circulates a single bit among the flip-flops to provide k distinguishable states. The number of sates can be doubled if the shift register is connected as a *switch-tail* ring counter. A switch-tail ring counter is a circular shift register with the complement of the last flip-flop being connected to the input of the first flip-flop. Figure 8.16 shows such a type

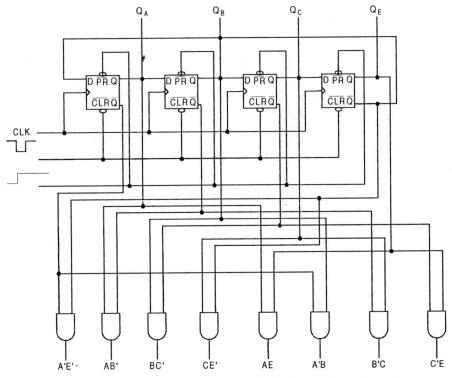

Figure 8.16 A 4-bit Johnson counter using D flip-flops and decoding gates.

of shift register. The circular connection is made from the complement of the rightmost flip-flop to the input of the leftmost flip-flop. The register shifts its contents once to the right with every clock pulse, and at the same time, the complement value of the E flip-flop is transferred into the A flip-flop. Starting from a cleared state, the switch-tail ring counter goes through a sequence of eight states as listed in Table 8.5. In general a k-bit switch-tail counter will go through 2k states. Starting with all 0s each shift operation inserts 1s from the left until the register is filled with all 1s. In the following sequences, 0s are inserted from the left until the register is again filled with all 0s.

A *Johnson* or *moebius counter* is a switch-tail ring counter with 2k decoding gates to provide outputs for 2k timing signals. The decoding gates are also shown in Figure 8.16. Since each gate is enabled during one particular state sequence, the outputs of the gates generate eight timing sequences in succession.

The decoding of a k-bit switch-tail ring counter to obtain 2k timing sequences follows a regular pattern. The all-0s state is decoded by taking the complement of the two extreme flip-flop outputs. The all-1s state is decoded by taking the normal outputs of the two extreme flip-flops. All other states are decoded from an adjacent 1, 0 or 0, 1 pattern in the sequence. For example, sequence 6 has an adjacent 0 and 1 pattern in flip-flops A and B. the decoded output is then obtained by taking the complement A and the normal of B, or the A'B.

Table 8.5 Count sequence of a 4-bit Johnson counter

Sequence number	Flip-flop outputs			
	A	B	C	E
1	0	0	0	0
2	1	0	0	0
3	1	1	0	0
4	1	1	1	0
5	1	1	1	1
6	0	1	1	1
7	0	0	1	1
8	0	0	0	1

One disadvantage of the circuit in Figure 8.16 is that, if it finds itself in an unused state, it will persist in moving from one invalid state to another and never find its way to a valid state. The difficulty can be corrected by modifying the circuit to avoid this undesirable condition. One correcting procedure is to disconnect the output from flip-flop B that goes to the D input of flip-flop C, and instead enable the input of flip-flop C by the function:

$$DC = (A + C)B$$

where DC is the flip-flop input function for the D input of the flip-flop C.

Johnson counters can be constructed for any number of timing sequences. The number of flip-flops needed is one-half the number of timing signals. The number of decoding gates is equal to the number of timing sequences and only 2-input gates are employed. Ring counter does not require any decoding gates, since in ring counter only one flip-flop will be in the set condition at any time.

8.9 SEQUENCE GENERATOR

A sequence generator is a circuit that generates a desired sequence of bits in synchronization with a clock. A sequence generator can be used as a random bit generator, code generator, and prescribed period generator. The block diagram of a sequence generator is shown in Figure 8.17.

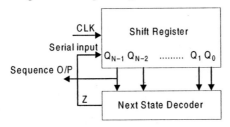

Figure 8.17 Block diagram of a sequence generator.

The sequence generator can be constructed using shift register and a next state decoder. The output of the next state decoder (Z) is a function of Q_{N-1}, Q_{N-2},......Q_1, Q_0 and is connected to the serial input of the shift register. This sequence generator is similar to a ring counter or a Johnson counter.

8.9.1 Design of a 4-bit Sequence Generator

We consider the design of a sequence generator to generate a sequence of 1001. The minimum number of flip-flops (n) required to generate a sequence of length N is given by

$$N \le 2^n - 1$$

Here N = 4 and hence, the minimum value of n to satisfy the above condition is 3, *i.e.*, three flip-flops are required to generate the given sequence. If the given sequence does not lead to four distinct states, then more than three flip-flops are required. The states of the given sequence generator are given in Table 8.6.

Table 8.6 State table for a 4-bit (1001) sequence generator

CLK	Flip-Flop outputs			Serial Input
	Q_2	Q_1	Q_0	Z
1	1	1	0	0
2	0	1	1	0
3	0	0	1	1
4	1	0	0	1
1	1	1	0	0
2	0	1	1	0
X	0	0	1	1
X	1	0	0	1

In Table 8.6, the given sequence (1001) is listed under Q_2 and the sequence under Q_1 and Q_0 are the same sequence delayed by one and two clock pulses respectively as indicated by arrow marks. Also, it is observed that all the four states are distinct and hence three flip-flops are sufficient to implement the sequence generator. The last column gives the serial input required at the shift register (*i.e.*, D_2 of MSB flip-flop), assuming D flip-flops are used and considering the output at Q_2. Now, the K-map for the serial input (Z) is shown in Figure 8.18.

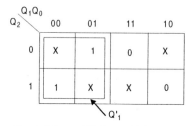

Figure 8.18 K-map of serial input (Z) for a 4-bit (1001) sequence generator.

From the K-map shown in Figure 8.18, the simplified expression for serial input Z can be written as

$$Z = Q'_1.$$

Therefore, using the simplified expression for Z, the logic diagram of a given 4-bit sequence generator can be drawn as shown in Figure 8.19.

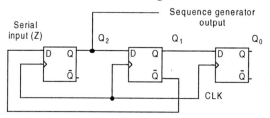

Figure 8.19 Logic diagram of a 4-bit (1001) sequence generator.

8.9.2 Design of a 5-bit Sequence Generator

We consider the design of a sequence generator to generate a sequence of 10011. The minimum number of flip-flops (*n*) required to generate a sequence of length N is given by the equation

$$N \le 2^n - 1.$$

Here N = 5 and hence, the minimum number of flip-flops required (*n*) to satisfy the above condition is 3. If the given sequence does not lead to five distinct states, then more than three flip-flops may be required. The states of the given sequence generator are shown in Table 8.7.

As explained in the previous section, the given sequence (10011) is listed under Q_2 and the sequence under Q_1 and Q_0 are the same sequence delayed by one and two clock pulses respectively as indicated by arrow marks. Also, it is observed that all 5 states are distinct and hence three flip-flops are sufficient to implement the sequence generator. The last column gives the serial input required at the shift register (*i.e.*, D_2 of MSB flip-flop), assuming D flip-flops are used and considering the output at Q_2. Now, the K-map for the serial input (Z) is shown in Figure 8.20.

Table 8.7 State table for a 5-bit (10011) sequence generator

CLK	Flip-Flop outputs			Serial Input
	Q_2	Q_1	Q_0	Z
1	1	1	1	0
2	0	1	1	0
3	0	0	1	1
4	1	0	0	1
5	1	1	0	1
1	1	1	1	0
2	0	1	1	0
X	0	0	1	1
X	1	0	0	1
X	1	1	0	1

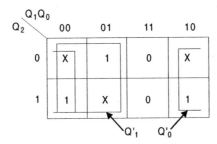

Figure 8.20 K-map of serial input (Z) for a 5-bit (10011) sequence generator.

From the K-map shown in Figure 8.20, the simplified expression for serial input Z can be written as

$$Z = Q'_1 + Q'_0$$

Therefore, using the simplified expression for Z, the logic diagram of a given 5-bit sequence generator can be drawn as shown in Figure 8.21.

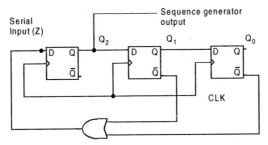

Figure 8.21 Logic diagram of a 5-bit (10011) sequence generator.

8.9.3 Design of a 6-bit Sequence Generator

We consider the design of a sequence generator to generate a sequence of 110101. The minimum number of flip-flops (n) required to generate a sequence of length N is given by the equation

$$N \le 2^n - 1.$$

Here N = 6 and hence, the minimum number of flip-flops required (n) to satisfy the above condition is 3. If the given sequence does not lead to six distinct states, then more than three flip-flops may be required. The states of the given sequence generator are shown in Table 8.8.

Table 8.8 State table for a 6-bit (110101) sequence generator

CLK	Flip-Flop outputs		
	Q_2	Q_1	Q_0
1	1	1	0
2	1	1	1
3	0	1	1
4	1	0	1
5	0	1	0
6	1	0	1
1	1	1	0
2	1	1	1

Table 8.9 Modified state table for a 6-bit (110101) sequence generator

CLK	Flip-Flop outputs				Serial Input
	Q_3	Q_2	Q_1	Q_0	Z
1	1	1	0	1	1
2	1	1	1	0	0
3	0	1	1	1	1
4	1	0	1	1	0
5	0	1	0	1	1
6	1	0	1	0	1
1	1	1	0	1	1
2	1	1	1	0	0
3	0	1	1	1	1
4	1	0	1	1	0
5	0	1	0	1	1
6	1	0	1	0	1

As explained in the previous section, the given sequence (110101) is listed under Q_2 and the sequence under Q_1 and Q_0 are the same sequence delayed by one and two clock pulses respectively as indicated by arrow marks. From Table 8.8, it is observed that all six states are not distinct, *i.e.*, 101 state occurs twice. Hence three flip-flops are not sufficient to generate the given sequence. Next, assuming $n = 4$, the modified state table for the given sequence generator can be shown in Table 8.9.

From Table 8.9, it is observed that all six states are distinct and hence four flip-flops are sufficient to implement the sequence generator. The last column gives the serial input required at the shift register (*i.e.*, D_3 of the MSB flip-flop), assuming D flip-flops are used and considering the output at Q_3. Now, the K-map for the serial input (Z) is shown in Figure 8.22.

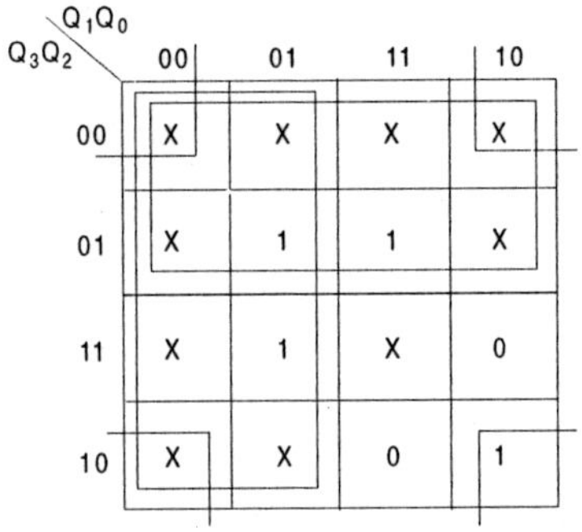

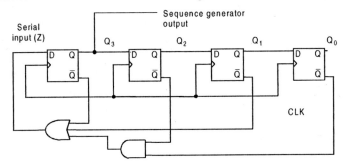

Figure 8.22 K-map of serial input (Z) for a 6-bit (110101) sequence generator.

From the K-map shown in Figure 8.18, the simplified expression for serial input Z can be written as

$$Z = Q'_3 + Q'_1 + Q'_2 Q'_0.$$

Therefore, using the simplified expression for Z, the logic diagram of a given 6-bit sequence generator can be drawn as shown in Figure 8.23.

Figure 8.23 Logic diagram of a 6-bit (110101) sequence generator.

8.10 SERIAL ADDITION

Serial addition is much slower than parallel addition, but requires less equipment. We now demonstrate the serial mode of operation.

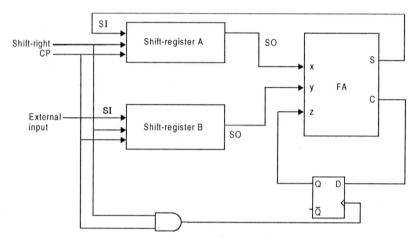

Figure 8.24 Serial adder

The two binary numbers to be added serially are stored in two shift registers. Bits are added one pair at a time, sequentially, through a single full-adder (FA) circuit shown in Figure 8.24. The carry out of the full adder is transferred to a D flip-flop. The output of this flip-flop is then used as an input carry for the next pair of significant bits. The two shift registers are shifted to the right for one word-time period. The sum bits from the S output of the full adder could be transferred into a third shift register. By shifting the sum into A while the bits of A are shifted out, it is possible to use one register to store both the augend and the sum bits. The serial input (SI) of register B is able to receive a new binary number while the addend bits are shifted out during the addition.

The operation of the serial adder is as follows. Initially the augend is in register A, the addend is in register B, and the carry flip-flop is cleared to 0. The serial outputs (SO) of A and B provide a pair of significant bits for the full-adder at x and y. Output Q of the flip-flop gives the input carry as z. The shift-right control enables both registers and the carry flip-flop. Hence at the next clock pulse, both registers are shifted once to the right, the sum bit from S enters the leftmost flip-flop of A, and the output carry is transferred into the flip-flop Q. The shift-right control enables the registers for a number of clock pulses equal to the number of bits in the registers. For each succeeding clock pulse, a new sum bit is transferred to A, a new carry is transferred to Q, and both registers are shifted once to the right. This process continues until the shift-right control gets disabled. Thus the addition is accomplished by passing each pair of bits together with the previous carry through a single full-adder circuit and transferring the sum, one bit at a time, into register A.

If a new number has to be added to the contents of register A, this number must be first transferred into register B. Repeating the process once more will add the second number to the previous number in A.

8.11 BINARY DIVIDER

We consider the design of a parallel divider for positive binary numbers. As an example we design a network to divide a 6-bit dividend by a 3-bit divisor to obtain a 3-bit quotient. The following example illustrates the division process:

```
              111          quotient
divisor   101 / 100110     (with a remainder of 4)
              101          dividend
              1001
              101
              1000
              101
              011          remainder.
```

Binary division can be carried out by a series of subtract and shift operations. To construct the divider, we will use a 7-bit dividend register and a 3-bit divisor register as shown in Figure 8.25. During the division process, instead of shifting the divisor right before each subtraction, we will shift the dividend to the left. Now an extra bit is required to the left end of the dividend register so that a bit is not lost when the dividend is shifted left. Instead of using a separate register to store the quotient, we will enter the quotient bit-by-bit into the right end of the dividend register as the dividend is shifted left.

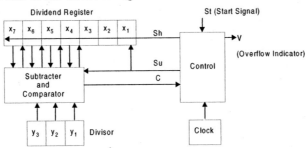

Figure 8.25 Block diagram for a parallel binary divider.

The preceding division example () is reworked below showing location of the bits in the registers at each clock time. Initially the dividend and the divisor are entered as follows:

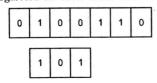

Subtraction cannot be carried out with a negative result, so we will shift before we subtract. Instead of shifting the divisor one place to the right, we will shift the dividend one place to the left:

```
1  0  0  1  1  0 | 0  ← Dividing line between dividend and quotient.
   1  0  1        |      Note that after the shift, the rightmost position in
                  |      the dividend is "empty." In effect, the quotient bit is
                  |      initially set to 0 and if subtraction occurs, it is changed
                  |      to 1.
```

Subtraction is now carried out and the first quotient digit of 1 is stored in the unused position of the dividend register:

0 1 0 0 1 0 |1   ◄——— first quotient digit.

Next we shift the dividend one place to the left:

1 0 0 1 0 |1 0
    1 0 1

Subtraction is again carried out and the second quotient digit of 1 is stored in the unused position of the dividend register:

0 1 0 0 0 |1 1

We shift the dividend one place to the left again:

1 0 0 0 |1 1 0
    1 0 1

A final subtraction is carried out and the third quotient bit is set to 1:

0 0 1 1 |1 1 1

remainder | quotient

The final result agrees with that obtained in the first example.

If, as a result of a division operation, the quotient would contain more bits than are available for storing the quotient, we say that an *overflow* has occurred. For the divider of Figure 8.25, an overflow may occur if the quotient is greater than 7, since only 3 bits are provided to store the quotient. It is not actually necessary to carry out the division to determine if an overflow condition exists, since an initial comparison of the dividend and divisor will tell if the quotient will be too large. For example, if we attempt to divide 38 by 4, the initial contents of the registers would be:

0 1 0 0 1 1 0
    1 0 0

Since subtraction can be carried out with a nonnegative result, we should subtract the divisor from the dividend and enter a quotient bit of 1 in the rightmost place in the dividend register. However, we cannot do this because the rightmost place contains the least significant bit of the dividend, and entering a quotient bit here will destroy that dividend bit. Therefore the quotient will be too large to store in the 3 bits we have allocated for it, and an overflow condition is detected. In general, for Figure 8.25, if initially $x_7 x_6 x_5 x_4 \geq y_3 y_2 y_1$ (i.e., if the left four bits of the dividend register exceeds or equal the divisor) the quotient will be grater than 7 and an overflow occurs. Note that if $x_7 x_6 x_5 x_4 \geq y_3 y_2 y_1$, the quotient is

$$\frac{x_7 x_6 x_5 x_4 x_3 x_2 x_1}{y_3 y_2 y_1} \geq \frac{x_7 x_6 x_5 x_4\ 000}{y_3 y_2 y_1} = \frac{x_7 x_6 x_5 x_4 \times 8}{y_3 y_2 y_1} \geq 8.$$

The operation of the divider can be explained in terms of the block diagram of Figure 8.25. A shift signal (Sh) will shift the dividend one place to the left. A subtract signal (Su) will subtract the divisor from the four leftmost bits in the dividend register and set the quotient bit (the rightmost bit of the dividend register) to 1. If the divisor is greater than the four leftmost dividend bits, the comparator output is C = 0; otherwise C = 1. The control circuit generates the required sequence of shift and subtract signals. Whenever C = 0, subtraction cannot occur without a negative result, so a shift signal is generated. Whenever C = 1, a subtract signal is generated and the quotient bit is set to one.

Figure 8.26 shows the state diagram for the control circuit. Initially, the 6-bit dividend and the 3-bit divisor are entered into the appropriate registers. The circuit remains in the stop state (S_0) until a start signal (St) is applied to the control circuit. If the initial value of C is 1, the quotient would require four or more bits. Since space is only provided for a 3-bit quotient, this condition leads to an overflow, so the divider is stopped and the overflow indicator is set by the V output. Normally, the initial value of C is 0, so a shift will occur first and the control circuit will go to state S_1. Then, if C = 1 subtraction takes place. After the subtraction is completed, C will always be 0, so that the next clock pulse will produce a shift. This process continues until three shifts have occurred and the control is in state S_3. Then a final subtraction occurs if necessary, and the control returns to the stop state.

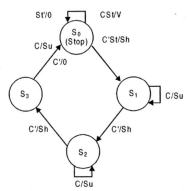

Figure 8.26 State graph for control circuit.

For this example, we will assume that when the start signal (St) occurs, it will be 1 for one clock time, and then it will remain 0 until the clock network is back in state S_0. Therefore, St will always be 0 in states S_1, S_2, and S_3. Table 8.10 gives the state table for the control circuit. Since we assumed that St = 0 in states S_1, S_2, and S_3, the next states and outputs are don't-cares for these states when St = 1. The entries in the output table indicate which outputs are 1.

Table 8.10 State table for Figure 8.26

AB	Present State	StC 00	01	11	10	Output 00	01	11	10
00	S_0	S_0	S_0	S_0	S_1	0	0	V	Sh
01	S_1	S_2	S_1	–	–	Sh	Su	–	–
11	S_2	S_3	S_2	–	–	Sh	Su	–	–
10	S_3	S_0	S_0	–	–	0	Su	–	–

For example, the entry Sh means Sh = 1 and the other outputs are 0. Using the state assignment shown in Table 8.10 for J-K flip-flops A and B, the following equations may be derived for the control circuit:

$$JA = BC', \qquad KA = B', \qquad JB = St \cdot C', \qquad KB = AC',$$

$$Sh = (St + B)C', \qquad Su = C(A + B), \qquad V = C \cdot St.$$

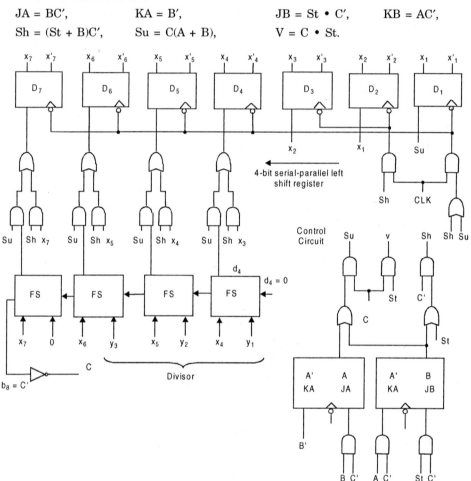

Figure 8.27 Logic diagram for binary divider.

Figure 8.27 shows a logic diagram for the subtractor/comparator, dividend register and control network. The subtractor is constructed using four full subtractors. When the numbers are entered into the divisor and dividend registers, the borrow signal will propagate through the full subtractors before the subtractor output is transferred to the dividend register. If the last borrow signal (b_8) is 1, this means that the result would be negative if the subtraction were carried out. Hence, if b_8 is 1, the divisor is greater than $x_7x_6x_5x_4$, and C = 0. Therefore, C = b'_8 and the dividend register so that if Sh = 1 a left shift will take place when the clock pulse occurs, and if Su = 1 the subtractor output will be transferred to the dividend register when the clock pulse occurs. For example,

$$D_4 = Su \bullet d_4 + Sh \bullet x_3 = x_4^+.$$

If Su = 1 and Sh = 0, and the subtracter output is transferred to the register of flip-flops. If Su = 0 and Sh = 1, and a left shift occurs. Since D_1 = Su, the quotient bit (x_1) is cleared when shifting occurs (Su = 0) and the quotient bit is set to 1 during subtraction (Su = 1). Note that the clock pulse is gated so that flip-flops x_7, x_6, x_5, x_4, and x_1 are clocked when Su or Sh is 1, while flip-flops x_3 and x_2 are clocked only when Sh is 1.

REVIEW QUESTIONS

8.1 A shift register has seven flip-flops. What is the largest binary number that can be stored in it? Octal number? Decimal number? Hexadecimal number?

8.2 What are the four basic types of shift registers? Draw a block diagram for each of them.

8.3 The hexadecimal number AC is stored in the IC 7491 shown in Figure 8.6. Show the waveforms at the output, assuming that the clock is allowed to run for eight cycles and that A = C = 0.

8.4 Why are shift registers considered to be basic memory devices?

8.5 Explain the workings of a serial-in–parallel-out shift register with logic diagram and waveforms.

8.6 Explain the workings of a serial-in–serial-out shift register with logic diagram and waveforms.

8.7 Describe a parallel-in–parallel-out shift register with a neat logic diagram.

8.8 Explain how a parallel-in–serial-out shift register works with a logic diagram.

8.9 Why does a Johnson counter have decoding gates, whereas a ring counter does not?

8.10 Construct a Johnson counter for twelve timing sequences.

8.11 Why are sequence generators used? Design a sequence generator to generate the sequence 111011.

8.12 Design a 6-bit ring counter using J-K flip-flops.

8.13 Determine the frequency of the pulses at points a, b, c, and d in the circuit of Figure P. 8.1.

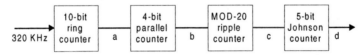

Figure P.8.1

8.14 The content of a 4-bit register is initially 1011. The register is shifted 7 times to the right with the serial input being 1010110. What is the content of the register after each shift?

8.15 Draw the circuit for a universal shift register and explain its operation.

8.16 (a) List the eight unused states in the switch-tail ring counter of Figure 8.16. Determine the next state for each unused state and show that, if the circuit finds itself in an invalid state, it does not return to a valid state.

(b) Modify the circuit as recommended in the text and show that:

(1) The circuit produces the same sequences as listed in Table 8.5, and

(2) The circuit reaches a valid state from any one of the unused states.

8.17 Write a short note on a Johnson counter.

❏ ❏ ❏

Chapter 9 — COUNTERS

9.1 INTRODUCTION

Counters are one of the simplest types of sequential networks. A counter is usually constructed from one or more flip-flops that change state in a prescribed sequence when input pulses are received. A counter driven by a clock can be used to count the number of clock cycles. Since the clock pulses occur at known intervals, the counter can be used as an instrument for measuring time and therefore period of frequency. Counters can be broadly classified into three categories:

(*i*) Asynchronous and Synchronous counters.

(*ii*) Single and multimode counters.

(*iii*) Modulus counters.

The asynchronous counter is simple and straightforward in operation and construction and usually requires a minimum amount of hardware. In asynchronous counters, each flip-flop is triggered by the previous flip-flop, and hence the speed of operation is limited. In fact, the settling time of the counter is the cumulative sum of the individual settling times of the flip-flops. This type of counters is also called *ripple* or *serial* counter.

The speed limitation of asynchronous counters can be overcome by applying clock pulses simultaneously to all of the flip-flops. This causes the settling time of the flip-flops to be equal to the propagation delay of a single flip-flop. The increase in speed is usually attained at the price of increased hardware. This type of counter is also known as a *parallel* counter.

The counters can be designed such that the contents of the counter advances by one with each clock pulse; and is said to operate in the *count-up* mode. The opposite is also possible, when the counter is said to operate in the *count-down* mode. In both cases the counter is said to be a *single mode counter*. If the same counter circuit can be operated in both the UP and DOWN modes, it is called a *multimode counters*.

Modulus counters are defined based on the number of states they are capable of counting. This type of counter can again be classified into two types: *Mod N* and *MOD < N*. For example, if there are n bits then the maximum number counted can be 2^n or N. If the counter is so designed that it can count up to 2^n or N states, it is called MOD N or MOD

2^n counter. On the other hand, if the counter is designed to count sequences less than the maximum value attainable, it is called a *MOD < N* or *MOD < 2^n* counter.

9.2 ASYNCHRONOUS (SERIAL OR RIPPLE) COUNTERS

The simplest counter circuit can be built using T flip-flops because the toggle feature is naturally suited for the implementation of the counting operation. J-K flip-flops can also be used with the *toggle* property in hand. Other flip-flops like D or S-R can also be used, but they may lead to more complex designs.

In this counter all the flip-flops are not driven by the same clock pulse. Here, the clock pulse is applied to the first flip-flop; *i.e.*, the least significant bit state of the counter, and the successive flip-flop is triggered by the output of the previous flip-flop. Hence the counter has cumulative settling time, which limits its speed of operation. The first stage of the counter changes its state first with the application of the clock pulse to the flip-flop and the successive flip-flops change their states in turn causing a *ripple-through* effect of the clock pluses. As the signal propagates through the counter in a *ripple* fashion, it is called a *ripple counter*.

9.2.1 Asynchronous (or Ripple) Up-counter

Figure 9.1 shows a 3-bit counter capable of counting from 0 to 7. The clock inputs of the three flip-flops are connected in cascade. The T input of each flip-flop is connected to a constant 1, which means that the state of the flip-flop will toggle (reverse) at each negative edge of its clock. We are assuming that the purpose of this circuit is to count the number of pulses that occur on the primary input called CLK (Clock). Thus the clock input of the first flip-flop is connected to the *Clock* line. The other two flip-flops have their clock inputs driven by the Q output of the preceding flip-flop. Therefore, they toggle their state whenever the preceding flip-flop changes its state from Q = 1 to Q = 0, which results in a negative edge of the Q signal.

Figure 9.1(*b*) shows a timing diagram for the counter. The value of Q_0 toggles once each clock cycle. The change takes place shortly after the negative edge of the *Clock* signal. The delay is caused by the propagation delay through the flip-flop. Since the second flip-flop is clocked by Q_0, the value of Q_1 changes shortly after the negative edge of the Q_0 signal. Similarly, the value of Q_2 changes shortly after the negative edge of the Q_1 signal. If we look at the values Q_2 Q_1 Q_0 as the count, then the timing diagram indicates that the counting sequence is 0, 1, 2, 3, 4, 5, 6, 7, 0, 1, 2, and so on. This circuit is a modulo-8 counter. Since it counts in the upward direction, we call the circuit an *up-counter*.

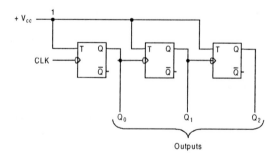

(a) Logic circuit diagram.

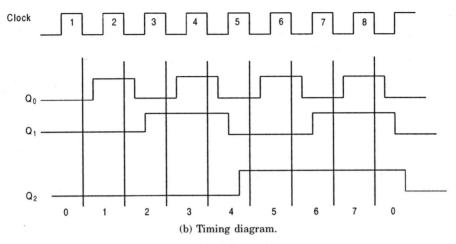

(b) Timing diagram.

Figure 9.1 A 3-bit asynchronous up-counter.

The counter in Figure 9.1(a) has three stages, each comprising of a single flip-flop. Only the first stage responds directly to the Clock signal. Hence we may say that this stage is synchronized to the clock. The other two stages respond after an additional delay. For example, when count = 3, the next clock pulse will change the count to 4. Now this change requires all three flip-flops to toggle their states. The change in Q_0 is observed only after a propagation delay from the negative edge of the clock pulse. The Q_1 and Q_2 flip-flops have not changed their states yet. Hence, for a brief period, the count will be $Q_2Q_1Q_0 = 010$. The change in Q_1 appears after a second propagation delay, and at that point the count is $Q_2Q_1Q_0 = 000$. Finally, the change in Q_2 occurs after a third delay, and hence the stable state of the circuit is reached and the count is $Q_2Q_1Q_0 = 100$.

Table 9.1 shows the sequence of binary states that the flip-flops will follow as ciock pulses are applied continuously. An n-bit binary counter repeats the counting sequence for every 2^n (n = number of flip-flops) clock pulses and has discrete states from 0 to 2^n-1.

Table 9.1 Count sequence of a 3-bit binary ripple up-counter

Counter State	Q_2	Q_1	Q_0
0	0	0	0
1	0	0	1
2	0	1	0
3	0	1	1
4	1	0	0
5	1	0	1
6	1	1	0
7	1	1	1

Figure 9.2 shows the 3-bit binary ripple counter with decoded outputs. It consists of the same circuit as shown in Figure 9.1 with additional decoding circuitry. In decoding the states of a ripple counter, pulses of one clock duration will occur at the decoding gate outputs as the flip-flops change their state.

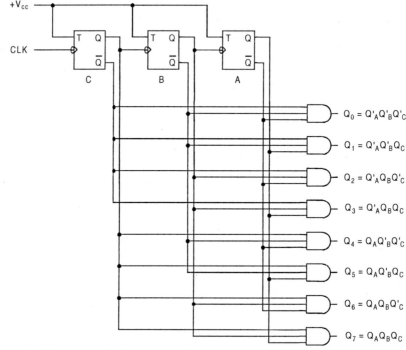

Figure 9.2 3-bit binary asynchronous counter with decoded outputs.

The decoding gates are connected to the outputs such that their outputs will be high only when the counter content is equal to the given state. For example, a decoding gate Q_6 connected in the circuit will decode state 6 (*i.e.*, $Q_A Q_B Q_C = 110$). Thus the gate output will be high only when $Q_A = 1$, $Q_B = 1$, and $Q_C = 0$. The remaining seven states of the 3-bit counter can be decoded in a similar manner using AND gates as Q_0, Q_1, Q_2, Q_3, Q_4, Q_5, and Q_7.

Now, theoretically each decoding output will be high only when the counter content is equal to a given state, and this state occurs only once during a cycle of 2^n states of the counter, where n is the number of flip-flops in the counter. But practically in an asynchronous counter, the decoding gate produces a high output more than once during the cycle of 2^n states. Such undesired high or low pulses of short duration, that appear at the decoding gate output at undesired time instants are known as *spikes* or *glitches*. The reason for these spikes is the cumulative propagation delay in the synchronous counter, which was already discussed in this chapter.

As TTL circuits are very fast, they will respond to even glitches of very small duration (a few nanoseconds). Therefore, these glitches should be eliminated. These can be eliminated by using any one of the following methods: (*i*) clock input to strobe the decoding gates, or (*ii*) using synchronous counters.

To understand the strobing of decoding gates with clock pulse input, we consider a four input AND gate to decode state 5 as shown in Figure 9.3. Here we are using the clock input as the strobe.

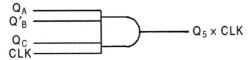

Figure 9.3 Decoding gate Q_5 with a clock as the strobe input.

When the clock input is used to strobe the decoding gate, as shown in Figure 9.3, it will produce the desired output only when the clock is *high*, resulting in perfect decoding of gate output ($Q_5 \times$ CLK) without any glitches. Thus, by strobing the decoding gates with the clock inputs, glitches can be completely avoided.

Modulus or MOD-Number of the Counter

The counter shown in Figure 9.2 has 8 different states. Thus it is a MOD-8 asynchronous counter. The Modulus (or MOD-number) of a counter is the total number of unique states it passes through in each of the complete cycles.

$$\text{Modulus} = 2^n$$

where n = Number of flip-flops.

The maximum binary number that can be counted by the counter is $2^n - 1$. Hence, a 3-flip-flop counter can count a maximum of $(111)_2 = 2^3 - 1 = 7_{10}$.

Frequency Division

Let us consider the counter shown in Figure 9.1. The input consists of a sequence of pulses of frequency, f. As already discussed, Q_0 changes only when the clock makes a transition from 1 to 0. Thus, at the first negative transition Q_0 changes from 0 to 1, and with the second negative transition of the clock Q_0 shifts from 1 to 0. Hence, two input pulses will result in a single pulse in Q_0. Hence the frequency of Q_0 will be $f/2$. Similarly, the frequency of Q_1 signal will be half that of Q_0 signal. Therefore its frequency is $f/4$. Similarly, the frequency of Q_2 will be $f/8$. Hence the circuit can be used to divide the input frequency. These circuits are called *frequency dividers*. If there are n flip-flops used in the circuit then the frequency will be divided by 2^n.

9.2.2 Asynchronous (or Ripple) Counter With Modulus < 2^n

The ripple counter shown in Figure 9.1 is a MOD N or MOD 2^n counter, where n is the number of flip-flops and N is the number of count sequences. This is the maximum MOD-number that is attainable by using n flip-flops. But in practice, it is often required to have a counter which has a MOD-number less than 2^n. In such cases, it is required that the counter will skip states that are normally a part of the counting sequences. A MOD-6 ripple counter is shown in Figure 9.4.

In the circuit shown in Figure 9.4(*a*), without the NAND gate, the counter functions as a MOD-8 binary ripple counter, which can count from 000 to 111. However, when a NAND gate is incorporated in the circuit as shown in Figure 9.4(*a*) the sequence is altered in the following way:

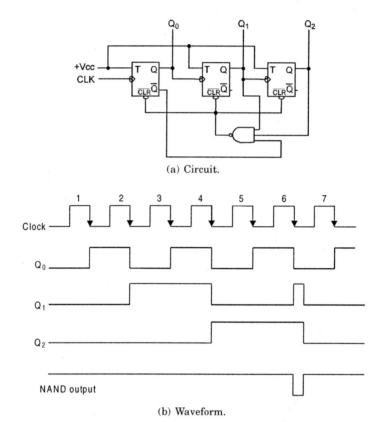

(a) Circuit.

(b) Waveform.

Figure 9.4 MOD-6 asynchronous counter.

1. The NAND gate output is connected to the *clear* inputs of each flip-flop. As along as the NAND gate produces a *high* output, it will have no effect on the counter. But when the NAND gate output goes *low*, it will clear all flip-flops, and the counter will immediately go to the 000 state.

2. The outputs Q_2, Q_1, and Q'_0 are given as the inputs to the NAND gate. The NAND output occurs *low* whenever $Q_2Q_1Q_0 = 110$. This condition will occur on the sixth clock pulse. The *low* at the NAND gate output will clear the counter to the 000 state. Once the flip-flops are cleared the NAND gate output goes back to 1.

3. Hence, again, the cycle of the required counting sequence repeats itself.

Although the counter goes to the 110 state, it remains there only for a few nanoseconds before it recycles to the 000 state. Hence we may say that the counter counts from 000 to 101, it skips the states 110 and 111; thus it works as a MOD-6 counter.

From the waveform shown in Figure 9.4(*b*), it can be noted that the Q_1 output contains a spike or glitch caused by the momentary occurrence of the 110 state before the clearing operation takes place. This glitch is essentially very narrow (owing to the propagation delay of the NAND gate). It can be noted that the Q_2 output has a frequency equal to 1/6 of the input frequency. So we may say that the MOD-6 counter has divided the input frequency by 6.

To construct any MOD-N counter, the following general steps are to be followed.

1. Find the number of flip-flops (n) required for the desired MOD-number using the equation

$$2^{n-1} \leq N \leq 2^n.$$

2. Then connect all the n flip-flops as a ripple counter.
3. Find the binary number for N.
4. Connect all the flip-flop outputs, for which $Q = 1$, as well as $Q' = 1$, when the count is N, as inputs to the NAND gate.
5. Connect the NAND gate output to the clear input of each flip-flop.

When the counter reaches the N-th state, the output of the NAND gate goes *low*, resetting all flip-flops to 0. So the counter counts from 0 through $N - 1$, having N states.

9.2.3 Asynchronous (or Ripple) Down-counter

A down-counter using n flip-flops counts downward starting from a maximum count of $(2^n - 1)$ to zero. The count sequence of such a 3-bit down-counter is given in Table 9.2.

Table 9.2 Count sequence of a 3-bit binary ripple down-counter

Counter State	Q_2	Q_1	Q_0
7	1	1	1
6	1	1	0
5	1	0	1
4	1	0	0
3	0	1	1
2	0	1	0
1	0	0	1
0	0	0	0

Such a down-counter may be designed in three different ways as follows:

Case 1. The circuit shown in Figure 9.1 can be kept intact; only the outputs of the counter may be taken from the complement outputs of the flip-flops, *i.e.*, Q', rather than from the normal outputs for each flip-flop as shown in Figure 9.5(a). The waveform is shown in Figure 9.5(b).

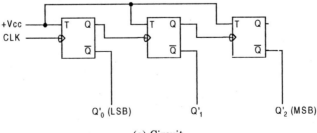

(a) Circuit.

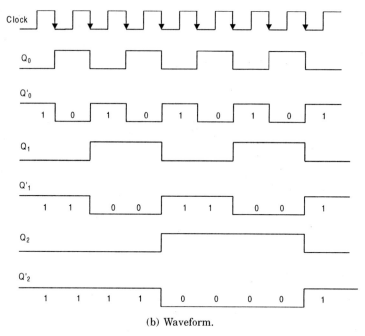

(b) Waveform.

Figure 9.5 A 3-bit asynchronous down-counter (taking outputs from
the complements of each flip-flop).

Here, since the outputs are taken from the complements of the flip-flops, the starting count sequence is $Q'_2Q'_1Q'_0 = 111$. With each negative edge of the clock Q_0 toggles its state. Similarly, with each negative transition of the output Q_0, the output Q_1 toggles and the same thing happens for Q_2, also. Hence the count sequences goes on decreasing from 7, 6, 5, 4, 3, 2, 1, 0, 7, and so on with each clock pulse.

Case 2. The circuit may be slightly modified so that the clock inputs of the second, third, and subsequent flip-flops may be driven by the Q' outputs of the preceding stages, rather than by the Q outputs as shown in Figure 9.6(a). The waveform is shown in Figure 9.6(b).

If the initial counter content is 000, at the first negative transition of the clock, the counter content changes to 111; at the second negative transition, the content becomes 110; at the third negative transition of clock, the content changes to 101, and so on. Thus, in the down-counter, the counter content is decremented by one for every negative transition of the clock pulse.

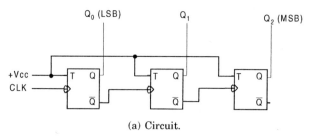

(a) Circuit.

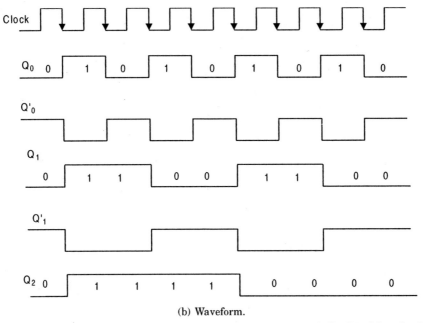

(b) Waveform.

Figure 9.6 A 3-bit asynchronous down-counter (clock inputs of each flip-flop driven by Q').

Case 3. The flip-flops used in the case of the up-counter shown in Figure 9.1, may be replaced by positive edge-triggering flip-flops as shown in Figure 9.7(*a*). The waveform is shown in Figure 9.7(*b*).

Here, since the flip-flops used for the circuit are all positive edge-triggering flip-flops, the flip-flops toggle their states with each positive edge transition of the clock pulse. If initially all the flip-flops are reset, with the first positive edge of the clock pulse Q_0 toggles to 1. Similarly, since Q_0 is driving the second flip-flop, Q_0 toggles from 0 to 1, and positive edge Q_1 also toggles from 0 to 1. A similar thing happens with Q_2. Hence, after the first clock pulse, the counter content becomes 111. Similarly, other count sequences also occur.

The major application of down-counters lies in situations where a desired number of input pulses that have occurred are found. In these situations, the down-counter is *preset* to the desired number and then allowed to countdown as the pulses are applied. When the counter reaches the zero state, it is detected by a logic gate whose output at that time indicates that the preset number of pulses are applied.

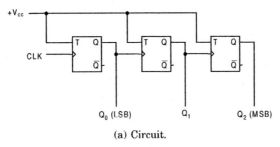

(a) Circuit.

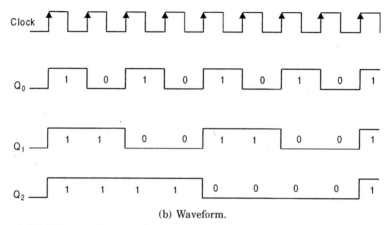

(b) Waveform.

Figure 9.7 A 3-bit asynchronous down-counter (using positive edge-triggering flip-flops).

9.2.4 Asynchronous (or Ripple) Up-down Counter

We have already considered up-counters and down-counters separately. But both of the units can be combined in a single up-down counter. Such a combined unit of up-down counter can count both upward as well as downward. Such a counter is also called a *multimode counter*. In the up-counter each flip-flop is triggered by the normal output of the preceding flip-flop; whereas in a down-counter, each flip-flop is triggered by the complement output of the preceding flip-flop. However, in both the counters, the first flip-flop is triggered by the input pulses.

A 3-bit up-down counter is shown in Figure 9.8. The operation of such a counter is controlled by the up-down control input. The counting sequence of the up-down counter in the two modes of counting is given in Table 9.3. From the circuit diagram we find that three logic gates are required per stage to switch the individual stages from count-up to count-down mode. The logic gates are used to allow either the noninverted output or the inverted output of one flip-flop to the clock input of the following flip-flop, depending on the status of the control input. An inverter has been inserted in between the count-up control line and the count-down control line to ensure that the count-up and count-down cannot be simultaneously in the HIGH state.

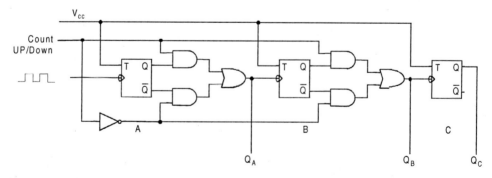

Figure 9.8 Asynchronous 3-bit up-down counter.

Table 9.3 Count sequence for a 3-bit binary ripple up-down counter

COUNT-UP Mode				COUNT-DOWN Mode			
States	Q_C	Q_B	Q_A	States	Q_C	Q_B	Q_A
0	0	0	0	7	1	1	1
1	0	0	1	6	1	1	0
2	0	1	0	5	1	0	1
3	0	1	1	4	1	0	0
4	1	0	0	3	0	1	1
5	1	0	1	2	0	1	0
6	1	1	0	1	0	0	1
7	1	1	1	0	0	0	0

When the count-up/down line is held HIGH, the lower AND gates will be disabled and their outputs will be zero. So they will not affect the outputs of the OR gates. At the same time the upper AND gates will be enabled. Hence, Q_A will pass through the OR gate and into the clock input of the B flip-flop. Similarly, Q_B will be gated into the clock input of the C flip-flop. Thus, as the input pulses are applied, the counter will count up and follow a natural binary counting sequence from 000 to 111.

Similarly, with count-up/down line being logic 0, the upper AND gates will become disabled and the lower AND gates are enabled, allowing Q'_A and Q'_B to pass through the clock inputs of the following flip-flops. Hence, in this condition the counter will count in down mode, as the input pulses are applied.

9.2.5 Propagation Delay in an Asynchronous Counter

Asynchronous counters are simple in circuitry. But the main disadvantage of these type of counters is that they are slow. In asynchronous counters each flip-flop is triggered by the transition of the output of the preceding flip-flop. Because of the inherent propagation delay time (t_{pd}), the first flip-flop only responds after a period of t_{pd} after receiving a clock pulse. Similarly, the second flip-flop only responds after a period of $2t_{pd}$ after the input pulse occurs. Hence, the nth flip-flop cannot change states for a period of $n \times t_{pd}$ even after the input clock pulse occurs. Therefore, to allow all the flip-flops in an n-bit counter to change states in response to a clock, the period of the clock T should be:

$$T \geq n \times t_{pd} .$$

Thus, the maximum frequency (f) that can be used in an asynchronous counter for reliable operation is given by:

$$\frac{1}{f} \geq n \times t_{pd}$$

$$\text{or, } \frac{1}{n \times t_{pd}} \geq f$$

Thus, f should be less than or equal to $\frac{1}{n \times t_{pd}}$. Hence, the maximum clock frequency that can be applied in an n-bit asynchronous counter is

$$f_{max} = \frac{1}{n \times t_{pd}}.$$

Example 9.1. *In a 5-stage ripple counter, the propagation delay of each flip-flop is 50 ns. Find the maximum frequency at which the counter operates reliably.*

Solution. The maximum frequency is

$$f_{max} = \frac{1}{5 \times 50\text{ns}}$$

$$= 4 \text{ MHz}.$$

9.3 ASYNCHRONOUS COUNTER ICs

The design of the asynchronous counter using flip-flops has been discussed above. Some asynchronous counters are available in MSI and are given in Table 9.4 along with some of their features. Depending on these features these ICs are divided into three groups A, B, and C. The group to which a particular IC belongs is indicated in the table. All these ICs consist of four master-slave flip-flops. The set, reset (clear), and load operations are asynchronous, *i.e.*, independent of the clock pulse.

9.3.1 Group *A* Asynchronous Counter ICs

Figure 9.9 shows the basic internal structure of IC 7490. IC 7490 is basically a BCD counter or decade counter (MOD 10), which consists of four master-slave flip-flops internally connected to provide a MOD-2 counter and a MOD-5 counter. The reset inputs R_1 and R_2 are connected to logic 1, to reset the counter to 0000, and the set inputs S_1 and S_2 are connected to logic 1 to set the counter to 1001. Since the output Q_A from flip-flop A is not internally connected to the succeeding stages, the counter can be operated in two count modes.

Table 9.4 Asynchronous counter ICs

IC No.	Description	Features	Group
7490, 74290	BCD counter	Set, reset	A
74490	Dual BCD counter	Set, reset	A
7492	Divide-by-12 counter	Reset	B
7493, 74293	4-bit binary counter	Reset	B
74390	Dual decade counters	Reset	B
74393	Dual 4-bit binary counters	Reset	B
74176, 74196	Presettable BCD counter	Reset, load	C
74177, 74197	Presettable 4-bit binary counter	Reset, load	C

1. When used as a BCD counter, the B input must be externally connected to the Q_A output. The incoming pulses are received by the input A, and a count sequence is obtained as the BCD output sequence as shown in Table 9.5. Two gated inputs are provided to reset the counter to 0. In addition, two more inputs are also provided to set a BCD count of 9 for 9's complement decimal applications.

2. When it is required to function as a MOD-2 counter and a MOD-5 counter, no external connections are necessary. Flip-flop A is used as a binary element for the MOD-2 function. The B input is used to obtain binary MOD-5 operation at the Q_B, Q_C, and Q_D outputs. In this mode, the two counters operate independently. But all four flip-flops are reset simultaneously.

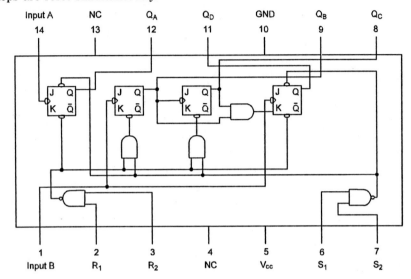

Figure 9.9 Internal structure of an IC 7490 ripple counter.

IC 74490 is a dual BCD counter consisting of two independent BCD counters. Each section consists of four flip-flops, all connected internally to form a decade counter. For each section there is a set (S) and a reset (R) input which are active high.

Table 9.5 Count sequence for IC 7490

Mode-1 (MOD-10)				Mode-2 (MOD-5)		
Q_D	Q_C	Q_B	Q_A	Q_D	Q_C	Q_B
0	0	0	0	0	0	0
0	0	0	1	0	0	1
0	0	1	0	0	1	0
0	0	1	1	0	1	1
0	1	0	0	1	0	0
0	1	0	1			
0	1	1	0			
0	1	1	1			
1	0	0	0			
1	0	0	1			

Example 9.2. *In a 7490 IC, if Q_D output is connected to A input and the pulses are applied at B input, find the count sequence of the Q outputs.*

Solution. If Q_D output is connected to A input and the pulses are applied at B input, we have the MOD-5 counter followed by the MOD-2 counter. The count sequence obtained is given in Table 9.6. Here the states of the MOD-5 counter change in a normal binary sequence and Q_A changes whenever Q_D goes from 1 to 0.

Table 9.6

Counter State	Flip-flop outputs			
	Q_D	Q_C	Q_B	Q_A
0	0	0	0	0
1	0	0	1	0
2	0	1	0	0
3	0	1	1	0
4	1	0	0	0
5	0	0	0	1
6	0	0	1	1
7	0	1	0	1
8	0	1.	1	1
9	1	0	0	1
10	0	0	0	0

The count sequence of this counter is different from that of a normal decade counter, although both are MOD-10 counters.

Example 9.3. *In a 7490 IC, if Q_A output is connected to B input and the pulses are applied at A input, find the count sequence of the Q outputs.*

Table 9.7

Counter State	Flip-flop outputs			
	Q_D	Q_C	Q_B	Q_A
0	0	0	0	0
1	0	0	0	1
2	0	0	1	0
3	0	0	1	1
4	0	1	0	0
5	0	1	0	1
6	0	1	1	0
7	0	1	1.	1
8	1	0	0	0
9	1	0	0	1
10	0	0	0	0

Solution. If Q_A output is connected to B input and the pulses are applied at A input, we have the MOD-2 counter followed by the MOD-5 counter. The count sequence obtained is given in Table 9.7. Here, when Q_A changes from 0 to 1, the state of MOD-5 counter does not change, whereas when Q_A changes from 1 to 0, the state of MOD-5 counter goes to the next state.

Example 9.4. *Design a MOD-5 counter using 7490 IC.*

Solution. First the counter is connected as a MOD-10 counter for normal binary sequence (as shown in Example 9.3). Then outputs Q_A and Q_C are connected to the reset inputs. Hence, as soon as Q_A and Q_C both become 1, the counter is reset to 0000. Figure 9.10 shows the MOD-5 ripple counter.

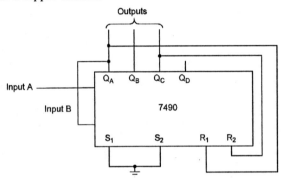

Figure 9.10 A MOD-5 ripple counter using IC 7490.

9.3.2 Group *B* Asynchronous Counter ICs

The basic structure of IC 7493 is shown in Figure 9.11. Basically the asynchronous counter ICs like 7492 and 74293 follows the same internal structure as IC 7493. The operation of these ICs is identical to the operation of IC 74990 except that the set inputs are not present.

IC 7493 is a 4-bit binary ripple counter that consists of four master-slave J-K flip-flops. These four flip-flops are internally connected to provide a MOD-2 and MOD-8 counter, the reset inputs R_1 and R_2 are used to reset the counter to 0000. Since the output Q_A from flip-flop A is not internally connected to the succeeding flip-flops, the counter may be operated in two independent modes as discussed below.

1. If the counter is to be used as a 4-bit ripple counter, output Q_A must be externally connected to input B. The input pulses are applied to input A. Simultaneous divisions of 2, 4, 8, and 16 are performed at the Q_A, Q_B, Q_C, and Q_D outputs respectively. The count sequence for this connection is given in Table 9.8.

2. If the counter is to be used as a 3-bit ripple counter, the input pulses are applied to the input B. Simultaneous frequency divisions of 2, 4, and 8 are performed at the Q_B, Q_C, and Q_D outputs. Independent use of flip-flop A is available if the reset function coincides with the reset of the 3-bit ripple counter.

Basically, these ICs are not used as counters but are used for frequency division. IC 7492 is a divide-by-12 counter, which consists of four master-slave J-K flip-flops. These four flip-flops are internally connected to provide a MOD-2 and MOD-6 counter. IC 74390 is a dual BCD counter consisting of two independent BCD counters similar to IC 7490. There is one *reset* (R) input for each section. IC 74393 is a dual 4-bit binary counter with one *reset* (R) input for each section that is active-high.

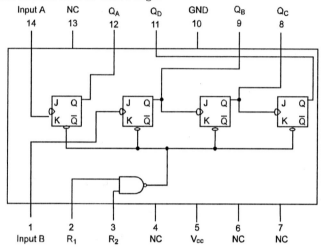

Figure 9.11 Internal structure of IC 7493—4-bit ripple counter.

Table 9.8 Count sequence for IC 7493—4-bit binary ripple counter

Mode-1 (MOD-16)				Mode-2 (MOD-8)		
Q_D	Q_C	Q_B	Q_A	Q_D	Q_C	Q_B
0	0	0	0	0	0	0
0	0	0	1	0	0	1
0	0	1	0	0	1	0
0	0	1	1	0	1	1
0	1	0	0	1	0	0
0	1	0	1	1	0	1
0	1	1	0	1	1	0
0	1	1	1	1	1	1
1	0	0	0			
1	0	0	1			
1	0	1	0			
1	0	1	1			
1	1	0	0			
1	1	0	1			
1	1	1	0			
1	1	1	1			
0	0	0	0			

Example 9.5. *If the output Q_A of a MOD-12 ripple counter 7492 IC is connected to the B input and the pulses are applied at the A input, find the count sequence.*

Solution. The count sequence is shown in Table 9.9. Here, it may be noted that simultaneous divisions of 2, 6, and 12 are performed at the Q_A, Q_C, and Q_D outputs respectively.

Table 9.9

Counter State	Flip-flop outputs			
	Q_D	Q_C	Q_B	Q_A
0	0	0	0	0
1	0	0	0	1
2	0	0	1	0
3	0	0	1	1
4	0	1	0	0
5	0	1	0	1
6	1	0	0	0
7	1	0	0	1
8	1	0	1	0
9	1	0	1	1
10	1	1	0	0
11	1	1	0	1

9.3.3 Group *C* Asynchronous Counter ICs

The basic internal structure of group C counter ICs is shown in Figure 9.12. IC 74196 and IC 74176 are both BCD counters with a difference in only maximum clock frequency specification. Similarly, IC 74197 and IC 74177 are both 4-bit binary counters with the same difference.

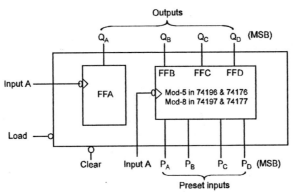

Figure 9.12 Basic internal structure of group C asynchronous counter ICs.

These counters are actually presettable versions of 7490 and 7493 counters respectively. The counter is cleared by connecting logic 0 to the clear input, which is active-low. The counter can be stopped any time and any binary number present at the preset inputs may be loaded into the counter by setting the load input to logic 0, while the clear input is at logic 1. For normal UP counting operation, both the load and clear inputs should be connected to logic 1.

The presettable 4-bit binary counters can be used as variable MOD-n counters in which the counter modulus is equal to 15–P, where P is the binary number connected at the preset input. In other words, for designing a MOD-n counter, the value of P is 15–n. When the counter output reaches the count 1111, the counter must be loaded again with P. This is made possible by using a four-input NAND gate between the Q outputs of the counter and the load input.

Example 9.6. *Design a divide-by-10 counter using IC 74177.*

Solution. The circuit of a divide-by-10 counter is shown in Figure 9.13. The value of P (= 15–n) is P = 1111–1010 = 0101. The counter is now loaded with 0101 as soon as the output reaches 1111.

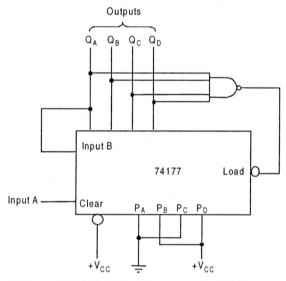

Figure 9.13 A divide-by-10 counter using IC 74177.

9.3.4 Cascading of Ripple Counter ICs

By cascading the ICs that were discussed above, we can construct ripple counters of any cycle length. The desired cycle length is decoded and used to reset all the counters to 0. The strobe should be used to eliminate false data.

The cascading arrangement for all the synchronous counter ICs is same where Q_D of the preceding stage goes to the clock input terminal of the succeeding stage. The load and clear inputs of all ICs are to be connected together.

Example 9.7. *Design a 2-decade BCD counter using IC 74390.*

Solution. The 74390 is a dual Decade counter and belongs to the Group B group of ICs. Hence only one IC is required to design a 2-decade BCD counter. The 2-decade counter is shown in Figure 9.14.

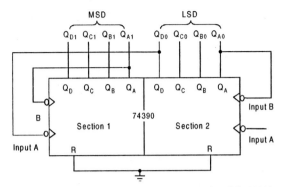

Figure 9.14 A 2-decade BCD counter using IC 74390.

9.4 SYNCHRONOUS (PARALLEL) COUNTERS

The ripple or asynchronous counter is the simplest to build, but its highest operating frequency is limited because of ripple action. Each flip-flop has a delay time. In ripple counters these delay times are additive and the total "settling" time for the counter is approximately the product of the delay time of a single flip-flop and the total number of flip-flops. Again, there is the possibility of glitches occurring at the output of decoding gates used with a ripple counter.

Both of these problems can be overcome, if all the flip-flops are clocked synchronously. The resulting circuit is known as a *synchronous counter*. Synchronous counters can be designed for any count sequence (need not be straight binary). These can be designed following a systematic approach. Before we discuss the formal method of design for such counters, we shall consider an intuitive method.

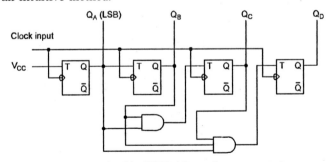

Figure 9.15 A 4-bit (MOD-16) synchronous counter.

A 4-bit synchronous counter with parallel carry is shown in Figure 9.15. In this circuit the clock inputs of all the flip-flops are tied together so that the input clock signal may be applied simultaneously to each flip-flop. Only the LSB flip-flop A has its T input connected permanently to logic 1 (*i.e.*, V_{CC}), while the T inputs of the other flip-flops are driven by some combination of flip-flop outputs. The T input of flip-flop B is connected to the output Q_A of flip-flop A; the T input of flip-flop C is connected with the AND-operated output of Q_A and Q_B. Similarly, the T input of D flip-flop is connected with the AND-operated output of Q_A, Q_B, and Q_C.

From the circuit, we can see that flip-flop A changes its state with the negative transition of each clock pulse. Flip-flop B changes its state only when the value of Q_A is 1 and a negative transition of the clock pulse takes place. Similarly, flip-flop C changes its state only when both Q_A and Q_B are 1 and a negative edge transition of the clock pulse takes place. In the same manner, the flip-flop D changes its state when $Q_A = Q_B = Q_C = 1$ and when there is a negative transition at clock input. The count sequence of the counter is given in Table 9.10.

Table 9.10 Count sequence of a 4-bit binary synchronous counter

State	Q_D	Q_C	Q_B	Q_A
0	0	0	0	0
1	0	0	0	1
2	0	0	1	0
3	0	0	1	1
4	0	1	0	0
5	0	1	0	1
6	0	1	1	0
7	0	1	1	1
8	1	0	0	0
9	1	0	0	1
10	1	0	1	0
11	1	0	1	1
12	1	1	0	0
13	1	1	0	1
14	1	1	1	0
15	1	1	1	1
0	0	0	0	0

9.4.1 Propagation Delay in a Synchronous Counter

Unlike asynchronous counters where the total propagation delay is given by the cumulative effect of the flip-flops, the total settling or response time of a synchronous counter is given as follows: the time taken by one flip-flop to toggle plus the time for the new logic levels to propagate through a single AND gate to reach the T inputs of the following flip-flop.

Total delay = Propagation delay of one flip-flop + Propagation delay of an AND gate.

Irrespective of the total number of flip-flops, the propagation delay will always be the same. Normally, this will be much lower than the propagation delay in asynchronous counters with the same number of flip-flops. Thus, the speed of operation of synchronous counters is limited by the propagation delays of an AND gate and a single flip-flop. Hence, the maximum frequency of operation of a synchronous counter is given by

$$f_{max} = \frac{1}{t_p + t_g}$$

where t_p is the propagation delay of one flip-flop and t_g is the propagation delay of one AND gate.

Because of common clocking of all the flip-flops, glitches can be avoided completely in synchronous counters.

9.4.2 Synchronous Counter with Ripple Carry

The 4-bit synchronous counter discussed in the previous section is said to be a *synchronous counter with parallel carry*. Moreover, in this type of counter, as the number of stages increases, the number of AND gates also increases, along with the number of inputs for each of those AND gates. This is a certain disadvantage for such type of circuits. Now this problem can be eliminated if we use the *synchronous counter with ripple carry* shown in Figure 9.16.

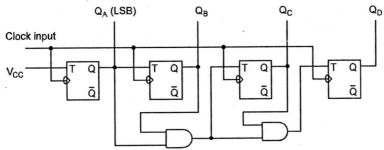

Figure 9.16 A 4-bit synchronous counter with ripple carry.

But in such circuits the maximum clock frequency of the counter is reduced. This reduction of the maximum clock frequency is due to the delay through control logic which is now 2_{tg} instead of t_g which was achieved with parallel carry. The maximum clock frequency for an n-bit synchronous counter with ripple carry is given by

$$f_{max} = \frac{1}{t_p + (n - 2)t_g}$$

where n = number of flip-flop stages.

9.5 SYNCHRONOUS DOWN-COUNTER

A parallel down-counter can be made to count down by using the inverted outputs of flip-flops to feed the various logic gates. Even the same circuit may be retained and the

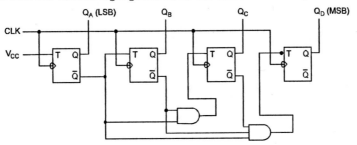

Figure 9.17 A 4-bit synchronous down-counter.

outputs may be taken from the complement outputs of each flip-flop. The parallel counter shown in Figure 9.17 can be converted to a down-counter by connecting the Q'_A, Q'_B, and Q'_C outputs to the AND gates in place of Q_A, Q_B, and Q_C respectively as shown in Figure 9.17. In this case the count sequences through which the counter proceeds will be as shown in Table 9.11.

Table 9.11 Count sequence of a 4-bit synchronous down-counter

State	Q_D	Q_C	Q_B	Q_A
15	1	1	1	1
14	1	1	1	0
13	1	1	0	1
12	1	1	0	0
11	1	0	1	1
10	1	0	1	0
9	1	0	0	1
8	1	0	0	0
7	0	1	1	1
6	0	1	1	0
5	0	1	0	1
4	0	1	0	0
3	0	0	1	1
2	0	0	1	0
1	0	0	0	1
0	0	0	0	0
15	1	1	1	1

9.6 SYNCHRONOUS UP-DOWN COUNTER

Combining both the functions of up- and down-counting in a single counter, we can make a synchronous up-down counter as shown in Figure 9.18. Here the control input (count-up/down) is used to allow either the normal output or the inverted output of one flip-flop to the T input of the following flip-flop. Two separate control lines (count-Up and count-

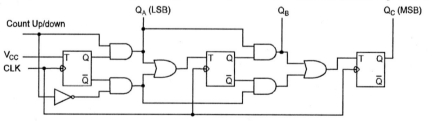

Figure 9.18 A MOD-8 synchronous up-down-counter.

down) could have been used but in such case we have to be careful that both of the lines cannot be simultaneously in the high state. When the count-up/down line is high, then the upper AND gates will be active and the lower AND gates will remain inactive and hence the normal output of each flip-flop is carried forward to the following flip-flop. In such case, the counter will count from 000 to 111. On the other hand, if the control line is low, then the upper AND gates remain inactive, while the lower AND gates will become active. So the inverted output comes into operation and the counter counts from 111 to 000.

9.7 DESIGN PROCEDURE OF A SYNCHRONOUS COUNTER

Following certain general steps, synchronous counters of any given count sequence and modulus can be designed. The steps are listed below:

Step 1. From the given word description of the problem, draw a state diagram that describes the operation of the counter.

Step 2. From the state table, write the count sequences in the form of a table as shown in Table 9.10.

Step 3. Find the number of flip-flops required.

Step 4. Decide the type of flip-flop to be used for the design of the counter. Then determine the flip-flop inputs that must be present for the desired next state from the present state using the excitation table of the flip-flops.

Step 5. Prepare K-maps for each flip-flop input in terms of flip-flop outputs as the input variables. Simplify the K-maps and obtain the minimized expressions.

Step 6. Connect the circuit using flip-flops and other gates corresponding to the minimized expressions.

9.7.1 Synchronous Counter with Modulus < 2^n

We have already discussed asynchronous (ripple) counters and different types of synchronous (parallel) counters, all of which have the ability to operate in either a count-up or count-down mode. But all of these counters progress one count at a time in a strict binary progression, and they all have a modulus given by 2^n, where n indicates the number of flip-flops. Such counters are said to have a "natural count" of 2^n.

A MOD-2 counter consists of a single flip-flop; a MOD-4 counter requires two flip-flops, and it counts four discrete states. Three flip-flops form a MOD-8 counter, while four flip-flops form a MOD-16 counter and so on. Thus we can construct counters that have a natural count of 2, 4, 8, 16, and so on. But it is often desirable to use counters having a modulus other than 2, 4, 8, 16, and so on. For example, a counter having a modulus of 3, 5, or 10 may be useful. A smaller modulus counter can always be constructed from a larger modulus counter by skipping state. Such counters are said to have a *modified count*.

9.7.2 Design of a MOD-3 Counter

A MOD-3 counter is a counter which has only three distinct states. To design a counter with three states, the number of flip-flops required can be found using the equation , where n is the number of flip-flops required and N is the number of states present in the counter. For N = 3, from the above equation, $n = 2$, *i.e.*, two flip-flops are required. Now we draw the state diagram as shown in Figure 9.19. Here it is assumed that the state transition from one state to another takes place only when a clock transition takes place. From the state diagram, we can form the state table for the counter as shown in Table 9.12. From the state table, along with the excitation table of the flip-flops, we can form the excitation table for the MOD-3 counter as shown in Table 9.13. Although any one of the four flip-flops, *i.e.*, S-R, J-K, T, and D can be used, we select here T flip-flops, which results in a simplified circuit.

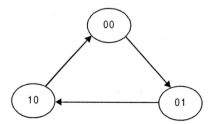

Figure 9.19 State diagram of a MOD-3 counter.

Table 9.12 State table for the counter

Present state	Next state
00	01
01	10
10	00

Table 9.13 Excitation table for the MOD-3 counter

Count Sequence		Flip-flop inputs	
A_1	A_0	TA_1	TA_0
0	0	0	1
0	1	1	1
1	0	1	0

Table 9.13 is the excitation table for the MOD-3 counter. The two flip-flops are given variable designations A_1 and A_0. The flip-flop excitation for the T inputs is derived from the excitation table of the T flip-flop and from inspection of the state transition from a given count (present state) to the next below it (next state). As an illustration, consider the count sequence for the row 00. The next state is 01. Comparing these two counts, we note that A_1 goes from 0 to 0; so TA_1 is marked with a 0 because flip-flop A_1 remains unchanged when a clock transition takes place. A_0 goes from 0 to 1; so TA_0 is marked with a 1 because flip-flop A_0 must be complemented in the next clock transition. The last row with the present state 10 is compared with the first count 00, which is its next state.

The flip-flop input functions from the excitation tables are simplified in the K-maps of Figure 9.20. The Boolean functions listed under each map specify the combinational-circuit part of the counter. Including these functions with the two flip-flops, we obtain the logic diagram of the counter as shown in Figure 9.21.

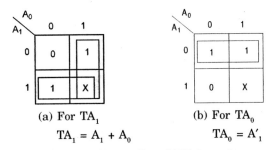

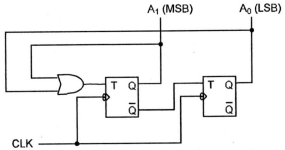

(a) For TA_1

$$TA_1 = A_1 + A_0$$

(b) For TA_0

$$TA_0 = A'_1$$

Figure 9.20 K-maps for a MOD-3 counter.

Figure 9.21 Logic diagram of a MOD-3 binary counter.

9.7.3 Design of a MOD-5 Counter

In order to design a MOD-5, which has five distinct states, the number of flip-flops required can be found using the equation, where n is the number of flip-flops required and N is the number of states present in the counter. For N = 5, from the above equation, $n = 3$, *i.e.*, three flip-flops are required. Now we draw the state diagram as shown in Figure 9.22. From the state diagram, we can form the state table for the counter as shown in Table 9.14. From the state table, along with the excitation table of the flip-flops, we can form the excitation table for the MOD-5 counter as shown in Table 9.15. Since we have already developed several counter circuits using T flip-flops, we select here J-K flip-flops, which also results in a simple circuit.

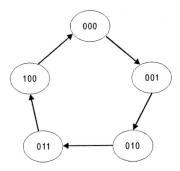

Figure 9.22 State diagram of a MOD-5 counter.

Table 9.14 State table for the counter

Present state	Next state
000	001
001	010
010	011
011	100
100	000

Table 9.15 Excitation table for the MOD-5 counter

Count Sequence			Flip-flop inputs					
A_2	A_1	A_0	JA_2	KA_2	JA_1	KA_1	JA_0	KA_0
0	0	0	0	X	0	X	1	X
0	0	1	0	X	1	X	X	1
0	1	0	0	X	X	0	1	X
0	1	1	1	X	X	1	X	1
1	0	0	X	1	0	X	0	X

Table 9.15 is the excitation table for the MOD-5 counter. The three flip-flops are given variable designations A_2, A_1, and A_0. The flip-flop excitations for the J and K inputs are derived from the excitation table of the J-K flip-flop and from inspection of the state transition from a given count (present state) to the next below it (next state).

The flip-flop input functions from the excitation tables are simplified in the K-maps of Figure 9.23. The Boolean functions listed under each map specify the combinational-circuit part of the counter. Including these functions with the two flip-flops, we obtain the logic diagram of the counter as shown in Figure 9.24.

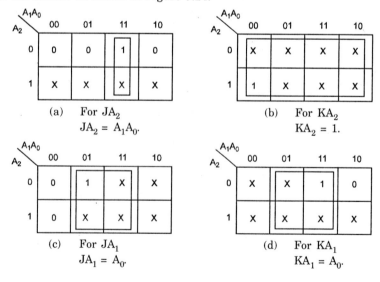

(a)   For JA_2
$JA_2 = A_1 A_0$.

(b)   For KA_2
$KA_2 = 1$.

(c)   For JA_1
$JA_1 = A_0$.

(d)   For KA_1
$KA_1 = A_0$.

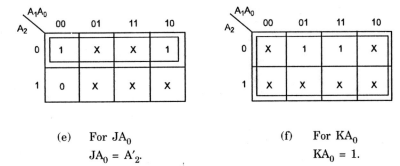

(e)   For JA_0

       $JA_0 = A'_2$.

(f)   For KA_0

       $KA_0 = 1$.

Figure 9.23 K-maps for a MOD-5 counter.

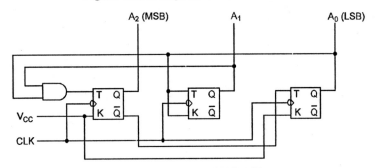

Figure 9.24 Logic diagram of a MOD-5 binary counter.

9.7.4 Design of a MOD-10 (or BCD or Decade) Counter

In order to design a MOD-10 or decade counter, which has ten distinct states, four flip-flops are required. Now we draw the state diagram as shown in Figure 9.25. From the state diagram, we can form the state table for the counter as shown in Table 9.16. From the state table and using the excitation table of the flip-flops, we can form the excitation table for the MOD-10 counter as shown in Table 9.17. We select T flip-flops to design the circuit.

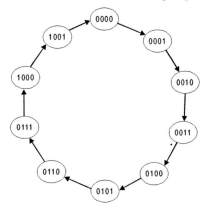

Figure 9.25 State diagram of a MOD-10 counter.

Table 9.16 State table for the counter

Present state	Next state
0000	0001
0001	0010
0010	0011
0011	0100
0100	0101
0101	0110
0110	0111
0111	1000
1000	1001
1001	0000

Table 9.17 Excitation table for the MOD-10 counter

Count Sequence				Flip-flop inputs			
A_3	A_2	A_1	A_0	TA_3	TA_2	TA_1	TA_0
0	0	0	0	0	0	0	1
0	0	0	1	0	0	1	1
0	0	1	0	0	0	0	1
0	0	1	1	0	1	1	1
0	1	0	0	0	0	0	1
0	1	0	1	0	0	1	1
0	1	1	0	0	0	0	1
0	1	1	1	1	1	1	1
1	0	0	0	0	0	0	1
1	0	0	1	1	0	0	1

Table 9.17 is the excitation table for the MOD-10 counter. The four flip-flops are given variable designations A_3, A_2, A_1, and A_0. The flip-flop excitations for the T inputs are derived from the excitation table of the T flip-flop and from inspection of the state transition from a given count (present state) to the next below it (next state).

The flip-flop input functions from the excitation tables are simplified in the K-maps of Figure 9.26. The Boolean functions listed under each map specify the combinational-circuit part of the counter. Including these functions with the two flip-flops, we obtain the logic diagram of the counter as shown in Figure 9.27.

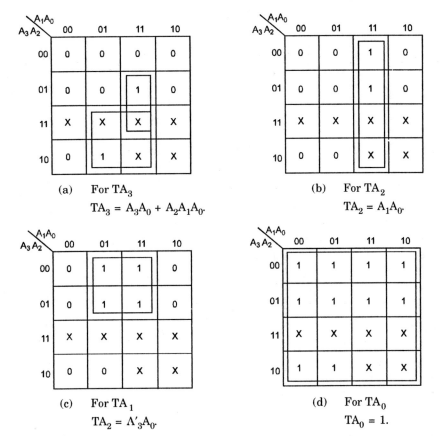

(a) For TA_3

$$TA_3 = A_3A_0 + A_2A_1A_0.$$

(b) For TA_2

$$TA_2 = A_1A_0.$$

(c) For TA_1

$$TA_2 = A'_3A_0.$$

(d) For TA_0

$$TA_0 = 1.$$

Figure 9.26 K-maps for a MOD-10 counter.

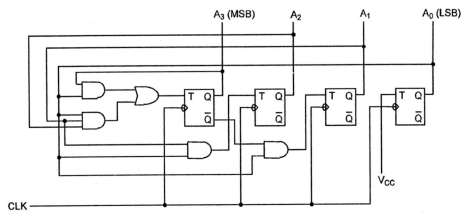

Figure 9.27 Logic diagram of a MOD-10 binary counter.

9.7.5 Lock Out

In the counters with modulus less than 2^n, it may happen that the counter by chance finds itself in any one of the unused states. For example, in the MOD-10 counter, logic states, $A_3A_2A_1A_0 = 1010, 1011, 1100, 1101, 1110,$ and 1111 are not used. Now, if by chance the counter enters into any one of these unused states, its next state will not be known. It may be possible that the counter might go from one unused state to another and never arrive at a used state. In such a situation the counter becomes useless for its intended purpose. A counter whose unused states have this feature is said to suffer from *lock out*. To make sure that at the starting point the counter is in its initial state or it comes to its initial state within a few clock cycles (count error due to noise), external logic circuitry is provided.

To ensure that lock out does not occur, we design the counter assuming the next state to be the initial state, from each of the unused states. Beyond this, the design procedure is the same as discussed earlier.

9.7.6 Design of an Irregular MOD-5 Counter

The counter circuit that follows irregular count sequences is called an irregular counter. It does not follow the natural binary sequence. The count sequences are prespecified, and the counter progresses according to that. To illustrate this, we can consider an irregular sequence and try to design the circuit for such a counter.

The count sequences are taken as: $000 \rightarrow 010 \rightarrow 100 \rightarrow 101 \rightarrow 110$. Now the state diagram is shown in Figure 9.28, and the excitation table is shown in Table 9.18. We can design the counter with any flip-flop. Here we consider the T flip-flop, which will lead to the simplest design.

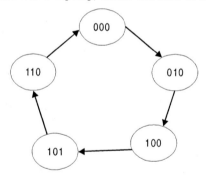

Fig 9.28 State diagram of the irregular MOD-5 counter.

Table 9.18 Excitation Table for the Counter

Count Sequence			Flip-flop inputs		
A_2	A_1	A_0	TA_2	TA_1	TA_0
0	0	0	0	1	0
0	1	0	1	1	0
1	0	0	0	0	1
1	0	1	0	1	1
1	1	0	1	1	0

Table 9.18 is the excitation table for the irregular counter. The three flip-flops are given variable designations A_2, A_1, and A_0. The flip-flop excitations for the T inputs are derived from the excitation table of the T flip-flop and from inspection of the state transition from a given count (present state) to the next below it (next state).

The flip-flop input functions from the excitation tables are simplified in the K-maps of Figure 9.29. The Boolean functions listed under each map specify the combinational-circuit part of the counter. Including these functions with the two flip-flops, we obtain the logic diagram of the counter as shown in Figure 9.30.

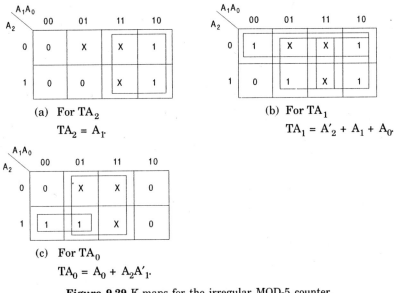

(a) For TA_2

$$TA_2 = A_1.$$

(b) For TA_1

$$TA_1 = A'_2 + A_1 + A_0.$$

(c) For TA_0

$$TA_0 = A_0 + A_2A'_1.$$

Figure 9.29 K-maps for the irregular MOD-5 counter.

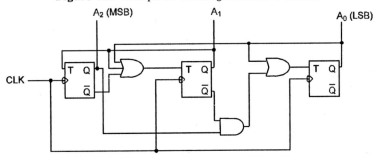

Figure 9.30 Logic diagram of the irregular MOD-5 binary counter.

9.7.7 Design of a MOD-8 Synchronous Counter Using S-R, J-K, and D Flip-flops

Normally, synchronous counters are designed using J-K or T flip-flops, since they lead to a simple design. However, synchronous counters can also be designed using S-R or D flip-flops. We have already designed a MOD-8 counter using T flip-flops. Now we want to show the design of the same counter using S-R, J-K, and D flip-flops.

To design a MOD-8 counter the number of flip-flops required is four. We now draw the state diagram of the counter. The state diagram is shown in Figure 9.31. Since we are using all three flip-flops for the counter design, the excitation table has entries for flip-flop inputs S_2R_2, S_1R_1, S_0R_0 (for design using S-R flip-flops); J_2K_2, J_1K_1, J_0K_0 (for design using J-K flip-flops) and D_2, D_1, D_0 (for design using D flip-flops). The table is shown in Table 9.19.

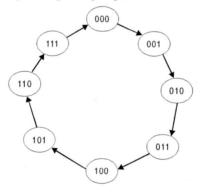

Figure 9.31 State diagram of a MOD-8 counter.

Table 9.19 Excitation table for the counter

Count Sequence			SR Flip-flop inputs						JK Flip-flop inputs						D Flip-flop inputs		
A_2	A_1	A_0	SA_2	RA_2	SA_1	RA_1	SA_0	RA_0	JA_2	KA_2	JA_1	KA_1	JA_0	KA_0	DA_2	DA_1	DA_0
0	0	0	0	X	0	X	1	0	0	X	0	X	1	X	0	0	1
0	0	1	0	X	1	0	0	1	0	X	1	X	X	1	0	1	0
0	1	0	0	X	X	0	1	0	0	X	X	0	1	X	0	1	1
0	1	1	1	0	0	1	0	1	1	X	X	1	X	1	1	0	0
1	0	0	X	0	0	X	1	0	X	0	0	X	1	X	1	0	1
1	0	1	X	0	1	0	0	1	X	0	1	X	X	1	1	1	0
1	1	0	X	0	X	0	1	0	X	0	X	0	1	X	1	1	1
1	1	1	0	1	0	1	0	1	X	1	X	1	X	1	0	0	0

Table 9.19 is the excitation table for the MOD-8 counter. The three flip-flops are given variable designations A_2, A_1, and A_0. The flip-flop excitations for the S-R, J-K, and D inputs are derived from the excitation table of the S-R, J-K, and D flip-flops and from inspection of the state transition from a given count (present state) to the next below it (next state).

The S-R flip-flop input functions from the excitation tables are simplified in the K-maps of Figure 9.32. Similarly, the J-K and the D flip-flop input functions from the excitation tables are simplified in the K-maps of Figure 9.33 and Figure 9.34 respectively. The Boolean functions listed under each map specify the combinational-circuit part of the counter. Including these functions with the two flip-flops, we obtain the logic diagram of the counter using S-R, J-K, and D flip-flops as shown in Figure 9.35, Figure 9.36, and Figure 9.37 respectively.

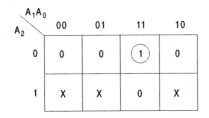

(a) For SA$_2$

SA$_2$ = A$'_2$A$_1$A$_0$.

(b) For RA$_2$

RA$_2$ = A$_2$A$_1$A$_0$.

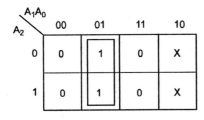

(c) For SA$_1$

SA$_1$ = A$'_1$A$_0$.

(d) For RA$_1$

RA$_1$ = A$_1$A$_0$.

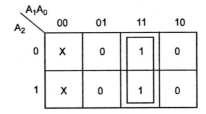

(e) For SA$_0$

SA$_0$ = A$'_0$.

(f) For RA$_0$

RA$_0$ = A$_0$.

Figure 9.32 K-maps for the MOD-8 counter using S-R flip-flops.

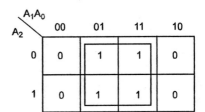

(a) For JA$_2$

JA$_2$ = A$_1$A$_0$.

(b) For KA$_2$

KA$_2$ = A$_1$A$_0$.

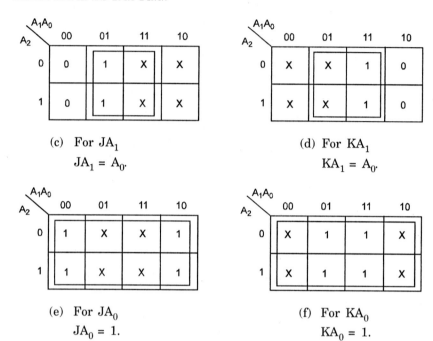

(c) For JA_1

$$JA_1 = A_0.$$

(d) For KA_1

$$KA_1 = A_0.$$

(e) For JA_0

$$JA_0 = 1.$$

(f) For KA_0

$$KA_0 = 1.$$

Figure 9.33 K-maps for the MOD-8 counter using J-K flip-flops.

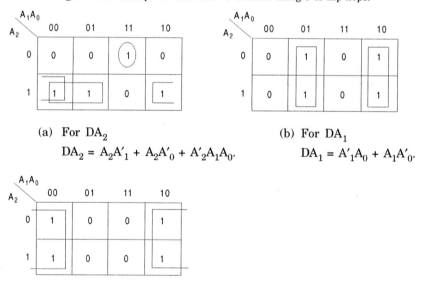

(a) For DA_2

$$DA_2 = A_2A'_1 + A_2A'_0 + A'_2A_1A_0.$$

(b) For DA_1

$$DA_1 = A'_1A_0 + A_1A'_0.$$

(c) For DA_0

$$DA_0 = A'_0.$$

Figure 9.34 K-maps for the MOD-8 counter using D flip-flops.

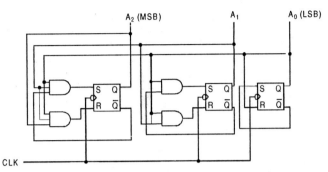

Figure 9.35 Logic diagram of the MOD-8 binary counter using S-R flip-flops.

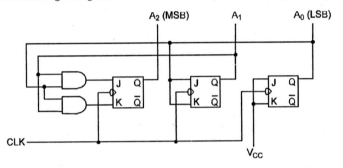

Figure 9.36 Logic diagram of the MOD-8 binary counter using J-K flip-flops.

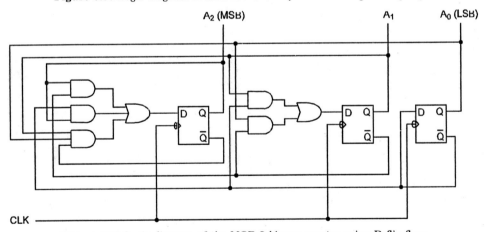

Figure 9.37 Logic diagram of the MOD-8 binary counter using D flip-flops.

9.8 SYNCHRONOUS/ASYNCHRONOUS COUNTER

This type of counter circuit may be formed by combining the synchronous and asynchronous counters. They represent a compromise between the simplicity of asynch-ronous counters and the speed of synchronous counters. In the BCD synchronous/asynchronous counter shown in Figure 9.38, the input pulses are only applied to flip-flop A in asynchronous counters, whereas the output Q_A drives the clock inputs of both B and D flip-flops so that they trigger simultaneously as in a synchronous counter. The operation of the counters is given below:

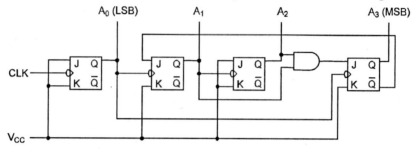

Figure 9.38 Synchronous/Asynchronous BCD counter.

1. Let us assume that the counter is initially in the 0000 state. Hence the J and K inputs of flip-flops A_0, A_1, A_2 are all HIGH. The flip-flop A_3 has J = 0 and K = 1 and so it will not be affected by any clock transition.

2. Flip-flops A_0, A_1, A_2 will function as a normal ripple counter for the first seven input pulses. When the present state is 0110, the input J becomes 1 for the flip-flop A_3, but since the flip-flop A_3 is active in the negative edge-transition of the clock pulse (which is coming from A_0), the flip-flop A_3 does not change its state.

3. In the 0111 state, the AND gate output is 1, so the flip-flop A_3 will have J = K = 1. When the eighth input pulse occurs, it will toggle flip-flop A_0 to the zero state, which in turn will toggle flip-flop A_1 LOW and flip-flop A_3 HIGH. The A_1 output transition will toggle A_2 LOW. Thus, the counter is now in the 1000 state.

4. The ninth input pulse simply toggles A_0 HIGH bringing the counter to the 1001 state.

5. In the 1001 state, the AND gate output is again 0, so flip-flop A_3 has J = 0, K = 1, which means it will go LOW on the next negative transition at its clock input. The flip-flop A_1 also has J = 0, K = 1 so it will remain in the LOW state. Thus, when the tenth pulse occurs, flip-flop A_0 will toggle LOW, which in turn will toggle A_3 LOW bringing the counter back to the 0000 state. The operation then returns to step 1 and repeats the sequence.

9.9 PRESETTABLE COUNTER

The up-counters generally start the count sequence from 00....0 state while down-counters start from 11....1 state. This is accomplished by applying a momentary pulse to all the flip-flop's CLEAR inputs before the counting operation begins. A counter can also be made to start counting in any desired state through the use of appropriate logic circuitry. Counters that have the capability to start counting from any desired state are called *presettable* or *programmable* counters.

A presettable MOD-16 ripple up-counter is illustrated in Figure 9.39. In this counter, the desired state is entered using the PRESET and CLEAR inputs irrespective of what is happening at the J and K or the clock inputs. The desired preset count is determined by the preset inputs P_A, P_B, P_C, and P_D whose values are transferred into the counter flip-flops when the PRESET LOAD input is momentarily pulsed to the LOW level. When the

PRESET LOAD input returns to HIGH, the NAND gates are disabled and the counter is free to count input clock pulses starting from the newly entered count that has been preset into the flip-flops.

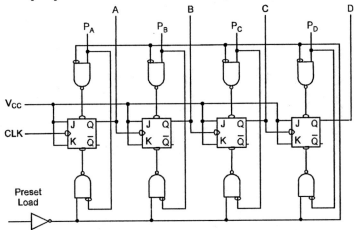

Figure 9.39 A presettable MOD-16 counter.

9.10 SYNCHRONOUS COUNTER ICs

Design of synchronous counters using flip-flops have been discussed above. Counters with any count sequence and modulus can be designed using these methods. Some synchronous counters are available as MSI ICs. They are listed in Table 9.20 along with some of their features. All these ICs are positive-edge-triggered, *i.e.*, the synchronous loading, clearing, change of states all take place on the positive going edge of the input clock pulse. Basically, these ICs can be divided into four groups—A, B, C, and D depending on their features.

Table 9.20 Synchronous counter ICs

IC No.	Description	Features	Group
74160	Decade Up-counter	Synchronous preset and asynchronous clear	A
74161	4-bit binary Up-counter	Synchronous preset and asynchronous clear	A
74162	Decade Up-counter	Synchronous preset and clear	A
74163	4-bit binary Up-counter	Synchronous preset and clear	A
74168	Decade Up-Down counter	Synchronous preset and no clear	B
74169	4-bit binary Up-Down counter	Synchronous preset and no clear	B
74100	Decade Up-Down counter	Asynchronous preset and no clear	C
74191	4-bit binary Up-Down counter	Asynchronous preset and no clear	C
74192	Decade Up-Down counter	Asynchronous preset and clear	D
74193	4-bit binary Up-Down counter	Asynchronous preset and clear	D

9.10.1 Group *A* Synchronous Counter IC

The block diagram and the function table of these types of ICs are given in Figure 9.40. There are two separate enable inputs in these types of ICs. They are ENT and ENP. The counting can be stopped asynchronously by setting either of these enable inputs to logic 0. Ripple carry (RC) output is normally at logic 0 and goes to logic 1 whenever the counter reaches its highest count (binary 9 for BCD counters and binary 15 for 4-bit counters). Setting ENT to logic 0 also inhibits RC changing from logic 0 to logic 1.

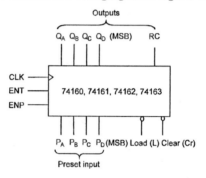

(a) Block diagram.

Load L	ENP	ENT	Cr	CLK	Mode
0	X	X	1	↑	Preset
1	0	1	1	X	Stop count
1	X	0	1	X	Stop count, disable RC
X	X	X	0	*	Reset to zero
1	1	1	1	↑	UP count

* X for 74160 and 74161

↑ for 74162 and 74163.

(b) Function table.

Figure 9.40 Group A synchronous counter.

Example 9.8. *Design a normal MOD-13 counter using IC 74161.*

Solution. The circuit is designed for normal up-counting. The Q_D, Q_C, and Q_A outputs are connected to the Cr terminal through a NAND gate, which clears the counter as soon as the output is 1101. The states of the counter are from 0000 to 1100. The MOD-13 counter is shown in Figure 9.41. In fact, using the above approach, the count can be terminated at any desired value and a counter with any modulus (less than 16 for binary and less that 10 for decade counter) can be obtained.

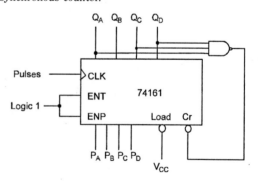

Figure 9.41 MOD-13 counter using IC 74161.

Example 9.9. *Design a divide-by-12 counter using IC 74163. Make use of the RC output and preset inputs.*

Solution. For obtaining a divide-by-12 counter using IC 74163, the counter is preset at binary 0100 (decimal 4). When the count reaches 1111, RC output goes to 1, which is used to load the data present at the preset inputs into the counter. The circuit of the counter is shown in Figure 9.42.

In general, for obtaining a divide-by-m counter, the preset input, P, is given by

$$P = 16-m \text{ for a 4-bit binary counter}$$
$$= 10-m \text{ for a decade counter}$$

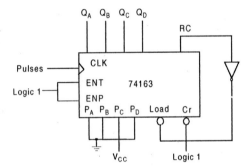

Figure 9.42 Divide-by-12 counter using IC 74163.

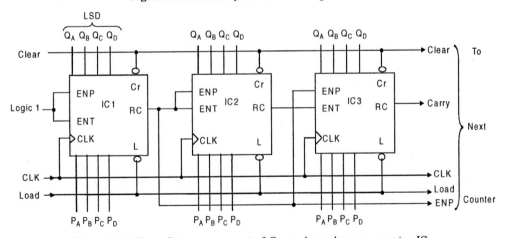

Figure 9.43 Cascading arrangement of Group A synchronous counter ICs.

9.10.2 Group *B* Synchronous Counter IC

The block diagram and the function table of group B synchronous counter ICs are given in Figure 9.44. The functions ENT and ENP are the same as in Group A ICs except that these are active-low. Ripple carry (RC) output is normally at logic 1 and goes to logic 0 whenever the counter reaches its highest count during up-counting, and when the count reaches minimum during down-counting. The value of the signal at the U/D' decides the direction of the counting operation. When U/D' = 1, up-counting takes place and when U/D' = 0, the down-counting happens.

Here the clear terminal in not available. Hence, if it is desired to terminate the count before it reaches the maximum value, a NAND gate is used to detect the count corresponding to the required number and its output is connected to the load input terminal. The preset input can be given corresponding to the required starting state of the counter.

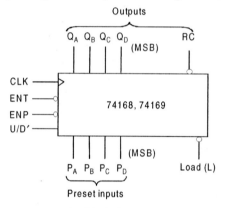

(a) Block diagram.

Load L	ENP	ENT	U/D'	CLK	Mode
0	X	X	X	↑	Preset
1	1	0	X	X	Stop count
1	X	1	X	X	Stop count, disable RC
1	0	0	1	↑	UP count
1	0	0	0	↑	DOWN count

(b) Function table.

Figure 9.44 Group B synchronous counter.

9.10.2.1 Frequency of Group B Synchronous Counter ICs

The frequency of the output waveform at RC (f_{out}) is related to the input clock frequency (f_{in}) as follows:

Decade counter 74168

$$f_{out} = \frac{f_{in}}{10 - N}, 0 \leq N \leq 8 \quad \text{(for up-counting)}$$

$$= \frac{f_{in}}{N + 1}, 1 \leq N \leq 9 \quad \text{(for down-counting)}$$

Binary counter 74169

$$f_{out} = \frac{f_{in}}{16 - N}, 0 \leq N \leq 14 \quad \text{(for up-counting)}$$

$$=\frac{f_{in}}{N+1}, 1 \leq N \leq 15 \qquad \text{(for down-counting)}$$

where N is the decimal equivalent of the preset input.

The cascading of the group B counter ICs is similar to that of group A counter ICs.

Example 9.10. *Design a counter with states 0011 through 1110 using the IC 74169 counter.*

Solution. The preset input is 0011 and as soon as the output reaches 1110, on the next pulse it should come back to its original state. Therefore, the number corresponding to the highest required state is to be detected for loading the counter. The counter is shown in Figure 9.45. If the counter is required to count up to the maximum/minimum value then the RC output is to be connected to the load input, for loading the initial count at the next pulse after maximum/minimum count has been reached.

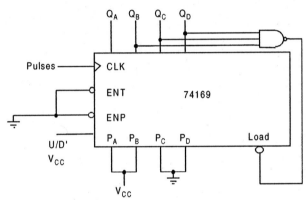

Figure 9.45 Counter for Example 9.10.

9.10.3 Group *C* Synchronous Counter IC

The block diagram and the function table of group C synchronous counter ICs are given in Figure 9.46. These ICs have only one enable input ENAB, which is active-low. MAX/MIN

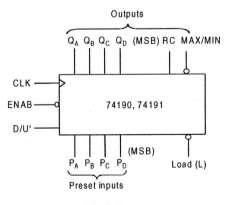

(a) Block diagram.

Load (L)	ENAB	D/U'	CLK	Mode
X	1	X	X	Stop count
0	0	X	X	Preset
1	0	0	↑	Up-count
1	0	1	↑	Down-count

(b) Function table.

Figure 9.46 Group C synchronous counter.

output is used to detect the maximum or minimum count of the counter. It is normally at logic 0 and goes to logic 1 when the count is maximum (1001 for 74190 and 1111 for 74191) for the up-counting and minimum (0000) for the down-counting. It serves the purpose of an underflow detector while down-counting and an overflow detector while up-counting. The RC output is normally at logic 1, and goes to logic 0 when the counter reaches a MAX/MIN point and the CLK input is low.

9.10.3.1 *Frequency Dividers*

These counters can also be used as programmable frequency dividers. By presetting any desired number into the counter and counting to the minimum (down-counting) or maximum (up-counting) count, division is achieved. The RC output is connected to the *load* input. The required output waveform is obtained at MAX/MIN output. The input clock frequency f_{in} and the frequency of the output waveform f_{out} are related as follows:

Decade Counter 74190

$$f_{out} = \frac{f_{in}}{N}, 1 \leq N \leq 9 \qquad \text{(for down-counting)}$$

$$= \frac{f_{in}}{9-N}, 1 \leq N \leq 8 \qquad \text{(for up-counting)}$$

Binary Counter 74191

$$f_{out} = \frac{f_{in}}{N}, 1 \leq N \leq 15 \qquad \text{(for down-counting)}$$

$$= \frac{f_{in}}{15-N}, 0 \leq N \leq 14 \qquad \text{(for up-counting)}$$

where N is the decimal equivalent of the preset input. For example, in the case of 74190, if the preset input is 0111 (decimal 7), and the clock frequency is 560 Hz, the frequency of the output waveform will be 280 Hz for up-counting and 80 Hz for down-counting. With the same values for the preset input and the clock frequency in the case of 74191, the frequency of the output waveform will be 70 Hz for up-counting and 80 Hz for down-counting.

9.10.3.2 *Cascading of Group C Counters*

These counters may be cascaded in three different ways:

1. *Synchronous counter ICs cascaded with ripple carry between stages.* The RC output of each stage is connected to the ENAB input of the succeeding stage. All CLK inputs are connected together and the clock pulses are applied at this common clock terminal.

2. *Synchronous counter ICs cascaded with parallel carry.* The speed of operation is maximum in this type of cascading. The number of stages that may be cascaded in

this manner may be restricted due to loading of MAX/MIN output by the external gating. A 3-decade synchronous counter with parallel carry is shown in Figure 9.47.

3. *Synchronous counter ICs cascaded as an asynchronous counter.* In this type of cascading, the RC output of each stage is connected to the CLK input of the succeeding stage and the clock pulses are applied at the CLK input of the first stage. In this, each IC is synchronous within itself, but between stages the overall system is a ripple counter.

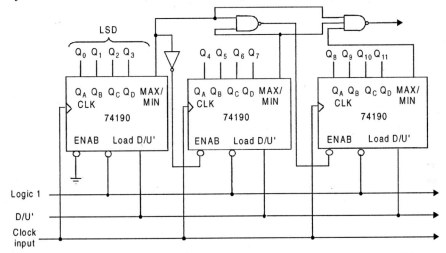

Figure 9.47 A 3-decade synchronous counter using 74190 counter ICs with parallel carry.

9.10.4 Group *D* Synchronous Counter IC

In the group D counters for down-counting, the clock is applied at the CLK-DOWN terminal with CLK-UP connected to logic 1, and for up-counting it is applied at the CLK-UP terminal with CLK-DOWN connected to logic 1. The block diagram of such a counter IC and its function table is given in Figure 9.48.

The carry and borrow outputs are normally at logic 1. The carry output drops to logic 0 when the counter shows its maximum count while up-counting and the CLK-UP input is at logic 0. The borrow output remains at logic 1 as long as the circuit is operating from the CLK-UP input. The function of borrow in down-counting is same as that of carry in up-counting.

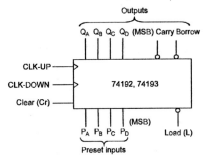

(a) Block diagram.

Load (L)	Clear (Cr)	CLK-UPC	LK-DOWN	Mode
X	1	X	X	Reset to Zero
1	0	↑	1	Up-count
1	0	1	↑	Down-count
0	0	X	X	Preset
1	0	1	1	Stop count

(b) Function table.

Figure 9.48 Group D synchronous counter.

9.10.4.1 Frequency Dividers

These counters can be used as programmable frequency dividers in a similar manner as the one used for group C counters except that carry (or borrow) output is to be connected to the load input for up (or down) counting.

The input clock frequency f_{in} and the frequency of the output waveform f_{out} are related as follows:

Decade Counter 74192

$$f_{out} = \frac{f_{in}}{N}, 1 \leq N \leq 9 \qquad \text{(for down-counting)}$$

$$= \frac{f_{in}}{9-N}, 1 \leq N \leq 8 \qquad \text{(for up-counting)}$$

Binary Counter 74193

$$f_{out} = \frac{f_{in}}{N}, 1 \leq N \leq 15 \qquad \text{(for down-counting)}$$

$$= \frac{f_{in}}{15-N}, 0 \leq N \leq 14 \qquad \text{(for up-counting)}$$

where N is the decimal equivalent of the preset input. For example, in the case of 74193, if the preset input is 1011 (decimal 11), and the clock frequency is 550 Hz, the frequency of the pulses at the borrow output will be 137.5 Hz for up-counting and 50 Hz for down-counting.

9.10.4.2 Cascading of Group D Counters

In order to cascade these counter ICs, the borrow and carry outputs of each stage are to be connected to the CLK_DOWN and CLK-UP inputs of the succeeding stage respectively. For steering the clock to CLK-DOWN or CLK-UP input the circuit shown in Figure 9.49 may be used.

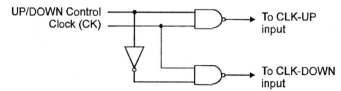

Figure 9.49 Circuit for steering clock pulses to CLK-UP (for up-count) or CLK-DOWN (for down-count).

9.11   COUNTER APPLICATIONS

There are many applications of the counters we have discussed. In this section we will discuss three representative applications.

9.11.1   Frequency Counter

A frequency counter is a circuit that can measure and display the frequency of a signal. The method for constructing a frequency counter is shown in Figure 9.50. The counter is driven by the output of an AND gate. The AND gate inputs are input pulses with unknown frequency, f_x, and a SAMPLE pulse that controls the time through which the pulses are allowed to pass through the AND gate into the counter.

The counter is usually made up of cascaded BCD counters, and the decoder/display unit converts the BCD outputs into a decimal display for easy monitoring.

The SAMPLE pulse goes HIGH from t_1 to t_2; this is called the *sampling interval*. During this sampling interval the unknown frequency pulses will pass through the AND gate and will be counted by the counter. After t_2 the AND gate output becomes LOW and the counter stops counting. Thus the counter counts the number of pulses that occur during the sampling interval. This is a direct measure of the frequency of the pulse waveform.

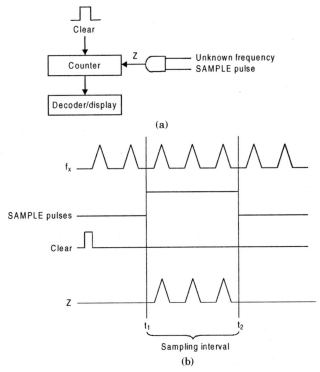

Figure 9.50 Basic frequency counter method.

The accuracy of this method depends almost entirely on the duration of the sampling interval, which must be very accurately controlled. A commonly used method for obtaining very accurate sample pulses is shown in Figure 9.51. A crystal controlled oscillator is used to generate a very accurate 100 kHz waveform, which is shaped into square pulses and fed to a series of decade counters that are being used to successively divide this 100 kHz frequency by 10. The frequencies at the output of each decade counter are as accurate (percentage wise) as the crystal frequency. These decade counters are usually binary or Johnson counters.

The switch is used to select one of the decade counter output frequencies to be fed to the clock input of a single flip-flop to be divided by 2.

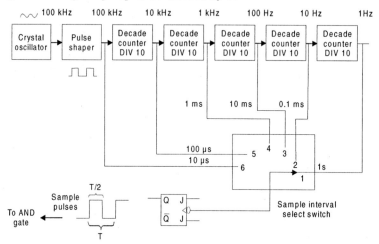

Figure 9.51 Method for obtaining accurate sampling intervals for frequency counters.

9.11.2 Measurement of Period

The principle of operation of the frequency counter can be modified and applied to the measurement of period rather than frequency. The basic idea is shown in Figure 9.52. An accurate 1 MHz reference frequency is gated into the counter/display for time duration equal to T_x, the period of the signal being measured. The counter counts and displays the values of T_x in units of μ's. For example, if T_x is 1.17 ms, the gate will allow 1170 pulses into the counter.

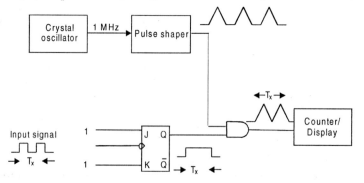

Figure 9.52 Measurement of period.

9.11.3 Digital Clock

A digital clock, which displays the time of day in hours, minutes, and seconds, is one of the most common applications of counters. To construct an accurate digital clock, a very highly controlled basic clock frequency is required. For battery-operated digital clocks (or watches) the basic frequency can be obtained from a quartz-crystal oscillator. Digital clocks operated from the AC power line can use the 50 Hz power frequency as the basic clock frequency. In either case, the basic frequency has to be divided down to a frequency of 1 Hz or pulse of 1 second (pps). The basic block diagram for a digital clock operating from 50 Hz is shown in Figure 9.53.

The 50 Hz signal is sent through a Schmitt trigger circuit to produce square pulses at the rate of 50 pps. The 50 pps waveform is fed into a MOD-50 counter, which is used to divide the 50 pps down to 1 pps. The 1-pps signal is then fed into the SECONDS section. This section is used to count and display seconds from 0 through 59. The BCD counter advances one count per second. After 9 seconds the BCD counter recycles to 0. This triggers the MOD-6 counter and causes it to advance one count. This continues for 59 seconds. At this point, the BCD counter is at 1001 (9) count and the MOD-6 counter is at 101 (5). Hence, the display reads 59 seconds. The next pulse recycles the BCD counter to 0. This, in turn, recycles the MOD-6 counter to 0.

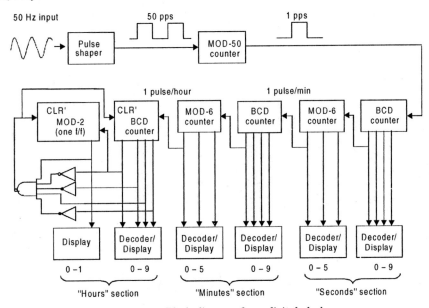

Figure 9.53 Block diagram for a digital clock.

The output of the MOD-6 counter in the SECONDS section has a frequency of 1 pulse per minute. This signal is fed to the MINUTES section, which counts and displays minutes from 0 through 59. The MINUTES section is identical to the SECONDS section and operates in exactly the same manner.

The output of the MOD-6 counter in the MINUTES section has a frequency of 1 pulse per hour. This signal is fed to the HOURS section, which counts and displays hours from

1 through 12. The HOURS section is different from the MINUTES and SECONDS section in that it never goes to the zero state. The circuitry in this section is different. When the hours counter reaches 12, it will be reset to zero by the NAND gate.

9.12 HAZARDS IN DIGITAL CIRCUITS

In asynchronous sequential circuits it is important that undesirable glitches on signals should not occur. The designer should be aware of the possible sources of glitches and ensure that the transitions in a circuit will be glitch free. The glitches caused by the structure of a given circuit and the propagation delays in the circuit are referred to as *hazards*.

If, in response to an input change and for some combination of propagation delays, a network output may momentarily go to 0 when it should remain a constant 1, we say the network has a *static 1-hazard*. Similarly, if the network output may momentarily go to 1 when it should remain a constant 0, we say the network has a *static 0-hazard*. If when, the output is supposed to change from 1 to 0 (or 0 to 1), the output may change three or more times, we say the network has a *dynamic hazard*. Figure 9.54 illustrates possible outputs from a network with hazards.

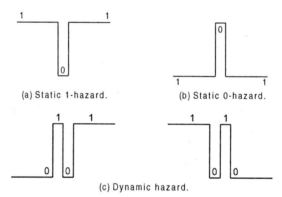

(a) Static 1-hazard. (b) Static 0-hazard.

(c) Dynamic hazard.

Figure 9.54 Types of hazards.

9.12.1 Static Hazards

A circuit with a static hazard is shown in Figure 9.55. Suppose that the circuit is in the state where $x_1 = x_2 = x_3 = 1$, in which case $f = 1$. Now let x_1 change from 1 to 0. Then the circuit is supposed to maintain $f = 1$.

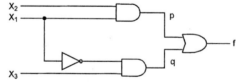

Figure 9.55 Circuit with a static 1-hazard.

Now we take into consideration the propagation delays through the gates. The change in x_1 will probably be observed at point p before it will be seen at point q. This is since the path from x_1 to q has an extra NOT gate in it. Hence the signal at p will become 0 before the signal at q becomes equal to 1. Thus, for a short time both p and q will be zero. This causes f to drop to 0 before it can recover back to 1. This gives rise to a static 1-hazard.

The circuit implements the function

$$f = x_1 x_2 + \overline{x_1} x_3 \ .$$

The corresponding Karnaugh map is shown in Figure 9.56. In the example, note that although in the steady state x_1 and x'_1 are complements, under transient conditions they are not. Thus, in the analysis of a network for hazards, we must treat a variable and its complements as if they were two independent variables. The hazard can be eliminated by including a redundant gate. Then the function would be implemented as

$$f = x_1 x_2 + \overline{x_1} x_3 + x_2 x_3 \ .$$

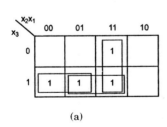

Figure 9.56 Karnaugh map for the circuit shown in Figure 9.55.

Now the change in x_1 from 1 to 0 would have no effect on the output f because the product term $x_2 x_3$ would be equal to 1 if $x_2 = x_3$, regardless of the value of x_1. The resulting hazard-free circuit is shown in Figure 9.57.

A potential hazard exists whenever two adjacent 1s in a Karnaugh map are not covered by a single product term. Therefore, a technique for removing hazards is to find a cover in which some product term includes each pair of adjacent 1s. Then, since a change in an input variable causes a transition between two adjacent 1s, no glitch can occur because both 1s are included in a product term.

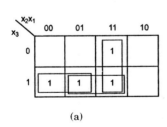

(a)

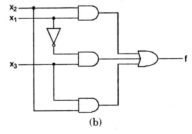

(b)

Figure 9.57 (a) Karnaugh map for a hazard-free circuit, (b) Hazard-free circuit.

From the previous example, it seems that static hazards can be avoided by including all prime implicants in a sum-of-products circuit that realizes a given function. This is indeed true. But it is not always necessary to include all prime implicants. It is only necessary to include product terms that cover the adjacent pair of 1s. There is no need to cover the don't-care vertices.

We consider the function in Figure 9.58. A hazard-free circuit that implements this function should include the encircled terms, which gives

$$f = \overline{x_1}x_3 + x_2 x_3 + x_3\overline{x_4}.$$

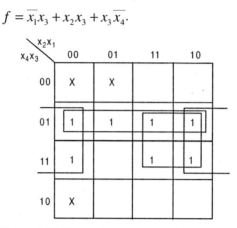

Figure 9.58 Karnaugh map for the above example.

The prime implicant $\overline{x}_2\overline{x}_1$ is not needed to prevent hazards, because it would account only for the two 1s in the left-most column. These 1s are already covered by $\overline{x}_1 x_3$.

Static hazards can also occur in other types of circuits. Figure 9.59(a) depicts a product-of-sums circuit that contains a hazard. If $x_1 = x_3 = 0$ and x_2 changes from 0 to 1, then f should remain at 0. However, if the signal at p changes earlier than the signal at q, then p and q will both be equal to 1 for a short time, causing a glitch $0 \to 1 \to 0$ on f.

In a POS circuit, it is the transitions between adjacent 0s that may lead to hazards. Thus, to design a hazard-free circuit, it is necessary to include sum terms that cover all pairs of adjacent 0s. In this example, we will get

$$f = (x_1 + x_2)(\overline{x}_2 + x_3) + (x_1 + x_3).$$

The circuit is shown in Figure 9.59(c).

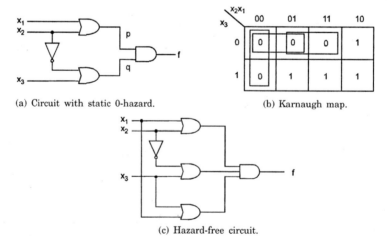

(a) Circuit with static 0-hazard.   (b) Karnaugh map.

(c) Hazard-free circuit.

Figure 9.59 Static 0-hazard in a POS circuit and removal of hazard.

In summary, the following procedure can be used to find all static 1- and 0-hazards in a network due to single input variable changes:

1. Determine the transient output function of the network F^t, and reduce F^t to SOP form treating each variable and its complement as separate variables.

2. Examine each pair of adjacent input states for which F^t is 1. If there is no 1-term that includes both input states of the pair, a 1-hazard is present. This is conveniently accomplished by plotting the 1-terms of F^t on a Karnaugh map and checking each pair of adjacent 1s on the map.

3. If the sum of products for F^t does not contain the product of a variable and its complement, no 0-hazards are present. If the sum of products for F^t contains the product of a variable and its complement, a 0-hazard may be present. To detect all 0-hazards,

   (a) Obtain the POS form for F^t by factoring or other means. x_i and x'_i are still treated as separate variables.

   (b) Examine each pair of adjacent input states for which F^t is 0. If there is no 0-term that includes both input states of the pair, a 0-hazard is present. This can be conveniently done by plotting the 0-terms of F^t on a Karnaugh map and checking each pair of adjacent 0s on the map.

Static 0- and 1-hazards can be eliminated by adding additional gates to the network to produce the missing 1-terms and 0-terms.

9.12.2 Dynamic Hazards

Dynamic hazards due to a change in an input variable x_i can only occur if there are three or more paths between x_i (and/or x'_i) input and the network output. This is necessary since a dynamic hazard involves a triple change in output, so the effect of the input change must reach the output at three different times. A network may have a dynamic hazard even if it is free of static hazards as illustrated by the example in Figure 9.60. For this network, the transient output function is

$$Y^t = (ac' + bc)(a' + c') = a'bc + ac' + bcc'$$
$$= (ac' + b)(ac' + c)(a' + c') = (a + b)(b' + c)(a + c)(c + c')(a' + c').$$

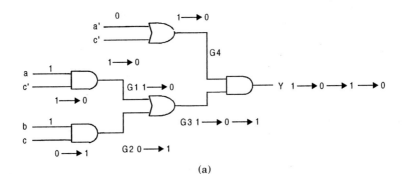

(a)

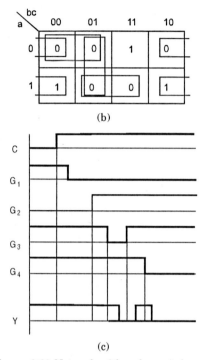

(b)

(c)

Figure 9.60 Network with a dynamic hazard.

Plotting the 1-terms and the 0-terms of Y^t on the map of Figure 9.60(b) reveals that there are no 1- or 0-hazards. Inspection of the network shows that the only input variable that could cause a dynamic hazard when it changes is c (since c is the only variable with three paths to the output). If we choose,, the effect of a change in c can propagate to the output along all three paths. If the gate outputs change in the order shown in Figure 9.60(c), the G_3 output undergoes a 1-0-1 change before the G_4 output changes 1 to 0, and the dynamic hazard shows up at the output. Dynamic hazards are not easy to detect nor easy to deal with. The designer can avoid dynamic hazards simply by using two-level circuits and ensuring that there are no static hazards.

9.12.3 Essential Hazards

Even though an asynchronous sequential network is free of critical races and the combinational part of the network is free of static and dynamic hazards, timing problems due to propagation delays may still cause the network to malfunction and go to the wrong state. For example, we consider the network of Figure 9.61. Clearly, there are no hazards in the combinational part of the network, and inspection of the flow table shows that there are no critical races. If we start in state "a" and change x to 1, the network should go to state "d." However, consider the following possible sequence of events:

1. x changes 0 to 1.
2. Gate 2 output changes 0 to 1.
3. Flip-flop f_1 output changes 0 to 1.

4. Gate 4 output changes 0 to 1.
5. Flip-flop f_2 output changes 0 to 1.
6. Inverter output x' changes 0 to 1.
7. Gate 1 output changes 0 to 1, gate 2 output changes back to 0, and gate 4 output changes back to 0.
8. Flip-flop output f_1 changes back to 0.

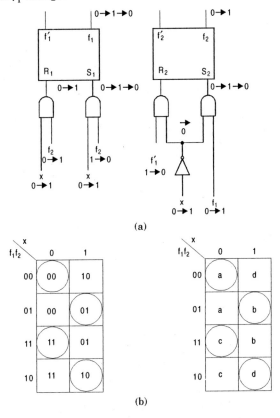

(a)

(b)

Figure 9.61 Network with essential hazards.

We may note that the final state of the network is "b" instead of "d." This came about because the delay in the inverter was longer than the other delays in the network. Hence, in effect, part of the network saw the value of $x = 1$ while the other part of the network had $x = 0$. The net result was that the network acted as if the input x had changed three times instead of once so that the network went through the sequence of states $f_1 f_2 = 00$, 10, 11, 01.

The malfunction illustrated in the previous example is termed an essential hazard. Essential hazards can be located by inspection of the flow table. Hence, we may define an essential hazard as follows:

A flow table has an *essential hazard* starting in stable total state "s" for input variable

x_i if, and only if, the stable total state reached after one change in x_i is different from the stable total state reached after three changes in x_i.

Essential hazards can be eliminated by adding delays to the network. For the example discussed above, if we add a sufficiently large delay to the output of flip-flop f_1, then the change in x will propagate to all of the gates before the change in f_1 takes place and the essential hazard is eliminated.

REVIEW QUESTIONS

9.1 Compare between a ripple and a synchronous counter.

9.2 Design a MOD-18 ripple counter using J-K flip-flops.

9.3 Design a MOD-32 parallel counter using S-R flip-flops.

9.4 Draw the gates necessary to decode all the stages of a MOD-16 counter using active-HIGH outputs.

9.5 What is the lock-out condition of a counter?

9.6 How can you convert an up-counter into a down-counter?

9.7 Design a MOD-16 up-down counter using D flip-flops.

9.8 Which flip-flop is best suited for designing a counter and why?

9.9 How does a digital clock operate? Explain.

9.10 What is meant when we say that a counter is presettable?

9.11 Explain why a ripple counter's maximum frequency limitation decreases as more flip-flops are added to the counter.

9.12 A certain J-K flip-flop has propagation delay of 12 ns. What is the largest MOD counter that can be constructed from these flip-flops and still operate up to 10 MHz?

9.13 What would the notation "DIV32" mean on a counter symbol?

9.14 Explain why the decoding gates for an asynchronous counter may have glitches on their outputs.

9.15 How does strobing eliminate decoding glitches?

9.16 Discuss the procedure for designing synchronous counters.

9.17 Design a synchronous counter that has the following sequence: 0010, 0110, 1000, 1001, 1100, 1101, and repeat. From the undesired states the counter must always go to 0010 on the next clock pulse.

9.18 Design a 4-bit counter with one control signals S. The counter should operate as
  (*a*) binary down-counter when S = 0,
  (*b*) binary up-counter when S = 1.

9.19 How many decade counters are necessary to implement a DIV1000 counter and DIV10000 counter?

9.20 Analyze the counter shown in Figure P.9.1 for a "lock-up" condition in which the counter cannot escape from an invalid state or states. An invalid state is one that is not in the counter's normal sequence.

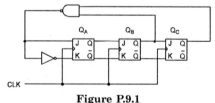

Figure P.9.1

❑ ❑ ❑

10 A/D AND D/A CONVERSION

10.1 INTRODUCTION

There are numerous advantages to processing signals using digital systems. And because of these advantages, digital systems are widely used for control, communication, computers, instrumentation, etc. In many such applications of digital systems, the signals are not available in the digital form. Therefore, to process these analog signals using digital hardware, they have to be converted into digital form. The process of conversion of analog signal to digital signal is referred as *analog-to-digital conversion*. The system that realizes the conversion is referred to as an *analog-to-digital converter* or *A/D Converter* or *ADC*.

The output of the system may be desired to be of analog form. Therefore, the output of the digital system is required to be converted back to the analog form. The process of converting the digital signal to analog form is called *digital-to-analog conversion* and the system used for this purpose is referred to as a *digital to analog converter* or *D/A converter* or *DAC*.

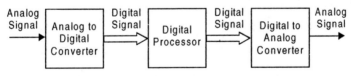

Figure 10.1

In the present trend of technology, most of the signal processing is based on digital systems. But the real-world signals are analog in nature. A/D converter and D/A converter are the bridge between the analog world and digital world. They find their applications in almost every system of signal processing. An elementary analog signal-processing system with the use of a digital processor is illustrated with the block diagram in Figure 10.1. In this chapter the D/A converter will be discussed first as it also serves as a sub-system of the A/D converter.

10.2 DIGITAL-TO-ANALOG CONVERTERS (DAC)

The input of a D/A converter is an *n*-bit binary signal, available in parallel form. Normally, digital signals are available at the output of latches or registers and the voltages correspond to logic 0 and logic 1. In general, the logic levels do not have precisely fixed voltages. Therefore, these voltages are applied directly to the converter for digital-to-analog computation, but they are used to operate digitally controlled switches. The switch is operated to one of the two positions depending upon the digital signal logic levels (logic 0 or logic 1) which connects precisely fixed voltages or voltage references V(1) or V(0) to the converter input, corresponding to logic 1 and logic 0 respectively.

The analog output voltage V_0 of an n-bit straight binary D/A converter can be related to the digital input by the equation

$$V_0 = K (2^{n-1} . b_{n-1} + 2^{n-2} . b_{n-2} + 2^{n-3} . b_{n-3} + \ldots\ldots + 2^2 . b_2 + 2 . b_1 + b_0). \tag{10.1}$$

Where   K = proportionality factor equivalent to step size in voltage,

$$b_n = 1, \quad \text{if the } n^{th} \text{ bit of the digital input is 1,}$$
$$= 0, \quad \text{if the } n^{th} \text{ bit of the digital input is 0.}$$

As an example, for a 4-bit D/A converter, there are 16 voltage levels and they are tabulated in Figure 10.2 assuming the voltage step size or the proportionality factor K = 1.

There are two types of commonly used D/A converters as mentioned below.

1. Weighted-resistor D/A converter, and
2. R-2R ladder D/A converter.

Digital Input				Analog Output
D_3	D_2	D_1	D_0	V
0	0	0	0	0
0	0	0	1	1
0	0	1	0	2
0	0	1	1	3
0	1	0	0	4
0	1	0	1	5
0	1	1	0	6
0	1	1	1	7
1	0	0	0	8
1	0	0	1	9
1	0	1	0	10
1	0	1	1	11
1	1	0	0	12
1	1	0	1	13
1	1	1	0	14
1	1	1	1	15

Figure 10.2

10.2.1 Weighted-resistor D/A Converter

Let us consider a resistor network that has N-bit straight binary inputs (through digitally controlled electronic switches), which produces a current I corresponding to logic 1 at the most significant bit, I/2 corresponding to logic 1 at the next lower bit, $I/2^2$ for the next lower bit and so on, and $I/2^{N-1}$ for logic 1 at the least significant bit position. The total current thus produced will be proportional to the digital inputs. This current can be converted to voltage with the help of a current-to-an voltage converter circuit by an using operational amplifier (OP AMP). The produced voltage is analog in nature and will be proportional to the digital inputs.

The scheme above for converting digital signals to analog voltage can be realized by the circuit diagram as shown in Figure 10.3. It may be observed in the circuit diagram that different values of resistances are used at the digital inputs and the resistance values are the multiple of the resistance corresponding to the most significant digital input to produce the currents I, I/2, $I/2^2$, ... $I/2^{N-1}$. Since the resistance values are weighted in accordance with the binary weights of the digital inputs, this circuit is referred to as a *weighted-resistor D/A converter*.

In the circuit in Figure 10.3, digitally controlled switches have two positions once they are connected to voltage V(1), which is equivalent to logic 1, and at other positions they are connected to V(0), which equivalent to logic 0.

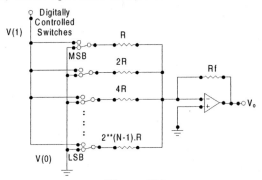

Figure 10.3

Current produced at the most significant bit through resistor R is I_{N-1}, at resistor 2R is I_{N-2}, at resistor 4R is I_{N-3}, so on, and at the least significant bit through resistor 2_{N-1}, R current is I_0. Then,

$$I_{N-1} = V_{N-1}/R$$
$$I_{N-2} = V_{N-2}/2R$$
$$I_{N-3} = V_{N-3}/2^2R$$
$$\vdots$$
$$\vdots$$
$$I_0 = V_0/2^{N-1}R. \tag{10.2}$$

Where V_n = V(1) if b_n = 1
$\qquad\qquad$ = V(0) if b_n = 0.

V_n is the voltage of the nth bit and b_n is the nth bit.

So, the total current at the input of OP AMP

$$I_i = I_{N-1} + I_{N-2} + I_{N-3} + \ldots + I_2 + I_1 + I_0. \tag{10.3}$$

Or, $I_i = V_{N-1}/R + V_{N-2}/2R + V_{N-3}/2^2R + \ldots + V_2/2^{N=3}R + V_1/2^{N-2}R + V_0/2^{N-1}R.$ (10.4)

If the output of OP AMP is considered as V_o, then I_i may be equated as $-V_o/R_F$, or

$$-V_o/R_F = V_{N-1}/R + V_{N-2}/2R + V_{N-3}/2^2R + \ldots + V_2/2^{N=3}R + V_1/2^{N-2}R + V_0/2^{N-1}R.$$

Or, $V_o = -R_F (V_{N-1}/R + V_{N-2}/2R + V_{N-3}/2^2R + \ldots + V_2/2^{N=3}R + V_1/2^{N-2}R + V_0/2^{N-1}R).$

Or, $V_o = -(R_F/2^{N-1}R)(2^{N-1}V_{N-1} + 2^{N-1}V_{N-2} + 2^{N-1}V_{N-3} + \ldots + 2^2V_2 + 2^1V_1 + 2^0V_0).$
$$\tag{10.5}$$

For straight binary inputs, $V(0) = 0$ and $V(1) = V_R$. Therefore V_o may be expressed as

$V_o = -R_F (V_R\, b_{N-1}/R + V_R\, b_{N-2}/2R + V_R\, b_{N-3}/2^2R + \ldots + V_R\, b_2/2^{N=3}R + V_R\, b_1/2^{N-2}R + V_R\, b_0/2^{N-1}R)$

$= -V_R(R_F\, b_{N-1}/R + R_F\, b_{N-2}/2R + R_F\, b_{N-3}/2^2R + \ldots + R_F\, b_2/2^{N=3}R + R_F\, b_1/2^{N-2}R + R_F\, b_0/2^{N-1}R).$
$$\tag{10.6}$$

From the above Expressions 10.5 and 10.6, we may conclude that output voltage V_o is the summation of input bits multiplied by the factor $R_F/2^{N-1}R$, where N is the bit position. These expressions are similar to the Expression 10.1 with the multiplying factor

$$K = (R_F/2^{N-1}R)\, V_R. \tag{10.7}$$

In this circuit the output swings in only one direction and therefore it is unipolar. Sometimes it may require that the output is desired with some offset voltage. This arrangement can be done with some modification of the circuit as in Figure 10.4.

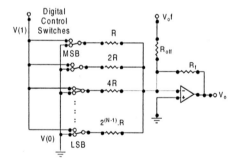

Figure 10.4

The offset produced in this circuit is $-V_{off}.R_F/R_{off}$, and the revised expression for the output will be

$$V_o = -V_{off}.R_F/R_{off} -(R_F/2^{N-1}R)(2^{N-1}V_{N-1} + 2^{N-1}V_{N-2} + 2^{N-1}V_{N-3} + \ldots + 2^2V_2 + 2^1V_1 + 2^0V_0). \tag{10.8}$$

For straight binary inputs where $V(1) = 1$ and $V(0) = 0$, the output expression will be

$V_o = -V_{off}.R_F/R_{off} -V_R(R_F\, b_{N-1}/R + R_F\, b_{N-2}/2R + R_F\, b_{N-3}/2^2R + \ldots + R_F\, b_2/2^{N=3}R + R_F\, b_1/2^{N-2}R + R_F\, b_0/2^{N-1}R).$
$$\tag{10.9}$$

Example 10.1. *(a) Consider a 4-bit D/A converter with V(1) = –1 V, V(0) = 0 V, and R_F = 8R. Obtain the analog voltage for each of the digital inputs from 0000 to 1111. (b) Adjust the offset voltage, using the circuit of Figure 10.4, such that V_o = 0 V for digital input 1000. Obtain analog voltage for each of the digital inputs with this offset. (c) Using the above offset voltage, obtain the analog voltages if MSB is complemented before applying to the D/A converter.*

Solution. *(a)* For a 4-bit D/A converter, the output voltage using Equation 10.6, V_o can be expressed as

$$V_o = -V_R(R_F\, b_{N-1}/R + R_F\, b_{N-2}/2R + R_F\, b_{N-3}/2^2R + R_F\, b_{N-4}/2^3R)$$
$$= -V_R(R_F\, b_3/R + R_F\, b_2/2R + R_F\, b_1/4R + R_F\, b_0/8R)$$
$$= 8b_3 + 4b_2 + 2b_1 + b_0\ \text{V}. \qquad \text{(Substituting } R_F = 8R \text{ and } V_R = -1 \text{ V.)} \quad (10.10)$$

The table for digital inputs 0000 to 1111 with their corresponding output voltages is similar to Figure 10.2 (notation D_3, D_2, D_2, D_0 are to be replaced by b_3, b_2, b_1, b_0).

Digital Input				Analog Output
b_3	b_2	b_1	b_0	V
0	0	0	0	-8
0	0	0	1	-7
0	0	1	0	-6
0	0	1	1	-5
0	1	0	0	-4
0	1	0	1	-3
0	1	1	0	-2
0	1	1	1	-1
1	0	0	0	0
1	0	0	1	1
1	0	1	0	2
1	0	1	1	3
1	1	0	0	4
1	1	0	1	5
1	1	1	0	6
1	1	1	1	7

Figure 10.5

(b) From the table in Figure 10.2, we can see that without applying the offset voltage, the output voltage is 8 V when digital input is 1000. So an offset voltage of –8 V must be produced. Therefore,

$$-V_{off} R_F/R_{off} = -8 \text{ V}.$$

We may use R_{off} = R and V_{off} = 1 V . The analog voltage corresponding to digital inputs with the offset voltage is shown in Figure 10.5.

(c) If MSB is complemented, *i.e.*, digital input b_3 is complemented to b_3', the analog voltage outputs are shown in the table in Figure 10.6, applying the same offset voltage.

You may notice from the table in Figure 10.6, that this is a D/A converter, which converts 2's complement format to an analog signal.

Digital Input				Analog Output
b_3	b_2	b_1	b_0	V
1	0	0	0	−8
1	0	0	1	−7
1	0	1	0	−6
1	0	1	1	−5
1	1	0	0	−4
1	1	0	1	−3
1	1	1	0	−2
1	1	1	1	−1
0	0	0	0	0
0	0	0	1	1
0	0	1	0	2
0	0	1	1	3
0	1	0	0	4
0	1	0	1	5
0	1	1	0	6
0	1	1	1	7

Figure. 10.6

Example 10.2. *Design a D/A converter for 4-bit digital inputs in 1's complement format.*

Solution. For 1's complement format, when MSB is 0 the data is positive and data is assumed as negative when MSB is 1. Therefore, 4-bit format data 0000 to 0111 represent positive numbers 0 to +7 and are the same as the representations of the unipolar binary numbers. No offset voltage is required for these inputs.

For negative numbers 1111 to 1000, the output analog voltage is offset by −15 V. This can be achieved by operating a switch with MSB of input to introduce the proper value of V_{off}. The circuit diagram is shown in Figure 10.7. The circuit components are

$$R_F = 8R, \ V(1) = -1 \ V, \ V(0) = 0 \ V.$$

The weighted-resistor D/A converter has the problem of having a wide range of resistance values of R to $2^{N-1}.R$. They are required to be very precise and also required to track over a wide temperature range to achieve analog voltage with high precision. It is very difficult to fabricate such a wide range of high-precision resistance values in a monolithic IC. This difficulty is eliminated by an R-2R ladder network type D/A converter as discussed next.

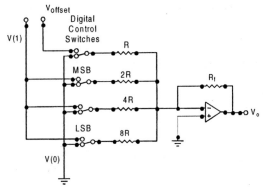

Figure 10.7

10.2.2 R-2R Ladder D/A Converter

The circuit diagram in Figure 10.8 illustrates an R-2R ladder D/A converter. It's comprised of only two types of resistor values, R and 2R. The inputs to the resistor network are applied through digitally controlled switches. A switch is in position 0 or 1 corresponding to digital input for that bit position being 0 or 1, respectively. To analyze the circuit, for simplicity, let us consider a 3-bit R-2R ladder network as in Figure 10.9(a) assuming the digital input as 001.

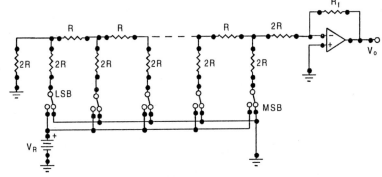

Figure 10.8

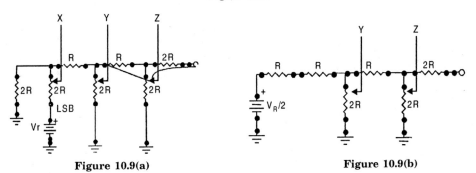

Figure 10.9(a)   **Figure 10.9(b)**

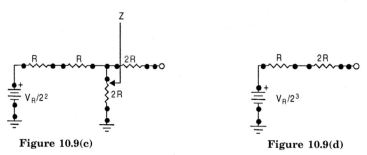

Figure 10.9(c)   **Figure 10.9(d)**

The circuit is simplified using Thevinin's theorem. Applying Thevnin's theorem at X, the circuit of Figure 10.9(a) may be modified to the circuit in Figure 10.9(b). The equivalent resistance before X is modified to R with the voltage source $V_R/2$. Similarly, applying Thevnin's theorem at Y of the circuit in Figure 10.9(b), an equivalent circuit can be reconstructed as in Figure 10.9(c).

Finally, an equivalent circuit is obtained in Figure 10.9(d), by applying Thevnin's theorem again at Z. You may notice in the circuit diagram in Figure 10.9(d) that equivalent resistance is R and equivalent voltage is $V_R/2^3$. Here LSB is assumed as 1.

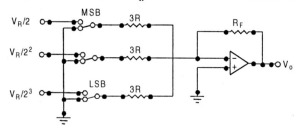

Figure 10.10

Similarly, for the digital inputs 010 and 100, the equivalent voltages will be $V_R/2^2$ and $V_R/2$ respectively. Now, by applying superposition theorem, all the digital inputs of the equivalent circuit can be reconstructed as in Figure 10.10, where effective resistance for each digital input is 3R. The output analog voltage V_O may be expressed as

$$V_O = -\{(R_F/3R).(V_R/2^3).b_0 + (R_F/3R).(V_R/2^2).b_1 + (R_F/3R).(V_R/2).b_2\} \qquad (10.11)$$

$$= -\{(R_F/3R).(V_R/2^3).(2^2b_2 + 2^1b_1 + 2^0b_0)\}. \qquad (10.12)$$

The above circuit may be expanded to any number of digital inputs. With the circuit analysis as above, it may be observed that for N^{th} digital input effective resistance is 3R and effective voltage applied is $V_R/2^N$. The generalized expression for analog output voltage is

$$V_O = -\{(R_F/3R).(V_R/2^N).(2^{N-1}b_{N-1} + 2^{N-2}b_{N-2} + ... + 2^2b_2 + 2^1b_1 + 2^0b_0)\}. \qquad (10.13)$$

Choosing $R_F = 3R$ and $V_R = -2^N$ V, the above expression for analog output voltage is reduced to

$$V_O = (2^{N-1}b_{N-1} + 2^{N-2}b_{N-2} + ... + 2^2b_2 + 2^1b_1 + 2^0b_0). \qquad (10.14)$$

The number of resistors required for an N-bit D/A converter is 2N in the case of an R-2R ladder type D/A converter, whereas the number of resistors required for weighted resistor type D/A converter is N. However, for an R-2R ladder network the type of resistance values is only two, which are required to be precise. On the other hand, if the number of bits N is large, the weighted-resistor type D/A converter needs wide-spread precision resistance values and hence it is not suitable for practical purpose. However, the weighted-resistor network can be modified in such a way that it can accommodate a large number of bits without consequent spread in resistance values.

10.2.3 Modified Weighted-resistor D/A Converter

Figure 10.11 illustrates one circuit example of modified type of weighted-resistor D/A converter. Here the bits are divided into groups of four. Each of the groups comprises a weighted-resistor type network. The most significant group of four bits is directly connected to the OP AMP input, but the least significant group is connected to the OP AMP input through an additional resistor R_x. The resistance R_x is introduced to produce input currents of OP AMP due to the least significant group of four bits and the most significant group of four bits in the ratio of 1:16. This means

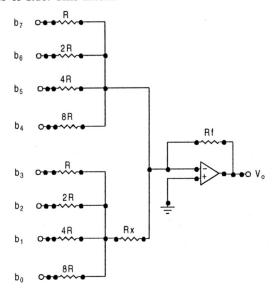

Figure 10.11

$$\frac{b_3}{b_7} = \frac{b_2}{b_6} = \frac{b_1}{b_5} = \frac{b_0}{b_4} = \frac{1}{16}. \tag{10.15}$$

The resistor R_x is determined by the following way. Let the bit b_3 be 1 and b_2, b_1 and b_0 are 0. The part of the circuit consisting of b_3, b_2, b_1 and b_0 is drawn at Figure 10.12(a) for simplification of analysis, and its further simplified equivalent circuit is drawn in Figure 10.12(b). From Figure 10.12(b), input current to OP AMP $I_{in}(1)$ is calculated, assuming OP AMP inputs are virtual grounded.

$$R_Y = 2R \parallel 4R \parallel 8R = R \cdot 8/7 \tag{10.16}$$

$$\text{So,} \quad I_{in}(1) = \frac{V_R}{R + \dfrac{R_x(R.8/7)}{R_x + R.8/7}} \times \frac{R.8/7}{R_x + R.8/7} = \frac{V_R}{R + R_x.15/8.} \tag{10.17}$$

Similarly, input current to OP AMP $I_{in}(2)$ is calculated considering b_7 bit is 1 and all other bits 0.

$$I_{in}(2) = V_R/R \qquad\qquad (10.18)$$

Now, $I_{in}(2) = 16\ I_{in}(1)$,   so, $V_R/R = 16.\ V_R/ (R + R_X \cdot 15/8)$.

After simplification, we get $R_X = 8R$. $\qquad\qquad (10.19)$

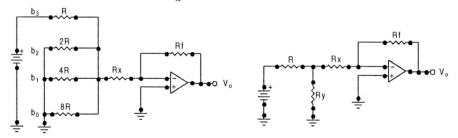

Figure 10.12(a)   **Figure 10.12(b)**

The relation $R_X = 8R$ may be verified for the other digital inputs b_2, b_1, and b_0, for which input current will be 1/16th of the currents due to b_6, b_5, and b_4 respectively. The output voltage V_0 of this modified D/A converter as shown in Figure 10.11 for $R_X = 8R$, can be expressed as follows

$$V_O = -\{(V_R/R)R_F b_7 + (V_R/2R)R_F b_6 + (V_R/4R)R_F b_5 + (V_R/8R)R_F b_4 + (V_R/16R)R_F b_3$$
$$+ (V_R/32R)R_F b_2 + (V_R/64R)R_F b_1 + (V_R/128R)R_F b_0\}$$
$$= -\{(V_R/2^7)\ (R_F/R)\ (2^7 b_7 + 2^6 b_6 + 2^5 b_5 + 2^4 b_4 + 2^3 b_3 + 2^2 b_2 + 2^1 b_1 + 2^0 b_0)\}. \quad (10.20)$$

Note that the analog voltage output derived in Expression 10.20 is proportional to the digital input. The number of resistors in this circuit is less and also the variety of resistor values is reduced. The circuit configuration can be extended to any number of bits.

Example 10.3. *Design a 2-decade BCD D/A converter.*

Solution. We have seen in the earlier chapters that each of the BCD numbers consists of four binary digits. Therefore, 2-decade numbers will have 8 binary digits. The modified weighted resistor D/A converter configuration may be used to design the 2-decade D/A converter. The binary inputs corresponding to the least significant digit are applied at b_3, b_2, b_1, and b_0 and the next significant digit is applied to b_7, b_6, b_5, and b_4. The circuit diagram is similar to Figure 10.11. The little difference for the BCD converter is that the value of R_X should be such that the input current to the OP AMP corresponding to the least significant digit is 1/10th of the input current due to the next significant digit. This means the equation at 10.15 is modified as

$$\frac{b_3}{b_7} = \frac{b_2}{b_6} = \frac{b_1}{b_5} = \frac{b_0}{b_4} = \frac{1}{10}. \qquad\qquad (10.21)$$

Using Equation 10.17, input current $I_{in}(1)$ due to b_3, when other bits b_2, b_1, and b_0 are grounded, is calculated as

$$I_{in}\ (1) = \frac{V_R}{R + R_X\ 15/8}. \qquad\qquad (10.22)$$

Similarly, input current to OP AMP $I_{in}(2)$ is calculated considering b_7 bit is 1 and all other bits 0.

$$I_{in}(2) = V_R/R \qquad\qquad (10.23)$$

Now, $\qquad\qquad I_{in}(2) = 10\ I_{in}(1), \qquad$ So, $V_R/R = 10.\ V_R/(R + R_X.15/8)$.

After simplification, we get

$$R_X = 4.8R. \qquad\qquad (10.24)$$

Therefore, by using resistor value R_X as 4.8R, the circuit diagram as illustrated in Figure 10.11 can be converted to a 2-decade BCD D/A converter.

10.3 SPECIFICATION OF D/A CONVERTERS

It is very important that the designers as well as the users be aware of the governing characteristics of D/A converters, as these characteristics play an important role to determine the stability and accuracy in analog output. The following characteristics of D/A converters are generally specified by the manufacturers.

1. Resolution.

2. Linearity.

3. Accuracy.

4. Settling time.

5. Temperature sensitivity.

10.3.1 Resolution

It is defined by the smallest possible change in the output voltage as a fraction or percentage of the full-scale output range. If an 8-bit D/A converter is considered for an example, there are 2^8 or 256 possible values of output analog voltage. Hence the smallest change in the output voltage is $1/255^{th}$ of full-scale output range. Therefore, the resolution is calculated as 1/255 or 0.4%. So a general expression of *resolution* for an N-bit D/A converter may be defined as below.

$$\text{Resolution} = \frac{1}{2^N - 1} \times 100\%. \qquad\qquad (10.25)$$

Alternatively, resolution is also defined by the number of bits accepted by the D/A converter. For example, a 12-bit D/A converter has 12-bit resolution.

10.3.2 Linearity

In D/A converters, it is desired that equal increments in the numerical significance of the digital inputs should result in equal increments in the analog output voltage. However, in practical circuits, due to an error in resistor values and potential loss at the switches, this type of linear input-output relationship is never achievable. The term *linearity* of a converter determines the measure of precision with which the linear input-output relationship is satisfied.

Figure 10.13 demonstrates the input-output relationship of a 3-bit D/A converter. The horizontal axis represents the input bit combinations with fixed interval separations in order of numerical significance and the vertical axis represents the output analog voltage. The output corresponding to each input is indicated by a dot. If the D/A converter is ideal or perfectly linear, the dots would be on the straight line.

However, in Figure 10.13, it has been shown that the dots are not in a single straight line. This is to show the practical nonlinearity behavior of the converter. The linearity error for a digital input is the difference between the actual voltage obtained corresponding to the dot and the ideal voltage expected. This is denoted by ε. The normal analog output voltage change corresponding to a digital input change equivalent to the least significant bit is indicated by Δ. This means the resolution of the converter is Δ.

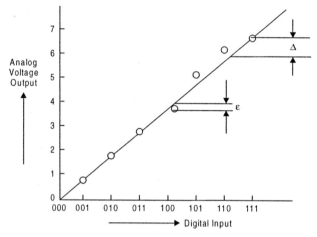

Figure 10.13

The linearity of a D/A converter is generally specified by comparing ε with Δ. For example, the linearity of a commercially available D/A converter is usually specified as *less than* ½ LSB. This implies that $|\varepsilon| < $ ½ Δ.

10.3.3 Accuracy

The *accuracy* of a D/A converter is determined by the measure of the difference between the actual output voltage and the expected output voltage. It is specified as the percentage of maximum output or the full-scale output voltage. For example, if a D/A converter is specified as the accuracy of 0.1%, with full-scale of maximum output voltage of 10 V, the maximum error at output voltage corresponding to any input combination will be $10 \times 0.1/100$ V = 10 mV.

10.3.4 Settling Time

It is one of the important governing factor of a D/A converter. For any change in digital input, the analog output voltage does not instantaneously attain its expected value corresponding to the digital input and takes some time to attain the steady state output. This is due to transients that appear at the output voltage and oscillation may occur because of the presence of switches, active devices, stray capacitance, and inductance associated with passive circuit components. A typical plot of change in analog output voltage with respect to time is shown in Figure 10.14.

The time required for analog output to settle to within ± ½ LSB of the final value due to a change in the digital input is usually specified by the manufacturers and is referred to as *settling time*. In Figure 10.14, it has been demonstrated that the final value of output voltage is settled within ± ½ LSB of the final value at time T, which is called settling time. Settling time limits the speed and operating frequency of a D/A converter. Operating frequency of the converter should be such that analog output must be settled to the correct voltage before any change in the digital input.

This means, if operating frequency is F_0, then $F_0 < 1/T$.

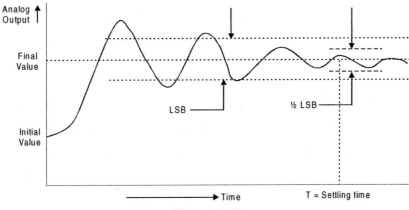

Figure 10.14

10.3.5 Temperature Sensitivity

The analog output voltage for any fixed digital input varies with temperature. This is due to the temperature sensitiveness of various active and passive components like reference voltage source, resistors, diodes, transistors, OP AMPs, etc. The temperature sensitivity is defined by the change in output voltage from its expected value in respect to temperature and is specified in terms of $\pm$ ppm/°C.

10.4 AN EXAMPLE OF A D/A CONVERTER

To understand the specification and application of a D/A converter, a typical commercially available D/A converter IC DAC 80 is being discussed here. The functional block diagram of this D/A converter IC is shown in Figure 10.15. It is a 12-bit D/A converter available in a 24-pin DIP package. It consists of matched bipolar switches, a precision resistor network, a low-drift high-stability voltage reference network with optional output amplifier. The options are available for 12-bit complementary binary (CBI) or three-digit BCD (complementary coded decimal—CCD) input codes, as well as the current or voltage output modes. The important performance characteristics are described below.

Most of the pins of the D/A converter are used in two options—voltage model and current model. The pins marked with dual functions have the first function for the voltage model and the second function for the current model. The table in Figure 10.16 describes the different input modes of operation of the D/A converter and the table in Figure 10.17 gives the various connections to be made to obtain different full-scale ranges (FSR).

The DAC 80 accepts complementary digital input codes in either binary (CBI) or decimal (CCD) format. The CBI code may be any one of these codes—complementary straight binary (CSB), complementary offset binary (COB), or complementary 2's complement. The 12-bit digital input is connected to pins B_1 to B_{12}, where B_1 is MSB and B_{12} is LSB.

Resolution	:	12 bits (binary) or 3 digits BCD
Linearity error	:	± 0.12 % (maximum)
Maximum gain drift	:	± 30 ppm/ °C
Power dissipation	:	925 mW (maximum)
Maximum conversion time	:	5 μs (voltage output)
(settling time to ± 0.01% of full scale range)		1 μs (current output)
Digital input format	:	12 bit CBI or 3 digit CCD
Analog output voltage ranges	:	± 2.5 V, ± 5 V, ± 10 V, 0 to +5 V,
		0 to +10 V (CBI), or 0 to +10 V (CCD)
Output current	:	± 5 mA (minimum)
Output impeadance (DC)	:	0.05 Ω
Analog output current ranges	:	± 1 mA, 0 to –2 mA (CBI)
		0 to –2 mA (CCD)
Output impedance	:	3.2 KΩ (bipolar) or 6.6 KΩ (unipolar)

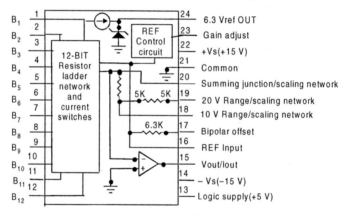

Figure 10.15

Digital input		Analog output			
Format	Code	Complementary straight binary (CSB)	Complementary offset binary (COB)	Complementary 2's complement (CTC)	Complementary coded decimal (CCD)
Binary	000000000000	+ Full scale	+ Full scale	1 LSB	–
	011111111111	+ ½ Full scale	Zero	– Full scale	–
	100000000000	Mid scale-1 LSB	– 1LSB	+ Full scale	–
	111111111111		– Full scale	Zero	–
BCD	011001100110	–	–	–	+ Full scale
	111111111111	–	–	–	Zero

Figure 10.16

Output range	Digital input codes	Connect pin 15 to pin	Connect pin 17 to pin	Connect pin 1 19 to pin
± 10 V	COB or CTC	19	20	15
± 5 V	COB or CTC	18	20	NC
± 2.5 V	COB or CTC	18	20	20
0 to +10 V	CSB	18	21	NC
0 to +5 V	CSB	18	21	20
0 to +10 V	CCD	19	NC	15

Figure 10.17

10.4.1 Calibration

Few external components are required for proper operation of the D/A converter and its offset and gain adjustment. This is demonstrated in Figure 10.18.

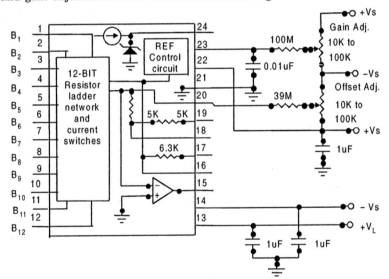

Figure 10.18

Offset Adjustment. Some external components are added as in Figure 10.18, to achieve the proper offset voltage. For unipolar operation, the digital code that should produce zero output voltage is applied and adjusts the offset potentiometer to produce zero output. For bipolar output configuration, digital input is applied in such a way that maximum negative output is produced and the offset potentiometer is also adjusted.

Gain Adjustment. Gain potentiometer, as indicated in Figure 10.18, is adjusted for maximum positive output of full-scale voltage after applying the digital input code that should produce maximum output.

The gain and offset adjustments for unipolar and bipolar modes of a D/A converter are shown in Figures 10.19(*a*) and 10.19(*b*).

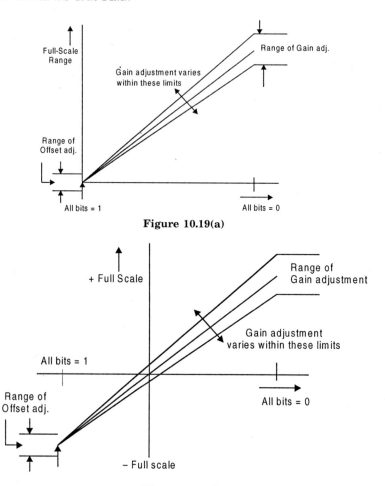

Figure 10.19(a)

Figure 10.19(b)

10.5 ANALOG-TO-DIGITAL CONVERTERS

An analog-to-digital converter, or A/D converter, is the reverse system of a D/A converter, which converts an analog signal to its digital form. In an analog-to-digital converter, the input analog voltage may have any value in a range and it will produce the digital output of 2^N number of discrete values for an N-bit converter. Therefore, the whole range of analog voltage is required to be represented suitably in 2^N intervals, and each of the intervals corresponds to a digital output.

Let us consider that an analog voltage range of 0 to V is represented by 3-bits digital output. Since a 3-bit digital system can generate $2^3 = 8$ different digital outputs, the full analog range will be divided into 8 intervals, and each interval of voltage of the size of V/8 is assigned unique digital value. This process is called *quantization*. The interval of analog voltage and their corresponding digital representations are tabulated in Figure 10.20.

It may be observed from the figure that a complete voltage interval is represented by a digital value, irrespective of any voltage value within the interval. Therefore, there involves always some error while converting any analog voltage interval to its digital value, which is referred to as *quantization error*.

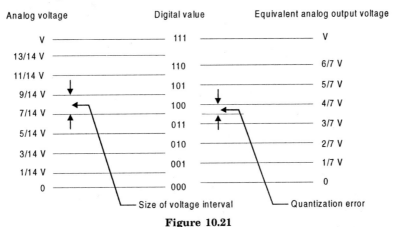

Figure 10.20

In Figure 10.20, the voltage interval 0 to 1/8 V is assigned to the digital value 000, voltage interval 1/8 V to 2/8 V is assigned to digital value 001, and so on. Also, it may be noted from the figure that the maximum quantization error is E = 1/8 V. This quantization error may be reduced if the analog voltage interval limits are considered a different way. Let us choose the size of voltage interval as 1/7 V, except the top and bottom intervals which are 1/14 V. The interval limits are at the middle of the voltage corresponding to the digital values shown in Figure 10.21.

Figure 10.21

Note that the quantization error in this case is 1/14 V and is equal to one half of the size of voltage interval or the quantization interval. This is less when compared to the earlier case in Figure 10.20. In practice, the quantization error is specified in terms of LSB. According to Figure 10.21, the maximum quantization error is ± LSB.

Example 10.4. *An analog voltage of the range of –V to +V is required to be converted into 3-bit 2's complement digital format. The digital value for 0 V should be 000 and the maximum quantization error should not exceed ± LSB. Determine the quqntization interval.*

Solution. If the quantization interval is denoted as S, then the digital value 000 should be assigned to the analog voltage interval 0V ± S/2. This is illustrated in Figure 10.22 along with the 2's complement representation corresponding to the analog voltage levels. Since in 2's complement representation, there is one more negative number than the positive number, the full analog voltage range from –V to +V is divided into seven intervals, each of size 2V/7, and one digital value is assigned to each interval. Therefore, the quntization interval is S = 2V/7. The extra digital value 100 may be assigned to represent the interval –V to –9/7 V.

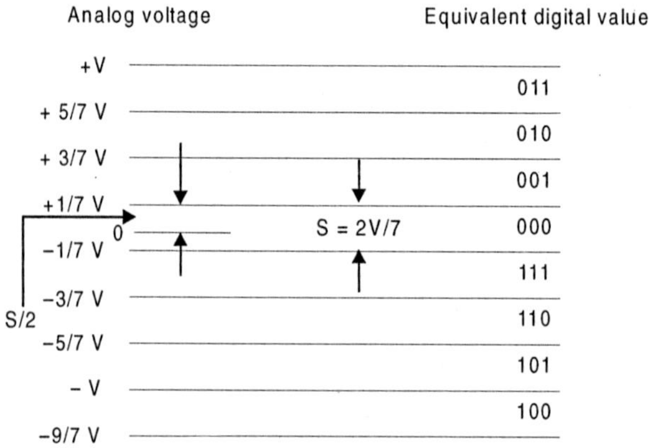

Figure 10.22

Some of the commonly used A/D converter techniques are discussed here.

10.5.1 Parallel Comparator A/D Converter

The schematic diagram of a 3-bit parallel-comparator type A/D converter is shown in Figure 10.23. V_a is the analog voltage that is converted to digital form. The corresponding full-scale voltage is V, and different reference voltage levels V_1 to V_7 have been generated with the help of a resistor network from the full-scale voltage V. The analog voltage V_a is simultaneously compared by seven comparators with reference voltage levels V_1 to V_7. Each of the outputs of the comparators is digital in nature and has only two levels. The seven outputs of the comparators are stored in latches. The seven output latches are converted to 3-bit binary format with the use of a decoder. The comparator outputs and corresponding digital output for each interval of analog voltage are given in the table in Figure 10.24.

The concept and principle of parallel comparator A/D conversion is the simplest as well as the fastest. Digital output with any number of bit system for an A/D converter can be realized by this simple concept of operation. However, as the number of bits for digital output increases, the number of comparators requirement increases. Here lies the main disadvantage of this type of A/D converter, because the number of comparators increases exponentially, as for N-bit A/D converters the number of comparators required is 2^N-1. Also, this increases the number of latches and complications of the decoder circuit.

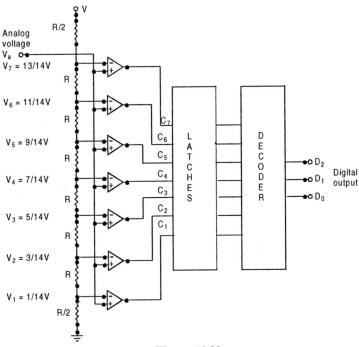

Figure 10.23

Analog Input	Comparator Outputs							Digital Output		
V_a	C_7	C_6	C_5	C_4	C_3	C_2	C_1	D_2	D_1	D_0
$0 < V_a < V_1$	0	0	0	0	0	0	0	0	0	0
$V_1 < V_a < V_2$	0	0	0	0	0	0	1	0	0	1
$V_2 < V_a < V_3$	0	0	0	0	0	1	1	0	1	0
$V_3 < V_a < V_4$	0	0	0	0	1	1	1	0	1	1
$V_4 < V_a < V_5$	0	0	0	1	1	1	1	1	0	0
$V_5 < V_a < V_6$	0	0	1	1	1	1	1	1	0	1
$V_6 < V_a < V_7$	0	1	1	1	1	1	1	1	1	0
$V_7 < V_a < V$	1	1	1	1	1	1	1	1	1	1

Figure 10.24

10.5.2 Successive Approximation A/D Converter

The essential elements of a successive approximation type A/D converter are a D/A converter and comparator. When any unknown analog voltage V_a is applied for A/D conversion, first it is compared with the analog voltage generated from the internal D/A converter, which is equivalent to ½ of the full-scale range. If the unknown voltage V_a is higher than ½ the full-scale range, then it is compared with ¾ of full-scale voltage generated from the internal

D/A converter, and if the analog voltage is less than the 1/2 full-scale voltage, then it will be compared with ¼ the full-scale voltage. Again, the analog voltage V_a will be compared with 1/8 the full scale if it is less than ¼ the full scale, or with 3/8 the full scale if it is higher than ¼ the ful scale, or with 5/8 the full scale if it is lower than ¾ the full scale, or with 7/8 the full scale if it is higher than ¾ the full-scale voltage. For an N-bit A/D converter, this comparison process is continued up to the lowest voltage segment of $1/2^N$ the full-scale range. The digital output is considered at the end of the comparison process and is determined by the last set input of a D/A converter. Since the digital output is determined by successive comparison technique, the process is referred to as successive approximation. The comparison process of a 3-bit successive approximation A/D converter is illustrated by the flow chart in Figure 10.25. An offset voltage is also introduced for calibration.

V_r = Reference analog voltage generated from a D/A converter

V_d = Digital output

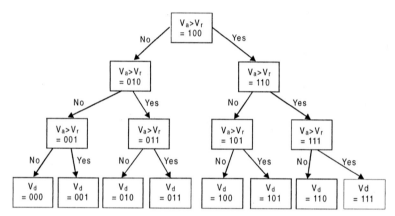

Figure 10.25

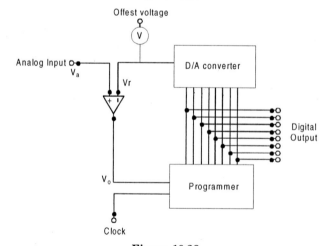

Figure 10.26

The schematic diagram of successive approximation type A/D converter is shown in Figure 10.26. In addition to the D/A converter and comparator, programmer and clock are also included to synchronize and monitor the comparison process. The programmer generated the digital reference which is converted to analog voltage and compared with the input unknown analog voltage. The comparator output sets or resets to generate the next digital data that is to be compared with the synchronization of the clock pulse. To start conversion, the programmer sets the MSB to 1 and all other bits to 0. If input analog voltage V_a is higher than V_r which is the analog equivalent of digital data set by the programmer, then comparator output is high, which sets next lower MSB to 1. On the other hand, if $V_a < V_r$, then comparator output is low and it reset MSB to 0 and set next lower MSB to 1. Thus, a 1 is tried in each of the D/A converters until the binary equivalent of analog input voltage is obtained.

Note that, unlike the parallel comparator A/D converter, the successive approximation type converter employs only one comparator, and associated hardware is much less. However, in successive approximation technique, the input analog voltage is compared N times for an N-bit A/D converter and N number of clock pulses are required to obtain the desired digital output. Hence it is slower than the parallel comparator type, but faster than other types of A/D converters and it is also very accurate. Therefore, it is very popular in practice.

10.5.3 Counting A/D Converter

The concept of a counting A/D converter is very similar to a successive approximation type A/D converter. The basic difference is that an up-counter is employed at counting type A/D converter in place of a programmer that is used in successive approximation type. The schematic diagram of a counting A/D converter is shown in Figure 10.27. The comparator output and clock are ANDed by an AND gate and applied to the clear input of the UP counter. The converter converts the counter output to analog voltage. To start the conversion, the counter is at the reset position, *i.e*, all the counter output bits are 0. So, D/A converter output is 0 and comparator output V_o is high because of the application of unknown analog input voltage. Therefore, clock is enabled and the counter starts counting upward. Since the number of clock pulses counted increases linearly with time, the D/A converter output voltage V_r increases, as shown in Figure 10.28. The counting process will stop when D/A converter output V_r is higher than analog input voltage V_a ($V_r > V_a$), and comparator output V_o is low to disable the AND gate. Since no clock pulse is now available, the counter will stop counting and at this instant digital output is available. Offset voltage may be applied at the input of the comparator for calibration purpose.

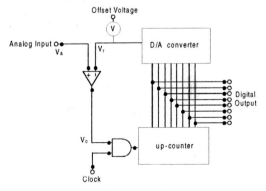

Figure 10.27

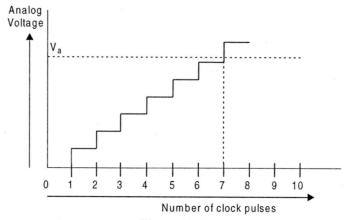

Figure 10.28

It may be noted that the conversion time for a counting A/D converter depends on counting the number of clock pulses. Therefore, the maximum conversion time for an N-bit converter is the time lapsed by 2^N number of clock pulses. Hence this type of A/D converter is slower than the previous two types of A/D converters.

10.5.4 Dual-Slope A/D Converter

A dual-slope A/D converter is one of the most commonly used types of converter. The schematic diagram of a dual-slope A/D converter is illustrated in Figure 10.29. It consists of the following major functional blocks.

1. An integrator.

2. A comparator.

3. A binary counter.

4. A switch driver.

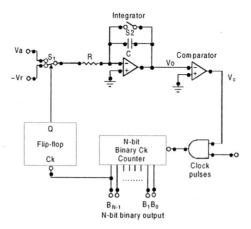

Figure 10.29

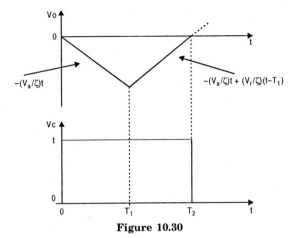

Figure 10.30

The principle of operation can be understood with the help of the timing diagram in Figure 10.30. Let us consider, the conversion process starts at time $t = 0$ with the switch S_1 connected to analog input voltage V_a and the N-bit counter is at cleared state, *i.e*, its output is all 0s. Analog voltage is passed through an integrator circuit with time constant $\zeta = RC$. The integrator output

$$V_o = -1/\zeta \int_0^t V_a dt = -(V_a/\zeta)t. \tag{10.26}$$

As V_o is negative, this will make the comparator output V_c HIGH, and enable the clock pulses to reach the clock input (Ck) of the N-bit binary counter. The counter will count from 00...00 to 11..11 when 2^N-1 clock pulses are applied. At the application of 2^{Nth} clock pulse at time T_1, the counter is cleared to all 0s output setting the flip-flop output Q to 1. Now the switch S_1 will connect reference voltage $-V_r$ to the integrator. The integrator output V_o starts to increase in a positive direction and at time instant T_2 it crosses zero voltage. The comparator output V_c remains HIGH up to time T_2 and becomes LOW after T_2. The counter will continue to count up to the time T_2, as comparator HIGH output enables the clock pulses to reach the counter clock input and thereafter the counter stops counting when V_c is low to disable the AND gate to prevent the clock pulses. The integrator output V_o and comparator output V_c behaviors are shown in Figure 10.30. Now, from the following derivation, it can be seen that the last counter reading where it stopped is directly proportional to the analog voltage.

From Equation 10.26, we obtain $V_o = -(V_a/\zeta)t$, while at negative slope. So at time T_1

$$V_o = -(V_a/\zeta)T_1. \tag{10.27}$$

When a switch is connected to reference voltage $-V_r$, the voltage expression of V_o can be written as

$$V_o = -(V_a/\zeta)T_1 + (V_r/\zeta)(t - T_1). \tag{10.28}$$

At time $t = T_2$, $V_o = 0$, as it reaches zero voltage, from Equation 10.28, we derive

$$0 = -(V_a/\zeta)T_1 + (V_r/\zeta)(T_2 - T_1).$$

Or, $T_2 - T_1 = (V_a/V_r)\, T_1.$ \hfill (10.29)

Now, if we assume that the duration of the clock pulse is T_c, then $T_1 = 2^N \cdot T_c$, as within the time T_1, the counter counted 2^N number of clock pulses. Again, if we assume that the counter stopped after n^{th} count at time instant T_2, then $T_2 - T_1 = n.T_c$, as the counter was reset at time instant T_1. So from Equation 10.28,

$$T_2 - T_1 = n.T_c = (V_a/V_r) \, T_1 = (V_a/V_r) \, 2^N.T_c.$$

Or, $n = (V_a/V_r) \, 2^N.$ (10.30)

From Equation 10.30, we observe that the count recorded at the counter is proportional to the input analog voltage V_a.

This type of A/D converter has very good conversion accuracy and is of low cost. This is often used in a digital voltmeter. However, the disadvantage of the dual-slope A/D converter is its slow speed.

10.5.5   A/D Converter Using Voltage-to-Frequency Conversion

An analog voltage can be converted into digital form by generating pulses whose frequency is proportional to the analog voltage. These pulses are then counted by a counter for a fixed time duration and the reading of the counter will be proportional to the frequency of the pulses and hence, to the analog voltage.

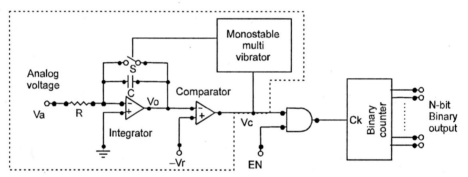

Figure 10.31 Voltage-to-frequency converter.

The concept of an A/D converter using a voltage-to-frequency converter is illustrated by the schematic diagram in Figure 10.31. The voltage-to-frequency converter section consists of an integrator, a comparator, and a monostable multivibrator. The analog input voltage is applied to the integrator whose output is applied to the inverting input terminal of the comparator. A reference voltage V_r is applied through the noninverting input terminal of the comparator. The comparator output activates a monostable multivibrator to produce a short pulse to control the active switch S of the integrator. The comparator will generate a pulse train. The operation can be explained with the help of a timing diagram as shown in Figure 10.32.

When the analog input is applied, at $t = 0$, integrator output starts decreasing with slope of time constant $\zeta = RC$. At this instant integrator output V_0 will be the function of analog input of the following relation.

$$V_0 = -1/\zeta \int_0^t V_a \ dt = -(V_a/\zeta)t$$ (10.31)

The integrator output will continue to decrease and the comparator output is LOW. At time instant $t = T$, the integrator output will cross the reference voltage $-V_r$, and the comparator output will be HIGH. The high output of the comparator activates the monostable multivibrator to produce a pulse to close the switch S for discharging the integrator capacitor. Monostable pulse width is very small compared to T, but sufficient to discharge the capacitor completely to make $V_o = 0$. Let the monostable pulse width be T_d, so the switch S is closed for time T_d and again V_o starts decreasing when S is open. Thus, a pulse train will be produced at comparator output V_c, as shown in Figure 10.32. Now we can write from Equation 10.31

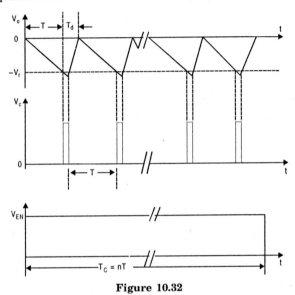

Figure 10.32

$$V_o = (V_a/\zeta)T = V_r.$$

Or, $$T = (V_r/V_a)\ \zeta. \tag{10.32}$$

If monostable pulse-width T_d is too small, such that $T \gg T_d$, then comparator output frequency F can be derived as

$$F = 1/(T + T_d) \approx 1/T = (1/\zeta)\ (V_a/V_r) \quad \text{[from equation 10.32]. } (10.33)$$

From Equation 10.33, you may notice that comparator output signal frequency F is proportional to the analog input voltage V_a.

Comparator output is now applied to the clock input Ck of an N-bit binary counter through an AND gate. An enable pulse of fixed duration T_C is applied to the other input of the AND gate. Therefore, clock pulses are available to the binary counter for a time duration of T_C and the counter will count the number of pulses for that duration. If the number of pulses recorded at the counter is n, then $T_C = n.T$.

So, $$T_C = n/F.$$

Or, $$n = F.\ T_C.$$

Or, $$n = (1/\zeta)\ (V_a/V_r).\ T_C. \tag{10.34}$$

From Expression 10.34, we can observe that the number of pulses counted in the counter is proportional to the analog input voltage and thus we obtain the equivalent digital form of the analog voltage. In this system, it is assumed the analog input voltage is constant over the time duration of T_C.

10.5.6 A/D Converter using Voltage-to-Time Conversion

The concept of an A/D converter using voltage-to-time conversion is very similar to the A/D converter with voltage-to-frequency conversion, as the digital output for both types of converters is derived by counting the number of pulses that is proportional to the analog input voltage. The difference is that an A/D converter using voltage-to-frequency conversion is based on the counting of pulses of variable frequency for a fixed time duration, whereas the A/D converter using voltage-to-time conversion counts the pulses of fixed frequency but of variable time. The schematic diagram of the A/D converter using voltage-to-time conversion is illustrated by Figure 10.33.

As shown in Figure 10.33, in this type of converter, a negative voltage reference $-V_r$ is integrated first by an integrator circuit with time constant $\zeta = RC$. At time $t = 0$, V_{EN} is LOW and switch S at integrator is open. The opening and closing of switch S is controlled by an ENABLE pulse control circuit. The output of the integrator will start increasing and it is compared with the analog input voltage. Initially, when analog voltage V_a is higher than the reference voltage (*i.e.*, $V_a > V_r$), then comparator output V_c is HIGH.

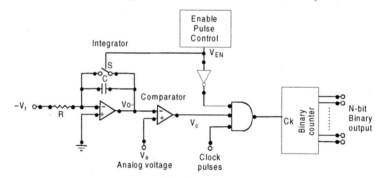

Figure 10.33

The comparator output is applied to one of the inputs of a 3-input AND gate, where the other two inputs of the AND gate complement V_{EN} and clock pulses time period T_C. Therefore the AND gate is enabled when comparator output is HIGH and V_{EN} is LOW, and clock pulses will reach the binary counter for count operation.

When integrator output will reach the analog input V_a, at $t = T$, the comparator output is LOW and the AND gate is disabled to stop counting at the counter. Now the digital output from the counter is proportional to the analog input and may be considered as the digital conversion of analog voltage. This can be proved by the following derivation.

$$\text{For}\ \ 0 < t < T, \qquad V_o = -1/\zeta \int_0^t V_r dt = -(V_r / \zeta)t. \tag{10.35}$$

At $t = T$, comparator output equals to V_a, so $V_a = (V_r / \zeta)\, T.$ $\hspace{2cm}$ (10.36)

Now, if the counter records the counting of n number of pulses and each pulse time period is T_C, then $\quad\quad\quad T = n \cdot T_C.$ $\quad\quad\quad\quad\quad\quad\quad\quad\quad\quad\quad\quad\quad$ (10.37)

Therefore, $\quad\quad V_a = (V_r/\zeta)T = V_a = (V_r/\zeta)\, n.T_C.$

Or, $\quad\quad\quad\quad\quad\quad n = (\zeta/T_C)(V_a/V_r).$ $\quad\quad\quad\quad\quad\quad\quad\quad$ (10.38)

Equation 10.38 indicates that counter reading is proportional to the analog input voltage V_a.

Though the digital output is obtained from the counter, the integrator output will continue to increase till V_{EN} is LOW. At $t = T_1$, switch S is closed as V_{EN} is HIGH and integrator capacitor discharges. At $t = T_2$, when the integrator capacitor is fully discharged, V_{EN} is made LOW to enable a fresh cycle of conversion operation.

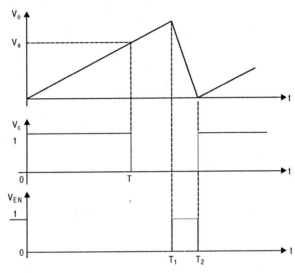

Figure 10.34

10.6 SPECIFICATION OF AN A/D CONVERTER

The A/D converters are usually specified by the following characteristics.

1. *Range of input voltage.* This is the factor that specifies the minimum and maximum analog input voltage that can be accepted by an A/D converter.

2. *Input impedance.* This is an important design criteria that limits the maximum input current to the A/D converter without deteriorating its performance or damage.

3. *Accuracy.* It is the error involved in the conversion process and is represented in %.

4. *Conversion time.* This characteristics specifies the maximum time required for the conversion process and is very critical while interfacing with other devices and synchronization with time. The output is considered only after the end of conversion.

5. *Format of digital output.* Digital output may be of various formats, like unipolar, bipolar, parallel, serial, etc. The information is essential while designing and interfacing with other networks.

10.7 AN EXAMPLE OF AN A/D CONVERTER IC

A/D converters are commercially available in monolithic IC packages. One of the A/D converters, ADC 80, is discussed here. It is a 12-bit successive approximation type A/D converter available in 32-pin DIP. The important performance characteristics are described below.

Linearity error	:	$\pm 0.012\%$
Maximum gain temperature coefficient	:	30 ppm/°C
Power dissipation	:	800 mW
Maximum conversion time	:	25 μs
Digital output format	:	Unipolar and bipolar, Parallel and serial
Output drive	:	2 TTL loads
Analog voltage range	:	± 2.5 V, ± 5 V, ± 10 V (bipolar) or
		0 to 5 V, 0 to 10 V (unipolar)

The functional block diagram of A/D converter ADC 80 is shown in Figure 10.35. It consists of a successive approximation register (SAR), 12-bit D/A converter (DAC), clock and control circuit, a reference generator, and comparator.

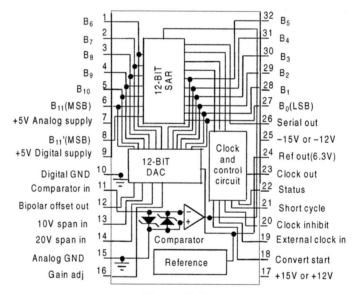

Figure 10.35

10.7.1 Operation

After receiving the convert start command, the A/D converter converts the voltage applied at its analog input terminal to an equivalent 12-bit binary number. How to connect the analog input will be discussed later in this chapter. The successive approximation register (SAR) generates a 12-bit binary number and it compared with the input analog voltage. During the conversion process, the status flag remains set. When the conversion is over, the status flag is reset and parallel output data becomes available. In-built parallel-to-serial register is provided in this A/D converter to make serial data available at output.

10.7.2 Analog input

The analog input is scaled as close to the maximum input signal range as possible in order to achieve the maximum resolution of the A/D converter. Different input signal ranges can be programmed by selecting appropriate external connections of input pins and span inputs as described in the table in Figure 10.36.

Input signal Ranges	Connect pin 12 To pin	Connect pin 14 To	Connect input Signal to pin
± 10 V	11	Input signal	14
± 5 V	11	Open	13
± 2.5 V	11	Pin 11	13
0 to +5 V	15	Pin 11	13
0 to +10 V	15	Open	13

Figure 10.36

The analog ground and digital ground must be connected together at one point, usually at the system's power supply ground.

10.7.3 Digital Output

The parallel digital data is available at pins B_{11} (MSB) to B_0 (LSB). For unipolar input ranges, the output is in the form of complementary straight binary code (CSB), whereas for bipolar input ranges the output is either in complementary offset binary (COB) format or in complementary 2's complement binary (CTC) format. For complementary offset binary (COB) format, pin 6 (B_{11}) is used as MSB, whereas for complementary 2's complement binary (CTC) format pin 8 (B_{11}') is used as MSB.

Serial data is available in CSB format for unipolar input ranges and in COB form for bipolar input ranges. The first bit at serial output is MSB and LSB comes out last.

The A/D converter can be used for 12-bit, 10-bit, or 8-bit resolution by connecting the short cycle terminal (pin 21) to pin 9, pin 28, or pin 30 respectively. The conversion time is reduced to 21 μs and 17 μs for 10-bit and 8-bit operations, respectively.

10.7.4 Calibration

For calibration of the device, zero adjustment and gain adjustment are performed by employing external potentiometers as shown in Figure 10.37. To prevent interaction of these two adjustments, zero is always adjusted first and then gain. Zero is adjusted with the analog input voltage near the most negative end of the analog voltage range (0 is for unipolar and full scale for bipolar input ranges). Gain is adjusted with the analog input voltage near the most positive end of the analog range. The analog input voltages to be used for these adjustments and the corresponding digital outputs are given in the table in Figure 10.38, for two ranges. Similarly, the corresponding analog input voltages can be determined for other ranges, also.

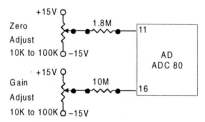

Figure 10.37

Input signal range	Set analog input voltage to	Adjust zero for digital output	Adjust gain for digital output
0 to +10 V	+1 LSB = 0.0024 V	111111111110	–
	+FSR – 2LSB = 9.9952 V	–	000000000001
	+5 V (Digital output = 011111111111)		
–10 V to +10 V	-9.9952 V	111111111110	–
	+9.9952 V	–	000000000001
	0.0 (Digital output = 011111111111)		

Figure 10.38

10.8 CONCLUDING REMARKS

Some of the commonly used techniques for digital-to-analog and analog-to-digital conversions are discussed in this chapter. The techniques are explained here with functional blocks for design concepts only. In actual practice, the circuits are more complex in nature.

Among the A/D converters discussed here, the fastest is the parallel comparator type. Therefore, this type of converter is the best choice where maximum speed of operation is required. The speed of the successive approximation type A/D converter is less than that of the parallel comparator type. But this requires less hardware and hence it is quite popular.

One of the most popular A/D converters is the dual-slope type and is widely used in instruments, such as digital voltmeters or digital multimeters, where conversion speed is not important.

The output voltage levels of A/D converters are often made compatible with different logic families such as TTL, CMOS, ECL, etc.

REVIEW QUESTIONS

10.1 What are D/A converters and A/D converters and what are their uses?

10.2 Refer to Figure 10.3. For an 8-bit weighted resistor D/A converter, what is the maximum resistor value if the value of the resistor for MSB is 2.2K ?

10.3 For question 10.2, what should the value of R_f be to make a full-scale analog output of 10V if $V(1) = -5$ V?

10.4 Verify equation 10.9 for bits b_2, b_1 and b_0.

10.5 Explain the terms linearity, resolution, and accuracy.

10.6 What is the conversion time of a D/A converter?

10.7 Explain the following codes and give examples using 4 bits.

   (*a*) CSB, (*b*) COB,   (*c*) CTC,   (*d*) CCD

10.8 What is quantization and what is quantization error?

10.9 What is the maximum quantization error for an 8-bit A/D converter if a full-scale analog input is 10 V?

10.10 An A/D converter has the conversion time of 20 μs. Find whether this A/D converter is suitable if the analog input is sampled at 100 kHz.

10.11 An 8-bit D/A converter provides an analog output that has a maximum value of 10 V. The output may have an error of ΔV due to drift in component values, temperature, etc. How large can ΔV be before the least significant bit would no longer be significant?

10.12 For the circuit diagram in Figure 10.39, find the analog output voltage for each of the digital inputs and show that this circuit can be used for converting digital signal in 1's complement format to analog input.

10.13 In the circuit in Figure P.10.1, if the offset switch is removed and the switch S_2 is replaced by a switch S_2' (with its resistor R/4 is replaced by a resistor R/3), verify that the operation of the circuit will not change.

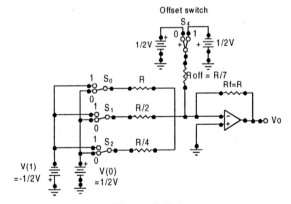

Figure P.10.1

10.14 Design a D/A converter circuit similar to the circuit in problem 10.2 for 4-bit input.

10.15 Design a D/A converter circuit similar to the circuit in problem 10.3 for 4-bit input.

10.16 Design a 3-bit parallel comparator A/D converter for 2's complement format.

10.17 A dual-slope A/D converter has a resolution of 12 bits. If the clock rate is 100 kHz, what is the maximum rate at which samples can be converted?

10.18 A D/A converter has a full-scale analog output of 10 V and accepts six binary bits as inputs. Find the voltage corresponding to each analog step.

10.19 Find the analog voltage corresponding to the LSB for each of the ranges of

(a) DAC 80

(b) ADC 80

10.20 How many bits are required at the input of a converter to achieve a resolution of 1mV, if full scale is + 5 V ?

10.21 What clock frequency must be used for a 10-bit dual-slope A/D converter if it is capable of making at least 7000 conversions per second?

10.22 What is the conversion time of a 12-bit successive approximation type A/D converter using a 1 MHz clock?

❑ ❑ ❑

11 LOGIC FAMILY

11.1 INTRODUCTION

Logic gates and memory devices are fabricated as integrated circuits (ICs) because the components used, such as resistors, diodes, bipolar junction transistors, and the insulated gate or metal-oxide semiconductor field-effect transistors are the integral parts of the chip. The various components are interconnected within the chip to form an electronic circuit during assembly. The chip is mounted on a metal or plastic package, and connections are welded to the external pins to form an IC. The ICs result in an increase in reliability and reduction in weight and size.

Small-scale integration (SSI) refers to ICs housing fewer than 10 gates in a single chip. Medium-scale integration (MSI) includes 11 to 100 gates, whereas large-scale integration (LSI) refers to more than 100 to 5000 gates in a single chip. Very large-scale integration (VLSI) devices contain several thousands of gates per chip.

Integrated circuits are classified into two general categories—(a) *linear* and (b) *digital*. Linear integrated circuits are operated with continuous signals and are used to construct electronic circuits such as amplifiers, voltage comparators, voltage regulators, etc. Digital circuits are operated with binary signals and are invariably constructed with integrated circuits. This chapter describes the basic internal structure of different types of logic families and analysis of their operation.

The various logic families can be broadly classified into categories according to the IC fabrication process—(a) bipolar and (b) metal oxide semiconductor (MOS). Integrated circuits are available in various types of packages as mentioned below.

 (*i*) Dual-in-Line Package or DIP

 (*ii*) Leadless Chip Carrier or LCC

 (*iii*) Plastic Leaded Chip carrier or PLCC

 (*iv*) Plastic QUAD Flat Package or PQFP

 (*v*) Pin Grid Array or PGA

According to the internal construction and fabrication process involved in the integrated circuits, they are placed in different logic families as follows.

RTL	Resistor-transistor logic
DTL	Diode-transistor logic
TTL	Transistor-transistor logic
ECL	Emitter-coupled logic
I^2L	Integrated-injection logic
MOS	Metal oxide semiconductor
CMOS	Complementary metal oxide semiconductor

The first two, RTL and DTL logic families have only historical significance, since they are seldom used in new designs. RTL was the first commercially available family to have been used extensively. It is included here because it represents a useful starting point to understand the basic operation of digital gates. A TTL circuit is the modification of a DTL and hence, DTL circuits have been gradually replaced by TTL. The operation of TTL will be easier to understand after DTL gates are discussed. These families have a large number of SSI circuits as well as MSI and LSI circuits. I^2L and MOS are mostly used for the construction of LSI functions.

The basic circuit in each digital IC logic family is either a NAND or a NOR gate. Combinational logic functions and more complex functions are generated using this basic circuit, which may be referred to as the primary building block. As an example, an RS latch is constructed from two NAND gates or two NOR gates connected back to back. A master-slave flip-flop is obtained from the interconnection of about ten logic gates. Each of the logic families provides numerous types of ICs that perform different types of logic functions. The differences in the logic functions available from different logic families are not so much in the function that they achieve, but they are different in specific characteristics of the basic building blocks from which the functions are constructed.

Inputs		Outnputs
A	B	F
L	L	H
L	H	H
H	L	H
H	H	L

NAND gate

(a) If *any* input is LOW, the output is HIGH.

(b) If *all* the inputs are HIGH, the output is LOW.

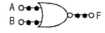

Inputs		Outnputs
A	B	F
L	L	H
L	H	L
H	L	L
H	H	L

NOR gate

(a) If *any* input is HIGH, the output is LOW.

(b) If *all* the inputs are LOW, the output is HIGH.

Figure 11.1

NAND and NOR gates are usually defined by the Boolean functions in terms of binary variables. While analyzing them as electronic circuits, it is more convenient to investigate their input-output relationship in terms of two voltage levels—*high level* (H) and *low level* (L) (refer to section 3.11). Binary variables use the logic values of 1 and 0. When positive logic is adopted, the high-voltage level is assigned the binary value of logic 1 and the low-voltage level is a binary 0. From the truth table of a positive logic NAND gate, we deduce its behavior in terms of high and low levels as indicated in Figure 11.1. The corresponding behavior of the NOR gate is also stated in the same figure. These statements must be remembered because they will be used during the analysis of all gates in this chapter.

11.2 CHARACTERISTICS OF DIGITAL IC

The various digital logic families are usually evaluated by comparing the characteristics of the basic gates of each family. The most important governing parameters or properties of various logic families are listed below.

1. Propagation delay (speed of operation).
2. Power dissipation.
3. Fan in.
4. Fan out.
5. Noise immunity.
6. Operating temperature.
7. Power supply requirement.
8. Current and voltage parameters.

11.2.1 Propagation Delay

Propagation delay is defined as the time taken for the output of a logic gate to change after the inputs have changed. It is the transition time for the signal to propagate from input to output. This factor governs the speed of operation of a logic circuit.

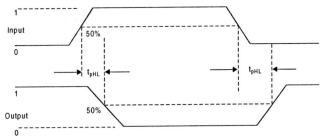

Figure 11.2

A logic signal always experiences a delay in going through a circuit. Two types of propagation delay times are explained by Figure 11.2, which are defined as

(a) t_{pLH} : It is the propagation delay time for a signal to change from logic LOW (0 state) to HIGH (1 state).

(b) t_{pHL} : It is the propagation delay time for a signal to change from logic HIGH (1 state) to LOW (0 state).

The delay times are measured by time lapsed between the 50% voltage levels of the input and the output waveforms while making the transition. In general, the two delays t_{pLH} and t_{pHL}, are not necessarily equal and will vary depending on load conditions. The average of the two propagation delays $(t_{pLH} + t_{pHL})/2$ is called the *average delay* and this parameter is used to rate the circuit. It depends on the switching time of the individual transistors or MOSFETs in the circuit.

11.2.2 Power Dissipation

Power dissipation is the measure of the power consumed by logic gates when fully driven by all inputs. The average power or the DC power dissipation is the product of DC supply voltage and the mean current consumed from that supply.

11.2.3 Fan In

The maximum number of inputs that can be connected to a logic gate without any impairment of its normal operation is referred to as fan in. For example, if the maximum of eight input loads is connected to a logic gate without any degradation of its normal operation, then its fan-in is 8. The parameter determines the functional capabilities of the logic circuit.

11.2.4 Fan Out

Fan out refers to the maximum number of standard loads that the output of the gate can drive without any impairment or degradation of its normal operation. A standard load is defined as the current flowing in the input of a gate in the same IC family. In a logic circuit a logic gate normally drives several other gates and the input current of each of the driven gates must be supplied from the driving gate. The driving gate must be capable of supplying this current while maintaining the required voltage level. Fan out depends on the output impedance of the driving gate and the input impedance of the driven gate. Usually the output impedance of a logic gate is made very low, while input impedance is made very high, so that a logic gate can drive many logic gates.

11.2.5 Noise Immunity or Noise Margin

The term noise denotes any unwanted signal, such as transients, glitches, hum, etc. Noise sometimes causes the change in the input voltage level, if it is too high, and leads to unreliable operation. Noise immunity or the noise margin is the limit of noise voltage that may appear at the input of the logic gate without any impairment of its proper logic operation. The difference between the operating input logic voltage level and the threshold voltage is the noise margin of the circuit.

Noise margin is related with the input-output transfer characteristics of a logic gate, which in turn depends on loading factors, power supply, operating temperature, fabrication process by the manufacturers, etc. These factors affect the signal values and sometimes they change the desired voltage levels causing unreliable operation. Considering these affecting factors, the input-output transfer characteristics of a logic gate must be kept within the maximum and minimum characteristics as specified by the manufacturer, which is shown in Figure 11.3.

Two circled points of the maximum and minimum transfer characteristics of Figure 11.3 are of notable interest. From the maximum characteristics, it is considered that any input voltage level less than $V_{IL}(max)$ is recognized by the logic gate as low-voltage level or logic 0. On the other hand, any input voltage level greater than $V_{IH}(min)$ is recognized as high level or logic 1. Regarding the minimum characteristics, the manufacture specifies that low level or logic 0 output voltage does not exceed $V_{OL}(max)$ and the high level or logic 1 output is always greater than $V_{OH}(min)$.

Hence, two types of noise margins may be derived. The worst-case low-level noise margin is V_{IL}(max) – V_{OL}(max) and the worst-case high-level noise margin is V_{OH}(min) – V_{IH}(min).

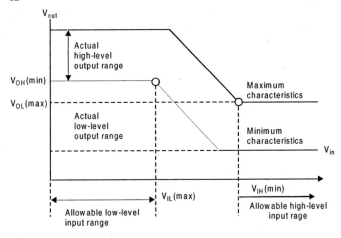

Figure 11.3

11.2.6 Operating Temperature

All the ICs are semiconductor devices and they are temperature sensitive by nature. The operating temperature ranges of an IC vary from 0°C to + 70°C for commercial and industrial application, and from –55°C to 125°C for military application.

11.2.7 Power Supply Requirements

The amount of power and supply voltage required for an IC is one of the important parameters for its normal operation. They are different for different logic families. The logic designer should consider these parameters while choosing the proper power supply.

11.2.8 Current and Voltage Parameters

All the IC manufacturers specify some voltage and current parameters for the input as well as for the output that are very important in designing the digital systems. A designer must consider these parameters as they directly affect the normal operation of the digital system. These are listed below.

High-level input voltage $(V_{IH})[V_{in(1)}]$. It is the minimum voltage level required for a logical 1 at an input.

Low-level input voltage $(V_{IL})[V_{in(0)}]$. It is the maximum voltage level required for a logical 0 at an input.

High-level output voltage $(V_{OH})[V_{out(1)}]$. It is the minimum voltage level available for a logical 1 at an output.

Low-level input voltage $(V_{OL})[V_{out(0)}]$. It is the maximum voltage level available for a logical 0 at an output.

High-level input current $(I_{IH})[I_{in(1)}]$. It is the current flow to an input when high-voltage level corresponding to a logical 1 is applied.

Low-level input current $(I_{IL})[I_{in(0)}]$. It is the current flow to an input when low-voltage level corresponding to a logical 0 is applied.

High-level output current $(I_{OH})[I_{out(1)}]$. It is the current available from the output when the output is considered to be of high-voltage level corresponding to a logical 1.

Low-level output current $(I_{OL})[I_{out(0)}]$. It is the current available from the output when the output is considered to be of low-voltage level corresponding to a logical 0.

11.3 BIPOLAR TRANSISTOR CHARACTERISTICS

It may not be irrelevant to review the characteristics of a bipolar transistor, which finds its application to form the internal structure of the logic gates. This section is devoted for this purpose. The information will be used for the analysis of basic circuits of five bipolar logic families.

A bipolar junction transistor (BJT) is the familiar *npn* or *pnp* junction transistor, and they are constructed either with germanium or silicon semiconductor material. IC transistors, however, are made with silicon. The operation of a bipolar transistor depends upon the flow of two types of carriers—*electrons* and *holes*. In contrast to that, the field effect transistor (FET) is said to be unipolar, as its operation depends on the flow of only one type of majority carrier which may be electrons (n-channel) or holes (p-channel). The first five logic families of the list of the previous section—RTL, DTL, TTL, ECL, and I²L use the bipolar transistors. The last two logic families of the list—MOS and CMOS employ a type of unipolar transistor called a metal-oxide semiconductor field effect transistor, abbreviated MOSFET or MOS in short.

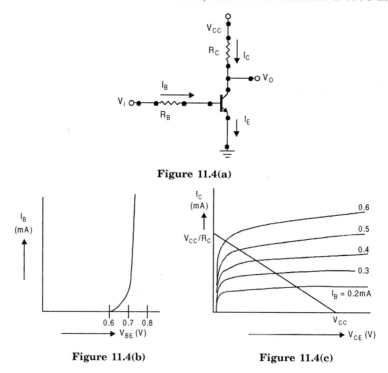

Figure 11.4(a)

Figure 11.4(b)   **Figure 11.4(c)**

The basic data needed for the analysis of digital circuits may be obtained from the typical characteristics curves of a common emitter *npn* silicon transistor, as shown in Figure 11.4. The circuit in Figure 11.4(*a*) consists of two resistors and a transistor, which performs as a logic inverter. The current marked I_C flows through resistor R_C and the collector of the transistor. Current I_B flows through resistor R_B and the base of the transistor. The emitter is connected to the ground and its current $I_E = I_C + I_B$. The supply voltage is between V_{CC} and the ground. The input is between V_i and the ground, and the output is between V_o and the ground.

The normal direction of currents of an *npn* transistor is indicated by the arrow marks in the figure. Collector current I_C and base current I_B are assumed to be positive when they flow into the transistor and the emitter current I_E is positive when it flows out of the transistor. The symbol V_{CE} stands for the voltage difference from collector to emitter and is always positive. V_{BE} denotes the voltage difference across the base to the emitter junction. This junction is termed to be forward biased when V_{BE} is positive and termed as negative biased when V_{BE} is negative.

The base-emitter characteristics of the *npn* transistor as biased as in Figure 11.4(*a*) are given in Figure 11.4(*b*). This is a plot of base current I_B with respect to V_{BE}. For a silicon-type transistor, if the base-emitter voltage is less than 0.6 V, the transistor is said to be *cut-off* and no base current flows. When the base-emitter junction is forward-biased with a voltage greater than 0.6 V, the transistor conducts and its base current I_B starts rising very fast with a very small rise of V_{BE}. The voltage V_{BE} across the base to the emitter of a conducting transistor seldom exceeds 0.8 V.

Figure 11.4(*c*) is the graphical representation of collector-emitter characteristics, together with the load line. When V_{BE} is less than 0.6 V, the transistor is at cut-off and no base current flows ($I_B = 0$) and a negligible current flows in the collector. The collector-to-emitter circuit then behaves like an open circuit. In an active region the collector-to-emitter voltage V_{CE} may be anywhere between 0.8 V to V_{CC}. Approximate collector current I_C in this region can be obtained by the relation $I_C = h_{FE}.I_B$, where h_{FE} is a transistor parameter called *DC current gain*. It should be noted that the maximum collector current does not depend on the I_B, but on the external circuit components connected to the collector. This is because V_{CE} is always positive and its lowest possible value is 0 V (practically minimum V_{CE} value is 0.2 V for silicon transistor). For example, in the inverter circuit shown in Figure 11.4(*a*), the maximum I_C obtained is V_{CC}/R_C, assuming $V_{CE} = 0$.

The collector current-to-base-current relationship $I_C = h_{FE}. I_B$ is valid only when the transistor is at *active* region. The parameter h_{FE} varies widely over the operating range of the transistor, but still it is useful to consider as an average value for the sake of analysis of transistor characteristics. In a typical operating range, h_{FE} may be of the value of 50, but under certain condition it may vary up to the value of 20. It may be observed that the base current may be increased to any desirable value, but the collector current is limited by the external circuit component, as in this case it is limited to V_{CC}/R_C. As a consequence, a situation can be reached when $h_{FE}.I_B$ is greater than I_C. This is the condition when the transistor is said to be in *saturation* region. Thus, the condition for saturation is determined by the relation

$$h_{FE}.I_B \geq I_{CS} \quad \text{or} \quad I_B \geq \frac{I_{CS}}{h_{FE}}$$

where I_{CS} is the maximum collector current flowing during saturation. V_{CE} attains its minimum value at saturation to 0.2V.

The typical values of basic parameters of the transistor characteristics are listed in the table in Figure 11.5, which are useful for the analysis digital circuit. In the cut-off region, V_{BE} is less than 0.6 V, V_{CE} is considered to be an open circuit and both base current and collector current are negligible. In the active region, V_{BE} is about 0.7 V, V_{CE} may vary over a wide range, and I_C can be calculated as a function of I_B. In saturation region V_{BE} hardly changes, but V_{CE} drops to 0.2 V and the base current must be large enough to satisfy the relation $I_B = I_C/h_{FE}$. V_{BE} is considered as 0.7 V when the transistor is in active or saturation region.

Operating Region	V_{BE} (V)	V_{CE} (V)	Current Relationship
Cut-off	< 0.6	Open circuit	$I_B = I_C = 0$
Active	0.6 to 0.7	> 0.2	$I_C = h_{FE}\,I_B$
Saturation	0.7 to 0.8	0.2	$I_B > I_{CS}/h_{FE}$

Figure 11.5

The analysis of digital circuits may be performed by following a prescribed procedure. For each transistor in the circuit where V_{BE} is less than 0.6 V, it is assumed that the transistor is at cut-off region and its collector to emitter is considered to be open circuit. If V_{BE} is greater than 0.6 V, the transistor may be in the active region or in saturation region. The base current is calculated considering $V_{BE} = 0.7$ V and the maximum possible value of the collector current I_{CS} is calculated as $I_{CS} = (V_{CC} - V_{CE})/R_C$, assuming $V_{CE} = 0.2$ V. If the base current is large enough to be $I_B \geq I_{CS}/h_{FE}$, we deduce that the transistor is at saturation region. However, if the base current is smaller and the above relationship is not satisfied, The transistor is said to be in the active region and I_C is recalculated using the relationship $I_C = h_{FE} \cdot I_B$.

To demonstrate with an example, let us consider the inverter circuit of Figure 11.4(a) with the following parameters:

$$R_C = 1K \qquad V_{CC} = 5 \text{ V (voltage supply)}$$
$$R_B = 22K \qquad H = 5 \text{ V (high voltage level)}$$
$$h_{FE} = 50 \qquad L = 0.2 \text{ V (low voltage level)}$$

If the input voltage $V_i = L = 0.2$ V, the transistor is at cut-off as $V_{BE} < 0.6$V. The collector-to-emitter circuit behaves like an open circuit and the voltage available at the output $V_o = H = 5$V.

When the input voltage $V_i = H = 5$ V, $V_{BE} > 0.6$ V. Assuming $V_{BE} = 0.7$V, base current is calculated as

$$I_B = \frac{V_i - V_{BE}}{R_B} = \frac{5 - 0.7}{22K} = 0.195 \text{ m A.} \tag{11.1}$$

Considering $V_{CE} = 0.2$ V, the maximum collector current

$$I_{CS} = \frac{V_{CC} - V_{CE}}{R_C} = \frac{5 - 0.2}{1K} = 4.8 \text{ mA.} \tag{11.2}$$

Now, $\quad I_{CS}/h_{FE} = 4.8 \text{ mA}/50 = 0.096 \text{ mA} \tag{11.3}$

Comparing Equations 11.1 and 11.3, we observe that 0.195 mA > 0.096 mA and hence,

$$I_B > I_{CS}/h_{FE}.$$

Therefore the transistor is at saturation region when V_i is high (H) and output

$V_o = V_{CE}(saturation) = 0.2 \text{ V} = \text{L}.$

Thus it is proved that the circuit behaves like an inverter.

It should be noted that the transistor in the inverter circuit above presents the desired output at only two of its operating conditions. Output is H, when input is L and transistor is at cut-off. On the other hand, output is L when input is H and the transistor is at saturation. The active region operation of the transistor is bypassed and hence this type of logic operation is referred to as *saturated logic*. Most of the bipolar logic families adopt the saturated logic operation. The procedure just described will be used extensively during the analysis of various digital circuits in the following sections.

There are occasions where not only transistors but also diodes are used in digital circuits. An IC diode is usually constructed from a transistor with its collector connected to the base as shown in Figure 11.6(a). The graphical symbol of a diode is shown in Figure 11.6(b). The diode behaves like a base-emitter junction of a transistor. A graphical plot of current I_D through the diode against its forward voltage drop V_D is shown in Figure 11.6(c), which is exactly identical to the transistor base characteristics. We can consider that a diode is off or nonconducting if its forward voltage V_D is less than 0.6V. When the diode conducts, current I_D flows in the direction as shown in Figure 11.6(b) and V_D stays at about 0.7 V. An external resistor must always be provided to limit the current in a conducting diode.

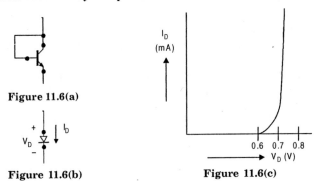

Figure 11.6(a)

Figure 11.6(b)

Figure 11.6(c)

11.4 RESISTOR-TRANSISTOR LOGIC (RTL)

Resistor-transistor logic or RTL is the first-generation digital logic circuit. The basic circuit of the RTL digital logic family is the NOR gate as shown in Figure 11.7. Each input is associated with one resistor and a transistor. The collectors of the transistors are tied together with a common resistor to the V_{CC} supply. The output is taken from the collectors joint. The voltage levels of the circuit are 0.2 V for low level and 1 to 3.6 V for high level.

Analysis of an RTL gate is simple. If any of the inputs is at high level, the corresponding transistor is at saturation. This causes the output at low irrespective of the conditions of other transistors, as all the transistors are connected in parallel. If all the inputs are at low level at 0.2 V, all the transistors are at cut-off condition because base-to-emitter voltage of all the transistors $V_{BE} < 0.6$ V, causing the output of the circuit at high level approaching the value of the supply voltage V_{CC}. Thus confirms the conditions of a NOR logic. Note that the noise margin for low signal input is $0.6 - 0.2 = 0.4$ V.

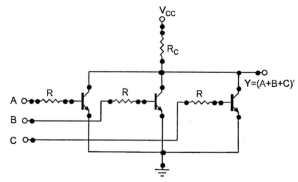

Figure 11.7

The RTL logic circuit has many drawbacks. The base current is practically independent of the emitter junction characteristics. The resistors increase the input resistance and reduce the switching speed of the circuit. This degrades the rise and fall times of any input pulse. Reduction in base resistors reduces the input resistance, increases power consumption, and decreases the fan in. An approach used in practice to increase the speed of RTL circuits is to connect speed-up capacitors parallel to the base resistors. In an RTL digital circuit the transistors are driven heavily to saturation resulting in long turn-off delays. The output high-voltage level reduces with the increase of load or the number of gates connected at output. Also, the collector reverse saturation current of a driver transistor at high temperature may become large enough to lower the already low output voltage.

The fan out of the RTL gate is limited by the value of the output voltage when high. As the output is loaded with inputs of other gates, more current is consumed by the load. This current must flow through R_C (typical value of R_C is 640 Ω). Assuming h_{FE} drops to the value of 20, with each of the base resistor value R = 450 Ω and V_{CC} = 3.6 V, the output voltage drops to 1 V when the fan out is 5. Any voltage below 1 V at the output may not drive the next transistor into saturation as required.

The following are the characteristics of the RTL family.

1. Speed of operation is low, *i.e.*, the propagation delay is high up to the order of 500 ns. It cannot operate at more than 4 MHz.

2. Fan out is 4 or 5 with a switching delay of 50 ns and fan in is 4.

3. Poor noise immunity.

4. High average power dissipation. Elimination of base resistors in RTL will reduce the power dissipation, which results in Direct-coupled Transistor Logic (DCTL).

5. Sensitive to temperature.

The RTL logic circuit of a NOR gate has been discussed here with three inputs. The working principle is similar for any number of inputs.

11.4.1 Resistor Capacitor Transistor Logic (RCTL)

RCTL circuit is an improvement over the RTL digital circuit, which employs a capacitor in parallel with each of the input resistors to increase speed and to improve noise immunity. The basic RCTL NOR gate circuit is shown in Figure 11.8.

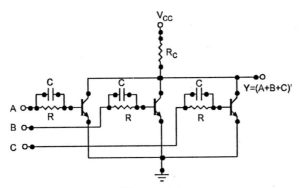

Figure 11.8

During the switching of logic levels the capacitor bypasses the resistor, resulting in the generation of base currents and the input capacitance discharges more quickly. The use of the capacitors allows employing higher values of resistors and thus lowering the power dissipation per gate. In comparison to the RTL family, the propagation delay of the RCTL circuit is less, although the fan in and fan out are the same. The main drawback of the RCTL circuit is that it is very difficult to fabricate the pn junction capacitor and it occupies a large area. Also, the RCTL circuit is not ideal for fabrication because it includes a high proportion of resistors and capacitors.

11.5 DIODE TRANSISTOR LOGIC (DTL)

The DTL family eliminates the problem of decreasing the output voltage with the increase of load. The basic circuit of a DTL NAND gate is shown in Figure 11.9. Each input is associated with one diode. The input diodes D_A, D_B, D_C, and resistor R_D form an AND gate. The transistor Q1 serves as a current amplifier as well as a digital INVERTER. The two voltage levels are 0.2 V for the low level and 4 V for the high level.

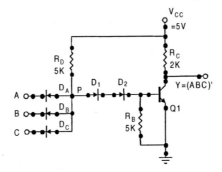

Figure 11.9

If any input of the gate is low to 0.2 V, the corresponding diode is forward biased and conducts current through V_{CC} and R_D. The voltage at point P is equal to the input voltage 0.2 V plus one diode drop of 0.7 V, for a total of 0.9 V. This is not sufficient to drive the transistor Q1 into conducting. In order for the transistor to conduct the voltage at P it must overcome a potential of one V_{BE} (0.6 V) drop in Q1 and two diode voltage drops (0.6 V each) across D1 and D2, or 0.6 + 2 × 0.6 = 1.8 V. Hence, the transistor Q1 remains at cut-off condition and its collector to emitter behaves like an open circuit. Therefore, the output voltage at collector of the transistor Y is high at 5 V (power supply V_{CC} is 5 V).

If all the inputs of the gate are high to 5 V, all the diodes are reverse biased and the current will flow through R_D, D1, D2, and the base of the transistor. The transistor is now driven to saturation region. The voltage at P is equal to one V_{BE} drop plus two diode drops across D1 and D2, or 3 × 0.7V = 2.1 V (V_{BE} drop is 0.7 V while conducting and forward biased diode drop becomes 0.7 V). This conforms that all the input diodes are reverse biased and off. With the transistor at saturated condition, the output drops to V_{CE}(sat) = 0.2 V, which is low level for the gate. Thus the DTL circuit operation as above conforms NAND logic gate behavior. The working principle is similar to any number of inputs.

The DTL family has better noise margin, higher fan out capability, and faster response than the RTL family. The DTL family has the following characteristics.

1. The propagation delay of a DTL circuit is in the order of 30 ns. The turn-off delay is considerably larger than the turn-on delay by a factor of 2 or 3.

2. Fan out is as high as 8, because the input impedance is high for DTL gates due to reverse biased input diodes at logic 1 states.

3. Fan in is at the order of 8.

4. Noise margin is high due to two diodes D1 and D2 connected in series. Typically the noise margin of a DTL NAND gate circuit is 0.8 V when the output is low and 3.4 V when the output is high.

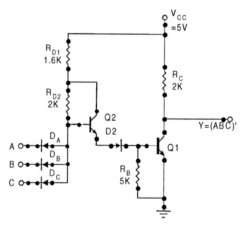

Figure 11.10

The fan out of the DTL circuit can be improved by replacing one of the diodes in the base circuit D1 with a transistor Q2 as shown in Figure 11.10. The biasing of the transistor Q2 is modified with resistors R_{D1} and R_{D2} as shown in the figure. The transistor Q2 is maintained at active region when the output transistor Q1 is at saturation. As a consequence, the modified circuit can supply a larger amount of base current to the output transistor, due to that R_{D1} is made smaller compared to R_D of a DTL circuit. The output transistor Q1 can draw a larger amount of collector current, before it goes to saturation. Part of the collector current comes from the conducting diodes of the loading gates, when

Q2 is in saturation. Thus, an increase in allowable collector current allows more loads to be connected at the output, and therefore increases the fan out capability of the gate.

11.5.1 High Threshold Logic (HTL)

There are several applications where the digital circuits operate in an environment that produces very high noise signals. For operation in such surroundings, a type of DTL gate is available that possesses a high threshold to noise immunity. This type of gate is called a *high threshold logic* or HTL gate.

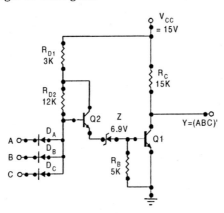

Figure 11.11

The HTL gate is the modification over the DTL gate of Figure 11.10. An HTL gate is shown in Figure 11.1. The normal diode D1, as used in DTL, has been replaced by a zenner diode (Z) in HTL. The supply voltage has been changed to 15 V and the resistor values have been modified accordingly to maintain an equal current with DTL. The zenner diode has the characteristics of maintaining a constant voltage of 6.9 V when reverse biased.

In order for output transistor Q1 to conduct, the emitter of Q2 must rise to a potential of one V_{BE} plus fixed zenner voltage of 6.9 V, that means a total of about 7.5 V. With a low-level input of 0.2 V, the base of Q2 is at 0.9 V and so Q1 is at cut-off. The noise signal must be higher than 7.5 V to change the state of Q1. With all inputs at high level to 15 V, sufficient voltage and current are available at the base of Q1 to drive it to saturation. On the other hand, the noise signal must be greater than 7.5 V in the opposite side to turn the transistor off. Thus the noise margin of the HTL gate is about 7.8V for both voltage levels.

HTL gates are quite useful in the industrial environment where the noise level is usually high due the presence of motors, high-voltage switches, relays, circuit breakers, etc.

11.6 TRANSISTOR TRANSISTOR LOGIC (TTL)

TTL is the most popular of all the logic families. The original basic TTL gate was a slight improvement over the DTL gate. As the TTL technology progressed, more and more additional improvements were carried out to make this logic family the most widely used type in the design of digital systems. Gates of this family possesses the highest switching speed when compared to other logic families that utilize the saturated transistors. TTL family, or the commercially available 74/54 series, evolved into five major divisions.

(*i*) Standard TTL (74/54 series)

(*ii*) High-speed TTL (74H/54H series)

(*iii*) Low-power TTL (74L/54L series)

(*iv*) Schottky diode clamped TTL (74S/54S series)

(*v*) Low-power schottky TTL (74LS/54LS series)

Although the high-speed and low-power TTL devices are designed for specific applications, all the groups of the family have several common features, and are compatible and capable of interfacing directly with one another. They have the following typical characteristics in common.

(*i*) Supply voltage is 5V.

(*ii*) Logic 0 output voltage level is 0 V to 0.4 V.

(*iii*) Logic 1 output voltage level is 2.4 V to 5 V.

(*iv*) Logic 0 input voltage level is 0 V to 0.8 V.

(*v*) Logic 1 input voltage level is 2 V to 5 V.

(*vi*) Noise immunity is 0.4 V.

But the five different TTL series as mentioned above differ from one another in terms of propagation delay and power dissipation values. Speed-power product is an important parameter for comparing the basic gates. This is a product of the propagation delay and the power dissipation measured in picojoules (pJ). A low value of this parameter is desirable in designing the digital circuit, because it indicates a low propagation delay without excessive power consumption or vice versa. Figure 11.12 provides a table comprising of the typical values of propagation delay, power consumption, speed-power product, maximum operating frequency, and fan out for different TTL series.

Name of the series	Abbreviation	Propagation delay (ns)	Power dissipation (mW)	Speed-power product (pJ)	Maximum clock rate (MHz)	Fan out
Standard TTL	TTL	10	10	100	35	10
Low-power TTL	LTTL	33	1	33	3	10
High-speed TTL	HTTL	6	22	132	50	10
Schottky TTL	STTL	3	19	57	125	10
Low-power Schottky TTL	LSTTL	9.5	2	19	45	10

Figure 11.12

The standard TTL gate was the first version of TTL family. The basic gate was constructed with different resistor values to produce gates with lower power dissipation or higher speed. The propagation delay of saturated logic family largely depends upon two factors—storage time and RC time constants. Reducing the storage time decreases the propagation delay and also reducing the resistor values in the circuit reduces the RC time constant, which in turn reduces the propagation delay. But reduction in resistor values causes higher power consumption. The speed of the gate is inversely proportional to the propagation delay.

All TTL versions are available in SSI packages and in more complex forms as MSI and LSI functions. The differences in the TTL versions are not in the digital functions that they perform, but rather in the values of resistors and type of transistor that are used to form their basic gates. In any case, TTL gates in all the versions come in three different types of output configurations as below.

1. Open collector output configuration.
2. Totem pole output configuration.
3. Tristate or three-states output configuration.

11.6.1 Standard TTL with Open Collector Output Configuration

The basic TTL NAND gate is shown in Figure 11.13, which is the modified circuit of a DTL gate. Q1 is a multiple emitters transistor and the logic inputs are applied to the emitters of Q1. These emitters behave like the input diodes in the DTL gate, as they form pn junction with their common base. The base-collector junction of Q1 acts like another pn junction diode, equivalent to the diode D1 of the DTL gate (refer to Figure 11.9). The transistor Q2 replaces the second diode D2 of the DTL gate. The output of the TTL gate is taken from the open collector of Q3. A resistor must be connected externally at the collector of Q3 to V_{CC}, to maintain the output voltage level to high when Q3 is at cut-off. The external resistor is termed as a *pull-up* resistor.

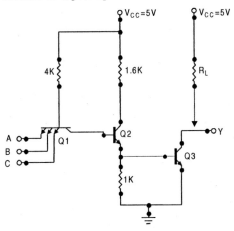

Figure 11.13

The two voltage levels of the TTL circuit are 0.2 V for the low level and 2.4 V to 5 V for the high level. If any input is low, the corresponding base emitter junction of Q1 becomes forward biased. The voltage level at the base of the transistor Q1 is 0.2 V plus one V_{BE} drop of 0.7 V, *i.e.*, 0.9 V. This voltage level is not sufficient to drive the transistor Q2 and Q3, and they are at cut-off condition. The voltage required at the base of Q1, to drive Q2 and Q3 into saturation should be V_{BE} of Q3 plus V_{BE} of Q2 plus one pn junction diode drop of Q1, *i.e.*, 0.7 V + 0.7 V + 0.7 V = 2.1V. When the output transistor Q3 cannot conduct, the output voltage level at Y will be high if any external resistor R_L is connected to V_{CC}.

When all the inputs are high, no base emitter junction of Q1 is forward biased and voltage at the base of Q1 is higher than 2.1 V. Hence transistor Q2 is driven to saturation as well as Q3, provided it has the current through the collector. The collector current may be available from the external pull-up resistance or from the connected loads at the output.

The output voltage at Y is V_{CE} (saturation) *i.e.*, 0.2 V. Thus the gate operation conforms the NAND function, as when any of the inputs is low, output Y is high and if all the inputs are high, output Y is low.

In the above analysis, it has been stated that the base-collector junction of Q1 acts like a pn diode junction. This is true in the steady state condition. However, during the turn-off transition, Q1 does exhibit transistor action resulting in a reduction in propagation delay. When one of the inputs is brought to low level from an all high inputs condition, both Q2 and Q3 are turning off. At this time the collector junction of Q1 is reverse biased and the emitter junction is forward biased. Hence the transistor momentarily goes into the active region. The collector current of Q1 comes from the base of Q2 and quickly removes the excess charge stored in Q2 during its previous saturation state. This results in the reduction in the storage time of the circuit as compared to the DTL type of input and also results in the reduction of the gate turn-off time of the gate.

You may notice that the TTL gate with an open collector output configuration can be operated without using any external resistor when connected to the inputs of other TTL gates. However, this is not recommended because noise immunity becomes low. Without an external resistor, the output of the gate will be an open circuit when Q3 is at cut-off. An open circuit to an input of a TTL gate behaves like a high-level input, but a very small amount of noise may change this to a low level. When Q3 conducts, its collector current will be available from the input of the loading gate through V_{CC}, the 4K resistor, and the forward-biased base-emitter junction.

The open collector output configuration has many useful applications. The output may be interfaced with another circuit that has a different supply voltage. The external resistor may be selected of a suitable value according to the supply voltage it is connected to. This facilitates to drive a lamp or a relay which may have a supply voltage other than 5 V used for TTL, directly from the open collector gate as shown in Figure 11.14. When the output transistor is off, no current flows through the lamp or relay and it remains off. When the output transistor Q3 is on, current path is available for the lamp or relay to make it on. Also, the open collector output gates can be used for interfacing with gates of another family like CMOS, where supply voltage varies from 3 V to 15 V.

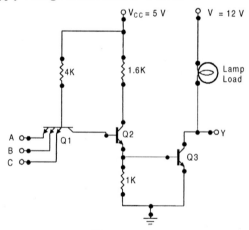

Figure 11.14

If the outputs of several open collector TTL gates are tied together with a single external pull-up resistor, a wired-AND logic is achieved. Remembered that a positive wire-AND function gives high level if all the outputs of the gates are high. This is due to that all the output transistors are at cut-off making the wired logic high. If any output transistor conducts, it forces the output to a low state.

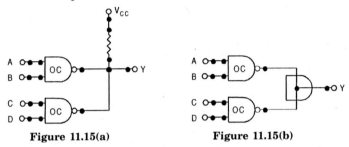

Figure 11.15(a)   **Figure 11.15(b)**

The wired logic function is illustrated in Figure 11.15. The physical wire connection is shown in Figure 11.15(*a*). Two open collector NAND gates are tied together and connected to a common resistor to V_{cc}. The graphic symbol of a wired-AND function is demonstrated in Figure 11.15(*b*). The AND gate drawn with the lines going through the center of the gate is distinguished from the conventional gate and symbolized as wired-AND. The Boolean expression obtained from the circuit of Figure 11.15 is the AND operation between the outputs of two NAND gates.

$$Y = (AB)'.(CD)' = (AB + CD)'$$

The second expression is preferred, as it is the commonly used AND-OR-INVERT expression.

Another important application of the open collector gates is to form a common bus by tying several such gates together. At any time, all the gates, except one, must be maintained in their output state. The selected gate output may be either high or low according to its logic information to be transmitted. A control circuit is provided to select the particular gate that drives the bus at any given time.

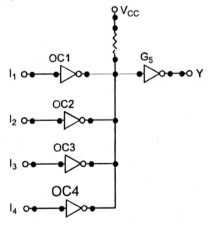

Figure 11.16

Formation of the common bus line with four sources tying together is demonstrated in Figure 11.16. (The control circuit mentioned above is not shown here.) Each of the four sources drives an open-collector inverter, and the outputs of the inverters are tied together with a common pull-up resistor to V_{CC} to form a single bus line. If any data is transmitted from the source I_1, then all the other sources (I_2 to I_4) are logic 0, so that they produce logic 1 on the bus line. The input I_1 can now transmit information through the common bus line to the output inverter G_5. Note that the AND operation is performed on the common bus by the wired logic. If I_1 is 0, its corresponding inverter OC1 produces logic 1 and so common bus has logic 1, because other open collector gates outputs are 1, hence output Y is 0. And if I_1 is logic1, the common bus is logic 0 and so Y is logic 1.

11.6.2 Standard TTL with Totem Pole Output Configuration

The TTL NAND gate with *totem pole* output configuration is shown in Figure 11.17. It is the same circuit as the open-collector gate, except for the output transistor Q4, a diode D1, and resistor 130Ω at the collector of Q4. It is called the *totem pole* output configuration, because the transistor Q4 sits upon Q3. The base of the transistor Q4 is driven from the collector of Q2, as shown in Figure 11.17.

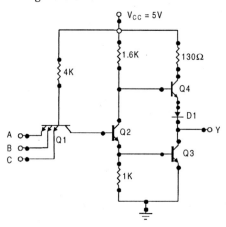

Figure 11.17

The two voltage levels of the TTL circuit are 0.2 V for the low level and 2.4 V to 5 V for the high level. If any input is low, the corresponding base emitter junction of Q1 becomes forward biased. The voltage level at the base of the transistor Q1 is 0.2 V plus one V_{BE} drop of 0.7 V, *i.e.*, 0.9 V. This voltage level is not sufficient to drive the transistor Q2 and Q3, and they are cut-off condition. The voltage required at the base of Q1, to drive Q2 and Q3 into saturation should be V_{BE} of Q3 plus V_{BE} of Q2 plus one pn junction diode drop of Q1, *i.e.*, 0.7 V + 0.7 V + 0.7 V = 2.1V. When Q2 and Q3 are off, high base current available for Q4 to operate and the output Y is logic high. The currents for the output loads or the gates connected at the output are supplied through transistor Q4 and its collector resistor 130Ω.

When all the inputs are high, no base-emitter junction of Q1 is forward biased and voltage at the base of Q1 is higher than 2.1 V. Hence, transistors Q2 and Q3 are driven to saturation. The output voltage at Y is V_{CE} (saturation) *i.e.*, 0.2 V. The voltage at the collector of Q2 is equal to one V_{BE} drop of Q3 plus one V_{CE} (saturation) drop of Q2, *i.e.*, 0.7 V + 0.2

V = 0.9 V. This voltage level is applied to the base of Q4 and is not sufficient to drive the transistor Q4. Since, to drive the transistor Q4, the voltage required at its base is one VCE (saturation) for Q3 plus one diode drop against D1 plus one V_{BE} drop of Q4, *i.e.*, 0.2 V + 0.7 V + 0.7 V = 1.6 V. Hence Q4 is at cut-off condition. While Q3 is in saturation, its collector current is available from the connected loads at the output. Thus the gate operation confirms the NAND function, as when any of the inputs is low, output Y is high and if all the inputs are high, output Y is low.

The output impedance of a gate is normally a resistive plus a capacitive load. The capacitive load consists of the capacitance of the output transistor, the capacitance of any of the fan out gates and any stray capacitance. When the output changes from low state to high state due to a change in the input condition, the output transistor goes from saturation to cut-off state. As soon as the transistor Q2 turns off and Q4 conducts because its base is connected to V_{CC} and the total load capacitance, C charges exponentially from the low- to high-voltage level with the time constant of RC, where R is the resistance value at the collector of the output transistor Q3 to supply voltage V_{CC}, which is also referred to as a pull-up resistor. For totem pole configuration as above, a network consisting of a transistor Q4, diode D1, and resistor 130Ω, is connected at the collector of the output transistor Q3 and is referred to as an *active pull-up* circuit. The active pull-up circuit offers low output impedance and hence the rise time of output is faster and in turn, the propagation delay is reduced to the order of 10 ns. Another advantage of the active pull-up circuit is that because of its low output impedance, it is capable of supplying more currents to the driven gate, and so the fan out capability increases.

As the capacitive load charges, the output voltage raises and the current in Q4 decreases, bringing the transistor into the active region. Thus, in contrast to the other transistors, Q4 is in the active region when a steady state condition is reached. The final value of the output voltage is 5V minus one V_{BE} drop of Q4, minus one diode drop of D1, *i.e.*, 5 V – 0.7 V – 0.7 V = 3.6 V. The transistor Q3 goes into cut-off very fast, but during the initial transition time both Q3 and Q4 are on and a peak current is drawn from the power supply. This current spike generates noise in the power supply distribution system. If the change of state is frequent, the transient current spikes increase the power supply current requirement and the average power dissipation of the circuit increases.

A wired logic connection like open-collector gates is not allowed with totem pole output configuration. When two totem poles are wired together with the output of one gate high and the output of other gate is low, an excessive amount of current will be drawn to produce heat and this may cause damage to the transistors in the circuit. Some TTL gates are constructed to withstand the amount of current that is produced under this condition. In any case, the collector current in the low gate may be high enough to move the transistor into the active region and produce an output voltage in the wired connection greater than 0.8 V, which is not a valid binary signal for TTL gates.

11.6.3 TTL Gate With Tri-state Output Configuration

As mentioned in the previous section that the outputs of the TTL gates with totem pole structures cannot be connected together in wired-AND fashion as in the open collector gates. There is, however, a special type of totem pole gate—the *tri-state gate*—that allows the wired connections at the outputs for the purpose of forming a common bus system.

We have seen that all the logic gates have two output states—logic 0 and logic 1. But the tri-state or three-state gate, as its name implies, has three output states as follows.

1. A low-level state or logic 0 state, when the lower transistor in the totem pole is on and the upper transistor is off.

2. A high-level state or logic 1 state, when the lower transistor in the totem pole is off and the upper transistor is on.

3. A third state when both transistors in the totem pole are off. This provides an open circuit or high impedance state which allows the direct wired connection of many outputs on a common line. Three states eliminates the need of open collector gates in common bus configurations.

Input	Enable	Output
1	0	1
0	0	0
X	1	Tristate

Figure 11.18(a) **Figure 11.18(b)**

Input	Enable	Output
1	0	0
0	0	1
X	1	Tristate

Figure 11.19(a) **Figure 11.19(b)**

The graphic symbol of a tri-state buffer gate and inverter are shown in Figures 11.18(a) and 11.19(a) respectively, and their truth tables are in Figures 11.18(b) and 11.19(b) respectively. Tri-state gates consists of an extra input called a control or enable input. Here, the control input is such that when it is logic 0, the gate performs its normal operation. When the control input is logic 1, the output of the gate goes to tri-state or high impedance state regardless of the value of input A. Other types of logic gates are also available with control input and tri-state output. Gates are also available with reverse control logic, *i.e.*, they perform their normal operation for control input of logic 1 and tri-state output for control input of logic 0.

The circuit diagram of a tri-state two-input NAND gate is shown in Figure 11.20. A and B are the data input and C is used as the control input. As compared to the standard TTL gate with totem pole output, the transistors Q6, Q7, and Q8 are introduced in a tri-state gate circuit, which is associated with the control input to form an open-circuit-type inverter and is connected to the collector of Q2 through diode D1. Transistors Q1 to Q5 associated with the data inputs form a totem pole NAND gate. When control input C is low, the base-emitter junction of Q6 is forward biased and sufficient base current as well as base voltage are not available to make Q7 on. This in turn makes Q8 off and no current will flow through the collector of Q8 and diode D1. Thus there is no potential effect at the collector of Q2 and the circuit performs a NAND operation on data from A and B.

When control input C is high, The base-emitter junction of transistor Q6 is reverse biased and the base-collector junction of it is forward biased. Thus voltage level at the base of Q7 is higher and sufficient current is available to make the transistor Q7 on and hence the transistor Q8 also becomes on and driven to saturation. The current will flow though the collector of Q8 and the diode D1. Therefore, the voltage level at the collector of Q2 is one diode drop plus one V_{CE} (saturation) drop of transistor Q7, *i.e.*, 0.7 V + 0.2 V = 0.9 V. This voltage level keeps the transistors Q5 and Q4 off, which requires at least two V_{BE}

drop *i.e.*, 1.4 V to turn them on. At the same time the saturation state of the transistor Q8 exhibits low input to one of the emitters of Q1 forcing the transistors Q2 and Q3 to turn off. As a result, both the totem pole transistors Q3 and Q4 are turned off and the output of the circuit behaves like an open circuit with a very high output impedance. The circuit operation may be tabulated to form a truth table as in Figure 11.21, which confirms the behavior of a NAND gate with a control input.

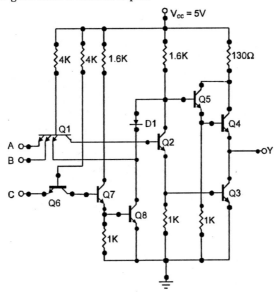

Figure 11.20

Control input	Data input		Output
C	A	B	Y
0	0	0	1
0	0	1	1
0	1	0	1
0	1	1	0
1	X	X	High impedance

Figure 11.21

Note that there is little difference between the standard totem pole TTL and the tri-state TTL at the output totem pole section of the circuit. In the standard totem pole configuration a transistor and a diode were used at the upper part, which have been replaced by two transistors Q4 and Q5 in darlington mode in tri-state output configuration. A darlington pair of transistors provides a very high current gain and extremely low impedance. This is exactly required during low- to high-voltage level swing at the output, resulting in a reduction of propagation delay. This circuit also takes care of the diode drop as well as increases in the current output capability.

A three-state bus can be created by wiring several three-state outputs together. At any given time, control input of only one single gate is enabled to transmit its information on the common output bus, and control inputs for all other gates must be disabled at that time instant to keep their outputs at a high impedance state. Extreme care must be taken to select the control input. If two or more control inputs are enabled at the same time, the undesirable condition of two or more active totem pole outputs will arise.

An important feature of most tri-state gates is that the output enable delay is longer than the output disable delay. If a control circuit enables one gate and disables another gate at the same time, then the disable gate enters the high impedance state before the enable gate comes to its action. This eliminates the undesirable situation of both gates being active at the same time.

There is a very small leakage current associated with the high impedance state condition in a tri-state gate. However, this current is so small that as many as 100 tri-state gates can be connected together to form a common bus line without degrading logic behavior of the gates.

11.6.4 Schottky TTL Gates

As mentioned earlier, a reduction in storage time results in the reduction of propagation delay. This is due to the time needed for a transistor to come out of its saturation condition delays the switching of the transistor from the on saturation condition to cut-off condition. Saturation condition can be eliminated by placing a *Schottky diode* between the base and the collector of each saturated transistor in the circuit as shown in Figure 11.22. The Schottky diode is formed by the junction of a metal and a semiconductor, in contrast to the conventional *pn* diode which is formed by the junction of *p*-type and *n*-type semiconductors. The forward-biased voltage across the Schottky diode is 0.4 V as compared to .7 V in a conventional diode. The presence of a Schottky diode between the base and collector prevents the transistor from driving into saturation. A transistor with a Schottky diode is referred to as a *Schottky transistor*. Figure 11.22 shows the graphic symbols of a Schottky diode and a Schottky transistor. The use of Schottky transistors in TTL circuits results in the reduction in propagation delay without sacrificing the power dissipation.

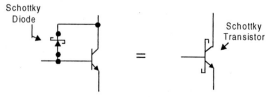

Figure 11.22

A two-input Schottky TTL NAND gate circuit is shown in Figure 11.23. With comparison to the standard TTL gate, all the transistors of the Schottky TTL circuit are of Schottky type, except Q4. Exception is made because the transistor Q4 does not go to the saturation region but remains at active region. It should be noted that the resistor values have been reduced to further decrease in the propagation delay.

In addition to employing the Schottky transistors and reducing the circuit resistor values, the Schottky TTL circuit in Figure 11.23 includes other modifications over the standard TTL circuit in Figure 11.17. Two new transistors, Q5 and Q6, have been introduced and Schottky diodes are provided at each of the input terminals to ground. The transistors

Q5 and Q4 are in darlington mode taking care of the diode V_{BE} drops, higher current gain and current output capability, low output impedance and reduction in propagation delay as already explained in the previous section.

The diodes at each input terminal as shown in the circuit are provided to prevent any ringing that may occur in the input lines. Under transient switching conditions, the signal lines appear inductive. This, along with the stray capacitance of the circuit, cause signals to oscillate or ring. When the output of a gate changes its level from high to low, the ringing waveform at the input of the connecting gate may have the excursions below ground as high as 2 to 3 V depending on the line length. The diodes connected to the ground help to clamp the ringing as they conduct when the negative voltage exceeds 0.4 V. When the negative excursion is limited, the positive swing also becomes limited and thus reduces the ringing as well as unwanted switching of the gate.

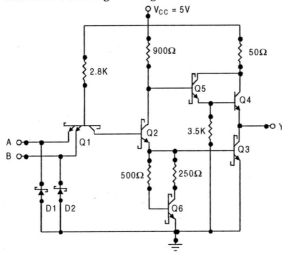

Figure 11.23

The emitter circuit of Q2 in Figure 11.17 has been modified in Figure 11.23 by a circuit consisting of a transistor Q6 and two resistors. The turn-off current spikes are reduced due to this modification, which also helps to reduce the propagation time of the gate.

11.6.5 TTL Parameters

TTL devices work reliably over a supply voltage range of 4.75 V to 5.25 V and within the temperature range of 0 to 70°C. However, some TTL devices are specially manufactured for the operating temperature range of –55°C to +125°C for military application.

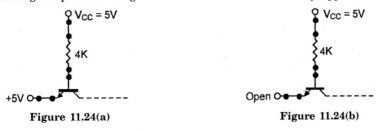

Figure 11.24(a)   **Figure 11.24(b)**

Floating inputs. When a TTL input is at high level or logic 1 (ideally +5V) as shown in Figure 11.24(a), the emitter current is zero. If the TTL input is unconnected or floating as shown in Figure 11.24(b), there is no flow of emitter current because of the open circuit. Hence, a floating input may be considered as high input.

Also, when the input terminal of a TTL gate is left open, it acts like a small antenna and picks up stray electromagnetic noise resulting in malfunctioning or undesirable operation of the gate. Therefore, it is mandatory to terminate the unused input terminals with either V_{CC} or ground depending on the gate function. For example, the unused terminals of an AND gate or NAND must be connected to V_{CC}, whereas the unused terminals of an OR gate or NOR gate should be connected to the ground.

Current sourcing and current sinking. When the output of a gate is high, it provides current to the input of the gate being driven, and in this situation, the output is said to act as the *current source*. For a TTL circuit the maximum current drawn by an input from a high output is 40 µA.

When the output of a TTL gate is at low level, it acts as a *current sink*, because it sinks current from the gate inputs, which are driven to low. In a standard TTL gate, when one of its inputs becomes low, the typical value of current that flows out of the device is 1.6 mA. (Refer to Figures 11.25(a) and 11.25(b).) Thus,

$$I_{IL} \text{ (maximum)} = -1.6 \text{ mA} \qquad \text{and}$$

$$I_{IH} \text{ (maximum)} = +40 \text{ µA}.$$

Here, the negative sign indicates that the current flows out of the device.

Standard loading. A TTL device can act as the source of the current when its output is high and it can sink the current when the output is low. The standard TTL datasheets specify that the TTL devices can sink up to 16 mA, denoted by $I_{OL}(\text{max}) = 16$ mA, and can source up to 400 µA, denoted by $I_{OH}(\text{max}) = -400\text{µA}$. The negative sign denotes that the current is flowing out of the device. Since the maximum output current capabilities—$I_{OH}(\text{max})$ and $I_{OL}(\text{max})$—are ten times larger than the input currents $I_{IH}(\text{max})$ and $I_{OH}(\text{max})$, up to 10 TTL gates can be connected at the output of a TTL gate.

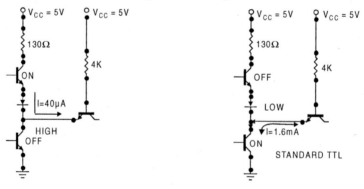

Figure 11.25(a)   Figure 11.25(b)

Fan out. The maximum number of TTL loads that can be driven by a TTL driver is called fan out. As discussed above 10 numbers of standard TTL loads can be connected to the output of a standard TTL gate. Thus the fan out of a standard TTL is 10. When the totem pole output of a standard TTL goes high, it reverse biases another gate input with the resulting current of 40 μA maximum, as shown in Figure 11.25(a). While standard TTL output goes low, it must sink the current of 1.6 mA from a standard TTL gate as shown in Figure 11.25(b). But in case of a low-power Schottky TTL the sink current is 0.36 mA. Using the standard unit as a reference, one unit load is then the same as the current of 1.6 mA into a low input. Since the standard TTL output drive is capable of sinking 16 mA of current, it can drive up to 10 loads.

For low-power TTL,

$$I_{IL}(\text{max}) = -0.18\text{mA}, \quad I_{IH}(\text{max}) = 10 \ \mu\text{A}$$
$$I_{OL}(\text{max}) = 3.6 \text{ mA}, \quad I_{OH}(\text{max}) = -200 \ \mu\text{A}.$$

Considering high-output state, $I_{OH}(\text{max})/I_{IH}(\text{max}) = 200 \ \mu\text{A}/10 \ \mu\text{A} = 20$.

Considering low-output state, $I_{OL}(\text{max})/I_{IL}(\text{max}) = 3.6\text{mA}/0.18\text{mA} = 20$.

Therefore, 20 numbers of low-power TTL gate can be connected to the output of another low-power TTL gate or fan out is 20.

For low-power Schottky TTL,

$$I_{IL}(\text{max}) = -0.36 \text{ mA}, \quad I_{IH}(\text{max}) = 20 \ \mu\text{A}$$
$$I_{OL}(\text{max}) = 8 \text{ mA}, \quad I_{OH}(\text{max}) = -400 \ \mu\text{A}.$$

Considering high-output state, $I_{OH}(\text{max})/I_{IH}(\text{max}) = 400 \ \mu\text{A}/20 \ \mu\text{A} = 20$.

Considering low-output state, $I_{OL}(\text{max})/I_{IL}(\text{max}) = 8 \text{ mA}/0.36 \text{ mA} = 22$.

Therefore, 20 numbers (whichever is less) of low-power Schottky TTL gate can be connected to the output of another low-power Schottky TTL gate or the fan out is 20.

Also, a particular type of TTL gate can be connected with other types of TTL. For example, if a standard TTL gate is connected with HTTL, the fan out is 8. Fan out will be 40, if LTTL is connected to the output of a TTL gate. A table is provided in Figure 11.26 for summarized data regarding fan out for different types of TTL loads.

TTL gates as driver	TTL loads				
	Standard TTL	High-speed TTL	Low-power TTL	Schottky TTL	Low-power Schottky TTL
Standard TTL	10	8	40	8	20
High-speed TTL	12	10	50	10	25
Low-power TTL	2	1	20	1	10
Schottky TTL	12	10	100	10	50
Low-power Schottky TTL	5	4	40	4	20

Figure 11.26

Switching speed. The TTL circuit has the fastest switching speed among all the saturated logic circuits. The propagation delay of the gate directly affects the switching speed. It is related with two switching parameters—propagation delay t_{pHL} during transition from logic 1 to logic 0 and propagation delay t_{pLH} during the transition from logic 0 to logic 1 at the output. It may be noted that t_{pHL} decreases with the increase in temperature, whereas t_{pLH} is independent of temperature. For a standard TTL gate, typical value of propagation delay is 10 ns.

Supply current characteristics. Power supply current requirements for all types of TTL families are specified as maximum current drain at maximum permissible power supply voltage V_{CC}. Maximum current drain I_{CCL} (for logic 0) per gate is specified as 5.5 mA and maximum current drain I_{CCH} (for logic 1) per gate is specified as 2.0 mA. At the nominal supply voltage of V_{CC} = 5V, typical I_{CCL} is 3 mA and I_{CCH} is 1 mA. Hence, I_{CCL} is three times higher than I_{CCH}.

Worst-case input and output voltages. For the TTL family, theoretically, the low state is 0 V and high sate is 5 V. But practically, this ideal situation never occurs and a range of voltage is defined to recognize each of the logic states. They are explained below.

V_{IL}(max) = 0.8 V. It means any voltage from 0 V to 0.8 V is recognized by the TTL gates as the low input level range without changing the output.

V_{IH}(min) = 2 V. This implies that any voltage from 2 V to 5 V is recognized by the TTL gates as the high input level range without changing the output.

A low voltage greater than 0.8 V and a high voltage lower than 2 V are not desirable at the input and lead to unpredictable output at the TTL gates. Similarly, worst-case output voltage ranges are defined below.

V_{OL}(max) = 0.4 V. This implies that logic low output voltage level must be within 0 V to 0.4 V.

V_{OH}(min) = 2.4 V. This means that logic high output voltage level must be within 2.4 V to 5 V.

Hence, as far as the TTL output is concerned, any voltage from 0.4 V to 2.4 V is undesirable and leads to an unpredictable output state.

Noise immunity. It is the maximum induced noise voltage a TTL device can withstand without any false change in the output state. This has been discussed in detail in earlier part of this chapter. The rating of the circuit depends on the smallest noise voltage that makes undesirable operation of the circuit. The noise immunity of TTL gates is much less. This can be understood from worst-case input/output voltage levels as discussed in the previous section.

The worst-case low voltages are

$$V_{IL}(\text{max}) = 0.8 \text{ V} \qquad \text{and} \qquad V_{OL}(\text{max}) = 0.4 \text{ V}.$$

The worst-case high voltages are

$$V_{IH}(\text{min}) = 2 \text{ V} \qquad \text{and} \qquad V_{OH}(\text{min}) = 2.4 \text{ V}.$$

Therefore, as defined in the earlier part of this chapter, the worst-case low-level noise immunity is

$$V_{IL}(\text{max}) - V_{OL}(\text{max}) = 0.8 \text{ V} - 0.4 \text{ V} = 0.4\text{V}.$$

And the worst-case high-level noise margin is

$$V_{OH}(\text{min}) - V_{IH}(\text{min}) = 2.4 \text{ V} - 2.0 \text{ V} = 0.4 \text{ V}.$$

Power dissipation. A standard TTL gate is operated with a power supply of 5 V, and its average current consumption is 2 mA. Therefore, average power dissipation of the TTL gate is 5 V × 2 mA = 10 mW.

Loading rules. A single TTL output may be connected to several TTL gates. When the gate output is low, the output transistor Q3 of the totem pole combination is at saturation condition and is acting as the current sink for all the currents from loading gates, as illustrated in Figure 11.27.

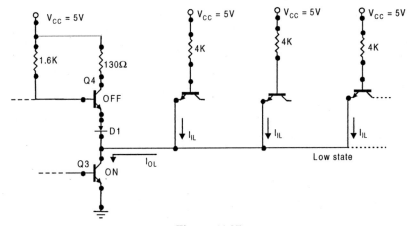

Figure 11.27

Although the transistor Q3 is saturated, its on-state resistance possesses some value other than zero. Therefore, the total low-level output sink current I_{OL}, which is the summation of all low-level input current of individual gate I_{IL}, produces an output voltage drop V_{OL}. The value of V_{OL} must not exceed 0.4 V for TTL, and this limits the value of I_{OL}, which in turn limits the number of loads to be connected.

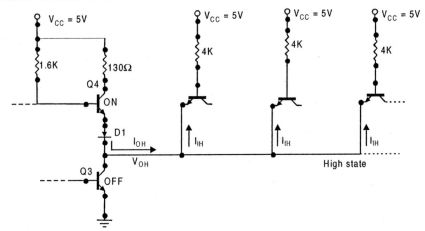

Figure 11.28

Figure 11.28 demonstrates the high state current situation. For high state output, the TTL gate behaves like a source and provides the currents to the external loads. At this condition, the transistor Q4 of the totem pole combination is in the active region and output source current I_{OH} passes through 130Ω resistor, Q4, and diode D1, and produces voltage drop across them. Output current I_{OH} increases with the number of gates driven and provides input high level current I_{IH} for each of the gates. Increase in I_{OH} results in increase in potential drop across Q4, D1, and 130Ω resistor network, and thus decrease in the output voltage level V_{OH}. But V_{OH} must not decrease below the minimum allowable voltage of 2.4 V and hence the number of loads is limited to this parameter.

Protective (clamping) diodes. The input signals of TTL circuits are always positive. If negative signals are inadvertently applied, excess input currents may flow through the input transistor, which might damage the circuit. Therefore, in general, protective diodes, called clamping diodes, are used at each of the input terminals of the TTL gate as shown in Figure 11.29.

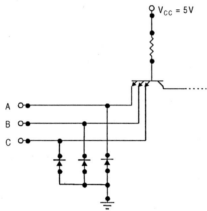

Figure 11.29

Also, these protective diodes suppress the oscillation or ringing at the input terminals, which are developed due to stray capacitance or large input cable length. This has been discussed in detail in earlier parts of this chapter.

11.6.6 Other TTL Gates

In the previous section, various configurations of TTL gates were discussed considering the NAND gate only, because it is the universal gate and any other logic gate can be realized on the basis of this universal gate. In this section, the internal circuits of other gates are also illustrated with basic totem pole configuration.

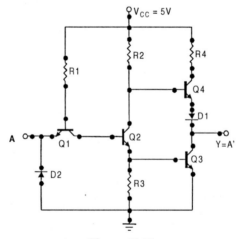

Figure 11.30

TTL INVERTER. The TTL INVERTER circuit diagram is shown in Figure 11.30. Q1 is the input coupling transistor forming two diodes from its base-emitter junction and base-collector junction. If input is low the base-emitter junction is forward biased and voltage level at the base of Q2 is much less to make Q2 and Q3 off. At this instant, the transistor Q4 receives sufficient base voltage and current to drive it to the active region. Therefore, output Y is logic high.

When the input of the gate is at high state, the base-emitter junction of Q1 is reverse biased and the base-collector junction is forward biased. Hence sufficient current and voltage is available at the base of Q2 making Q2 as well as Q3 turn on. At the same time, Q4 is at cut-off. So voltage level of output Y is equal to the V_{CE} (saturation) or low-level state. Thus, the circuit conforms the INVERTER operation.

TTL NOR gate. The TTL NOR gate circuit is shown in Figure 11.31. Here the base of two transistors Q3 and Q4 are connected to the collectors of Q1 and Q2 respectively. When both the inputs of the circuit A and B are at low level, the base-emitter junctions of both the transistors are forward biased, providing no current at the bases of the transistors Q3 and Q4. Hence, both Q3 and Q4 are at cut-off, and also the transistor Q5 is at cut-off. At this time Q6 receives sufficient current to drive into the active region making output Y at high state.

If either or all the inputs of A and B are high, the base-collector junction of either or both the transistors Q1 and Q2 are forward biased, making either or both of Q3 and Q4 turn on. So base voltage and current for the transistor Q5 is available to drive it to saturation. At this time, base of Q6 is too low to turn on the transistor Q6. Therefore, output Y is driven to low state, conforming the NOR logic function.

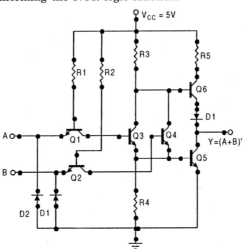

Figure 11.31

TTL AND gate. The TTL NAND gate circuit has already been discussed in detail in this chapter. As AND logic is the complement of NAND logic function, an AND circuit is almost similar to the NAND circuit. The TTL AND gate is shown in Figure 11.32. The main difference from the NAND circuit is that a transistor and a diode have been introduced after the input circuit to obtain the inverter action. When either or both the inputs are at logic 0, one or both of the base-emitter junctions of Q1 are forward biased making Q2 and Q3 turn off. Now base current for transistor Q4 is available to drive it into saturation. Hence, the transistor Q6 is at cut-off and transistor Q5 is at saturation. Therefore, output Y is at low state.

If both inputs A and B are high, the base-collector junction is forward biased, driving the transistor Q2 and Q3 into saturation. Hence, the base voltage of transistor Q4 is low enough to keep it at cut-off. This, in turn, makes the transistor Q5 remain at cut-off. At this time, sufficient base current is available for the transistor Q6 to drive it to the active region. So, output Y goes to high state, conforming the AND logic operation.

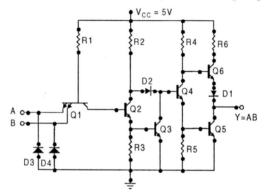

Figure 11.32

TTL OR gate. As OR function is the complement of NOR function, their circuits are similar. The circuit of a TTL OR gate is illustrated in Figure 11.33.

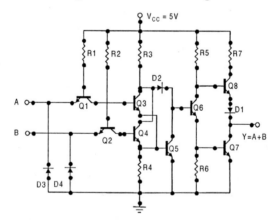

Figure 11.33

When both inputs are at logic 0, base-emitter junctions of both the transistors Q1 and Q2 are forward biased, keeping the transistors Q3 and Q4 at cut-off. Base current is available for Q6, driving it into saturation, in turn the transistor Q7 is also driven to saturation. At this time upper transistor Q8 of the totem pole is at cut-off, and output Y becomes low state.

When either or both of the inputs are high, the base-collector junction of either or both of the transistors Q1 and Q2 are forward biased, driving either or both the transistors Q3 and Q4 into saturation, also the transistor Q5. So the transistors Q6 as well as Q7 are at cut-off. Now sufficient base current is available to drive the transistor Q8 to the active region. Hence, output Y becomes logic high, thus conforming the OR logic operation.

11.6.7 Other TTL Series

The standard TTL with its various output configurations as well as Schottky TTL series have been discussed in detail in this section, as they form the basic structures of all TTL series. However, for critical design application, some special types of TTL series components are used. They are a little variation over the standard TTL series or Schottky TTL series as mentioned below.

Low-power TTL. The basic circuit is similar to the standard TTL series, except that all the resistor values are increased. The increase in resistor values results in the reduction in power dissipation. The power requirements of low-power gates are less than one-tenth of those of standard TTL gates. 74L/54L series devices have power dissipation in order of 1 mW per gate. But propagation delay for low-power TTL gates increases and they are used at low power and low frequency application, like battery-operated systems, calculators, etc.

High-speed TTL. The basic circuitry for this series of ICs is essentially the same as the standard TTL series, except that smaller resistor values are used and the upper transistor Q4 of the totem pole section (refer to Figure 11.17) is replaced with a darlington pair of transistors. This arrangement provides higher speed of operation, but with the sacrifice of power consumption. Flip-flpos, counters, high-speed data transfer network, etc., are the fields of application of 74H/54H series high-speed TTL gates.

Low-power Schottky TTL. With a little increase of internal resistor values and using Schottky devices, a compromise has been made between the speed of operation and power dissipation. Thus, ICs of this series (74LS/54LS) are faster as well as consume less power with comparison to the standard TTL gates.

11.7 EMITTER-COUPLED LOGIC (ECL)

Emitter-coupled logic (ECL) is a current mode logic (CML) or nonsaturated digital logic family. We have seen in earlier sections that the time taken for a transistor to come out of its saturation delays the switching of the transistor from on condition to the off condition and it directly affects the speed of operation of the logic gates. For the gates of the ECL family, the transistors never go to saturation and operate within the cut-off region to the active region. That is why the ECL family is called a nonsaturated logic family. Since the transistors do not saturate in the ECL family, it is possible to achieve propagation delay as low as 2 ns and even below 1 ns. The ECL family has the lowest propagation delay of all the families and is used mostly in systems requiring a very high speed of operation. Its noise immunity and power dissipation are, however, the worst of all the logic families.

A typical basic circuit of the ECL family is shown in Figure 11.34. The outputs provide both the OR and NOR functions. Each input is connected to the base of a transistor. In contrast to other logic families, negative power supply of –5.2 V to ground is used in the case of the ECL family. However, positive logic is employed here, *i.e.*, the lowest voltage level of –5.2 V is equivalent to logic 0 and the highest voltage level of ground level or 0 V is equivalent to logic 1. Practically, a voltage level higher than –0.8 V is considered high-level logic and voltage level below –1.8 V is considered low-level logic. The ECL circuit consists of a differential amplifier, a temperature- and voltage-compensated bias network, and an emitter follower output. The emitter outputs require pull-down resistors for current to flow. This is obtained from the input resistor, any of R1 to R4 of another similar gate connected, or from an external resistor connected to the negative power supply.

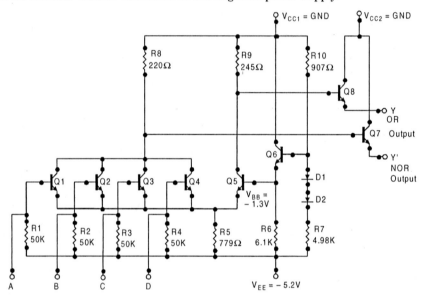

Figure 11.34

The temperature- and voltage-compensated bias network is formed by the transistor Q6, diodes D1, D2, and resistors R6, R7, and R10. The bias circuit provides a reference voltage of V_{BB} = –1.3 V to the differential amplifier, which is set at the midpoint of the signal logic swing. The bias voltage V_{BB} is constant despite any change in temperature or supply voltage.

If any input of the ECL gate is high (–0.8V assumed for worst case), the corresponding transistor is turned on and its emitter voltage level is –1.6 V. Hence the emitter of the transistor Q5 possesses the voltage level of –1.6 V. This forces the transistor Q5 to cut-off, as its base voltage is maintained at –1.3 V, which is not sufficient to turn it on (V_{BE} drop must be greater than 0.6 V to make a transistor on). Therefore, the base current for transistor Q8 is available through the resistor R9, driving Q8 into the active region, provided a load resistor is connected at the emitter of Q8. The voltage across R9 is nominal and the OR output is almost equal to –0.8 V, which is equivalent to high logic state. At this time, current through the resistor R8 is such that the voltage drop across R8 is 1 V below ground level. So, NOR output voltage level is –1 V minus one VBE drop of the transistor Q7, *i.e.*, –1 V –0.8 V = –1.8 V, which is considered to be of low logic level.

If all the inputs are low level (below -1.8 V), all the input transistors Q1 to Q4 turn off, and Q5 conducts. The voltage at the common emitter node is V_{BB} minus one V_{BE} drop for Q5, *i.e.*, –1.3 V –0.8 V = –2.1 V. Each of the input transistors Q1 to Q4 has the input voltage of −1.8 V and each base-emitter junction has the voltage difference of 0.3 V, which is not sufficient to turn them on. The voltage drop across R9 is about 1 V below ground level. The OR output is equal to −1 V minus one V_{BE} drop for Q8, or −1 V − 0.8 V = −1.8 V, which is considered to be of low state. On the other hand, as all the transistors from Q1 to Q4 are off, transistor Q7 receives base current through the resistor R8 to conduct. Current through R7 is so low that it produces nominal voltage drop against R8 and may be neglected. So, the NOR output will be one V_{BE} drop below ground or −0.8 V, which is equivalent to high state. This verifies the OR and NOR operations of the circuit.

As no transistor of the ECL circuit is driven to saturation, the minimum propagation delay is achieved in order of 2 ns. The power dissipation is 25 mW. This gives the speed-power product of 50, which is about the same as for Schottky TTL. The noise margin of an ECL gate is about 0.3 V, which is not as good as in the TTL gate. An ECL circuit has high input impedance due to the presence of a differential amplifier and also low output impedance due to an emitter follower at the output, making it possible to achieve high fan out. It may be noted that a power supply with any end grounded may be used for an ECL circuit, however, a positive grounded power supply is recommended for better noise margin.

The graphic symbol for the ECL gate is shown in Figure 11.35(*a*). Two outputs are available—one for the OR function and other for the NOR function. Wired logic is possible for ECL gates by tying two or more gate outputs together. As shown in Figure 11.35(*b*), an external wired connection of two OR outputs produces a wired-AND logic, and an external wired connection of two NOR outputs generates a wired-OR function. This property of ECL gates may be utilized to form the OR-AND-INVERT and the OR-AND functions.

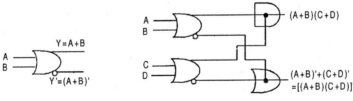

Figure 11.35(a)   **Figure 11.35(b)**

11.7.1 ECL Characteristics

An ECL family possesses the least propagation delay, and it is used in high-speed application. The characteristics of an ECL circuit are summarized as below.

1. The logic levels are nominally −0.8 V for high state and −1.8 V for low state.

2. The transistors never saturate, so storage delay in an ECL circuit is eliminated, and switching speed is high. Typical propagation delay is 2 ns, which makes it faster than advanced Schottky TTL devices (74AS series).

3. Because of very low noise margin in the order of 0.3 V only, ECL circuits are not suitable in heavy industrial environments.

4. An ECL gate is generally provided with true output and its complement, thus eliminating the need of an INVERTER.

5. Power supply requirement is −5.2 V.

6. Fan out is typically 25.

7. Typical power dissipation is 25 mW, which is a little higher in comparison with TTL gates. This is because the transistors of an ECL circuit are operating in the active region.

8. The total current flow in an ECL circuit remains relatively constant regardless of its logic state. This helps to maintain a constant current drain from the power supply, even during the switching and transitions. Thus, unlike TTL circuits, no noise spikes are internally generated for ECL.

ECL devices are not widely used as TTL and MOS families, except in applications where speed is critical. Their low noise margin and high power consumption are disadvantages in comparison to other logic families. Also, the use of negative power supply makes them incompatible to other logic families and difficult to interface. Because of the extreme high-speed of operation, external wires act like transmission lines. Except for very short wires of a few centimeters, ECL outputs must be used with coaxial cables with a resistor termination to reduce line reflections and interferences.

11.8 INTEGRATED-INJECTION LOGIC (I²L)

Integrated-injection logic, or IIL, or I²L, is latest generation LSI technique, also called Merged Transistor Logic (MTL), that uses both *npn* and *pnp* bipolar junction transistors to form a large number of logic gates on a single chip. It reduces the number of metal connections. This allows more circuits to be placed in a chip to form complex digital functions. It also eliminates all the resistors in the circuit, thus increasing the speed as well as reducing power dissipation.

The I²L basic gate is similar to the RTL gate, with a few major differences.

(a) The base resistor used in an RTL gate is removed altogether in the I²L gate.

(b) The collector resistor used in an RTL gate is replaced by a *pnp* transistor that acts as a load for the I²L gate.

(c) I²L transistors use multiple collectors instead of individual transistors as employed in RTL.

The schematic diagram of a basic I²L inverter gate is shown in Figure 11.36. It has a multiple collector transistor Q1 and a *pnp* transistor Q2 at the base of Q1. The emitter of Q2 is connected to the supply voltage V_{BB} and its base is grounded. Q2 acts as a current source and active pull-up, and the multiple collector *npn* transistor Q1 operates as an inverter when one or more collectors are connected with other gates. Most of the current leaving from Q2 is injected directly to the base of Q1, and hence the emitter of Q2 is known as the *injector* and the integrated structure is called the integrated injection logic.

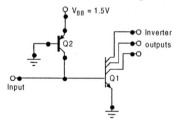

Figure 11.36

The operation of the basic gate can be best analyzed when it is connected to other gates. Figure 11.37 demonstrates a NOR logic function implemented with I²L basic gates.

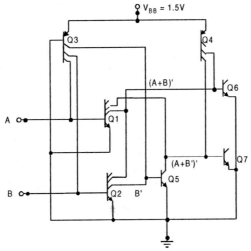

Figure 11.37

In Figure 11.37, transistors Q1 and Q2 are multiple collector transistors acting as inverters. Their base currents are injected through the multiple collector *pnp* transistor Q3, which is also acting as an active pull-up at the bases of Q1 and Q2. It is also acting as the active pull-up of the collector of transistor Q2. Collectors of Q1 and Q2 are connected with active pull-up *pnp* transistor Q4, producing a NOR function (A+B)′. Input signal B is complemented by the transistor Q2, which is connected to the base of the transistor Q5. One of the collectors of Q5 is connected with one of the collectors of Q1 and also to the active pull-up Q4, producing another NOR function (A + B′)′. OR functions may be available by the transistors Q6 and Q7, if they are connected to a pull-up circuit or other gates. Thus we can see that two NOR functions are realized just with a few number of transistors or basic gates. The graphic symbol of the functions as produced by the figure is shown in Figure 11.38.

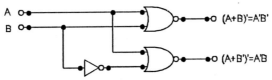

Figure 11.38

Use of Merged Transistor Logic (MTL) with multiple collectors and *pnp* transistors turned out to be most efficient method of construction of ICs of the I²L family, since it reduces the required chip area and increases the package density. The typical value of package density is 1500 gates per square mm. Complex digital circuits may be constructed with less numbers of active devices. I²L devices also consume less power. At high speeds, when propagation delay is 5 ns, it dissipates only 5 mW per gate. The typical values of parameters of I²L devices are as follows.

Packing density:	1500 gates/sq. mm
Gate delay:	25 to 250 ns
Power dissipation per gate:	5 mW to 75 mW
Supply voltage:	1 to 15 V
Logic voltage swing:	0.6 V

Because of high speed, low power consumption, and high package density, the I^2L devices find their application mostly in LSI functions and large computers. It is not available in SSI packages containing individual gates. Its range of application includes microprocessor and microcontroller chips, memory devices, video games, watches, television tuning and control, etc.

11.9   METAL OXIDE SEMICONDUCTOR (MOS)

The field-effect transistor (FET) is a unipolar transistor, since its operation depends on the flow of only one type of carrier—either holes or electrons. There are two types of field-effect transistors—junction field-effect transistors (JFET) and metal oxide semiconductor field-effect transistors (MOSFET). The JFETs are used in linear circuits, whereas the MOSFETs are employed in developing digital circuits.

11.9.1   Construction of a Mosfet

MOS technology derives its name from the basic structure of a metal electrode on an oxide insulator over a semiconductor substrate. The basic structure of the MOS transistor is shown in Figure 11.39(a). The p-channel MOS consists of a lightly doped substrate of n-type silicon material. Two regions are heavily doped by diffusion with p-type impurities to form the *source* and *drain*. The region between the two heavily doped areas of p-sections serves as the *channel*. A metal plate is placed on the channel area with a separation layer consisting of insulating dielectric of silicon dioxide. The metal plate serves as a *gate*. When a negative voltage with respect to the substrate is applied to the gate, it causes an induced electric field in the channel, which attracts p-type carriers from the substrate. As the magnitude of the negative voltage on the gate increases, the region below the gate accumulates more p-type carriers, conductivity in the channel region increases, and current can flow from source to drain, provided a voltage difference is maintained between these two terminals.

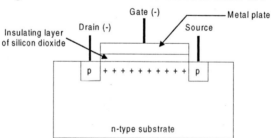

Figure 11.39(a) *p*-channel MOS.

There are four basic types of MOS structures. The channel can be either *p-type* or *n-type*. If the majority of carriers is positive type or holes, it is called a *p*-type MOS, and it is called an *n*-type MOS when the majority of carriers are negative type or electrons. The basic structure of an *n*-channel MOS is shown in Figure 11.36(*b*). The mode of operation can be enhancement type or depletion type, depending on the state of the channel region at zero voltage. If the channel is initially doped lightly with impurities (diffused channel), a conducting channel exists at zero gate voltage, and the device is called a *depletion* type. In this mode, current flows unless the channel is depleted by an applied gate field. If the channel region is initially uncharged, the gate field induces a conducting channel before current can flow. Thus the current is enhanced with the application of gate voltage and such a device is called an *enhancement* type.

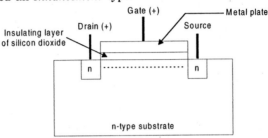

Figure 11.39(b) *n*-channel MOS.

The terminal through which the majority of carriers enter the semiconductor bar is called the source, and through drain, the majority carriers of leave the bar. For a *p*-channel MOS, the source terminal is connected to the substrate and a negative voltage is applied to the drain terminal. When the gate is above threshold voltage (V_T) of about –2 V, no current will flow in the channel, and the drain to the source path is like an open circuit. When the gate voltage is sufficiently negative below V_T, a channel is established and *p*-type carriers flow from source to drain. The *p*-type carriers are positive and hence the corresponding positive current flow from source to drain. The *p*-channel MOSs are normally referred to as PMOS.

In the *n*-channel MOS, the source is connected to the substrate and positive voltage is applied to the drain terminal with respect to the source or the substrate. When the gate voltage is below the threshold voltage V_T (about 2 V), no current flows through the channel. If sufficiently large positive gate voltage is applied above the threshold voltage, *n*-type carriers will flow from source to drain. As *n*-type carriers are negative in nature, positive current will flow from drain to source. The threshold voltage may vary from 1 V to 4 V, depending on the particular process of fabrication used. The *n*-channel MOSs are generally referred as NMOS.

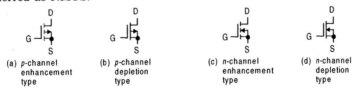

Figure 11.40

The graphic symbols of MOS transistors are shown in Figure 11.40. G, D, and S represent the gate, drain, and source respectively. Enhancement types of MOSFETs are symbolized by the broken lines between source and drain. In these symbols substrates can be identified and are shown connected to the source. The arrow indicates the direction of the flow of carriers.

Because of the symmetrical construction of the source and drain in the MOS transistors, they can be operated as bilateral devices. Although in normal operation carriers are allowed to flow from source to drain, in certain circumstances it is convenient to allow the carriers to flow from drain to source.

One advantage of the MOS device is that it can be used not only as transistor, but also as a resistor. A resistor may be constructed with technology by permanently biasing the gate terminal for conduction. The value of resistance is determined by the ratio of the source-drain voltage to the channel current. Different resistor values may be realized during construction by fixing the channel length and width of the MOS device.

11.9.2 Mosfet Logic Gates

Three logic gates using MOS devices are shown in Figure 11.41. For an n-channel MOS, positive supply voltage V_{DD} (about + 5 V) is applied between the drain and source with substrate connected to the source to allow positive current flow from drain to source. The n-channel gates generally employ positive logic and the two voltage levels are the function of the threshold voltage V_T. Any voltage from 0 V to V_T is considered low logic level and high-level ranges from V_T to V_{DD}. Negative supply voltage for V_{DD} is used with p-channel MOS devices to allow positive current flow from source to drain. It is convenient to use a negative logic system with p-channel MOS circuits.

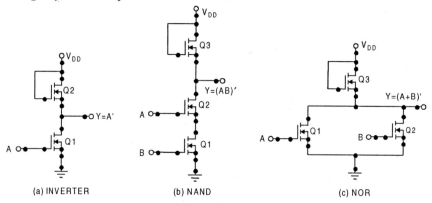

Figure 11.41

Figure 11.41(a) shows an INVERTER logic circuit with the use of MOS devices. Two MOSFETs, Q1 and Q2, are used here. Q1 acts as a normal MOS transistor, but Q2 behaves like a load resistor as its gate is biased with V_{DD} and is always in the conduction state. When a low-level logic (below V_T) is applied at the gate of Q1, it remains at cut-off. Since Q2 is at conduction state, the output voltage of Y is at about V_{DD} or high logic state. When input A is of high-level logic (above V_T), MOS transistor Q1 turns on and current flows from V_{DD} through load resistor Q2 and into Q1. The output Y becomes low, as its voltage level goes below the threshold voltage V_T. This verifies the INVERTER action. Note that resistance of Q1 while conducting must be much less compared to that of Q2. This means $R_{DS(ON)}$ of Q1 $\ll$ $R_{DS(ON)}$ of Q2, where $R_{DS(ON)}$ is on-state resistance of MOS transistor.

A NAND gate circuit has been shown in Figure 11.41(b), three MOS transistors are used. Q1 and Q2 are the normal MOS transistors connected in series, and Q3 acts as a load resistor. When either or all the inputs A and B are low, the corresponding transistor(s) will be at the cut-off state preventing any current through Q1 or Q2. So output Y is at high state. When both the inputs A and B are high, both the transistors Q1 and Q2 turn on, producing a current path through Q1 and Q2. Hence the voltage level of output drops below the threshold voltage and becomes low logic level. Again, the condition that must be fulfilled is $R_{DS(ON)}$ of Q1 + $R_{DS(ON)}$ of Q2 << $R_{DS(ON)}$ of Q3.

Similarly, a NOR gate circuit has realized with MOS devices as in Figure 11.41(c). Here two MOS transistors, Q1 and Q2, are active devices connected in parallel and Q3 behaves like a load resistor for both Q1 and Q2. When both the inputs A and B are at low logic state, none of Q1 and Q2 is operating, hence output Y is at high state. If either or all of the inputs are high, the corresponding transistor(s) turns on and the output Y becomes low. Thus conforms the NOR logic. The similar condition of the above regarding on-state resistance of MOS must be maintained.

MOSFET logic gates consume much less power and possess higher noise margin. It has high fan out capabilities, because of the extremely high input resistance at each of the inputs. It is also very simple to fabricate as it does not require resistors or diodes, etc., that leads to high package density. Its low power dissipation makes it ideally suited for LSI packages, this is where the MOS logic has made its greatest impact in the digital field. However, MOS gates are slower than the TTL gates.

11.9.3 Complementary MOS (CMOS) Logic

Complementary metal oxide semiconductors or CMOS circuits take advantage of the fact that both n-channel and p-channel devices can be fabricated on the same substrate. In CMOS circuits, both types of MOS devices are interconnected to form the logic functions. The power consumption of CMOS devices is extremely low. They have enhanced noise immunity, high fan out capability, and simpler interfacing with other logic circuits.

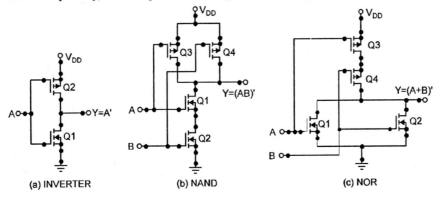

(a) INVERTER   (b) NAND   (c) NOR

Figure 11.42

The CMOS inverter circuit can be made with only one n-channel MOS and one p-channel MOS transistor as shown in Figure 11.42(a). The source terminal of the p-channel device is connected at V_{DD} and the source terminal of the n-channel device at ground. The value of V_{DD} may be anywhere from +3 V to +18 V. The two logic voltage levels are 0V for low-level logic and V_{DD} for high-level logic. To understand the operation of the CMOS circuit, the following behaviors of the CMOS devices must be remembered.

1. The n-channel MOS conducts when its gate-to-source voltage is positive.

2. The p-channel MOS conducts when its gate-to-source voltage is negative.

3. Either type of device is turned off when gate-to-source voltage is zero.

In Figure 11.42(a), Q1 is in the n-channel MOS and Q2 is the p-channel MOS. Input A is connected to the gates of both Q1 and Q2. When a low-level input is applied at A, gates of both the transistors Q1 and Q2 are at zero potential. This keeps Q1 at off condition, but Q2 turns on as its gate-to-source voltage is effectively negative. Under this condition, a low impedance path from V_{DD} to output is available, and output terminal to ground becomes a very high impedance path and hence output Y becomes logic high. On the other hand, when input A is at high state, gate-to-source voltage of Q1 is high and it turns on. Whereas gate-to-source voltage of Q2 is effectively zero and it turns off. As a result, the output Y approaches to low level to 0 V.

Figure 11.42(b) represents a CMOS two-input NAND gate circuit. Here, the transistors Q1 and Q3 form the complementary pair where Q1 is n-channel MOS and Q3 is p-channel MOS, and one of the inputs A is connected to their gates. Similarly, Q2 is an n-channel MOS and Q4 is a p-channel MOS making another complementary pair, and their gates are connected to another input B. The n-channel MOS transistors Q1 and Q2 are in series and p-channel MOS transistors are in parallel. When both the inputs A and B are at low logic level, both n-channel transistors Q1 and Q2 are at off condition, and both p-channel transistors Q3 and Q4 are at conduction as their base voltages are negative with respect to sources, producing a low impedance path from V_{DD} to output Y. Therefore output Y becomes high level. If input A is low and B is high, Q1 is off and Q2 is on. Their respective complementary transistors Q3 is on and Q4 is off. Hence, a low impedance path from V_{DD} to Y is available through Q3, one of the transistors of parallel combination, and Y to ground level is at high impedance as Q1, one of the transistors of the series combination, is off. This makes output Y at high level. Similarly, if A is high and B is low, output Y remains high. Because this time, transistor Q2 of the series combination is at off state, producing a high impedance path for Y to ground, and transistor Q4 of the parallel combination is at conduction state providing a low impedance path from V_{DD} to output Y. When both the inputs A and B are at high state, both the transistors Q1 and Q2 of the series combination are conduction state whereas the transistors Q3 and Q4 of parallel combination are at off condition. So a high impedance path is produced for V_{DD} to Y and low impedance path is available from Y to ground. Therefore Y becomes low and the above input-output combinations verify the function of a NAND logic gate.

The two-input NOR circuit in Figure 11.42(c) can be explained in similar fashion. Here the n-channel MOS transistors Q1 and Q2 are employed in parallel, and p-channel transistors Q3 and Q4 are used in series. When both the inputs A and B are at low level, both the transistors Q1 and Q2 of parallel combination are at off state, and both the transistors Q3 and Q4 of series combination are at on conduction, producing a high impedance path for Y to ground and a low impedance path for V_{DD} to Y. Hence output Y becomes high. When either or both the inputs are high level, one or both the transistors of the parallel combination of Q1 and Q2 conducts producing a low impedance path for Y to ground. At the same time one or both transistors of the series combination of Q3 and Q4 is at cut-off state, preventing any low impedance path from V_{DD} to Y. Therefore, output Y remains at low level, which are conditions for a NOR logic function.

11.9.3.1 Characteristics of CMOS Logic Gates

One great advantage of the CMOS logic gate is that it consumes extremely low DC power usually in the order of 10 nW and high package density. This is why this technology is becoming more and more popular and finds its wide application in the fabrication of LSI chips as well as SSI and MSI chips. Other advantages include higher noise immunity, high fan out capability, and wide range of supply voltage. The characteristics of the CMOS family are described below.

Power dissipation. We have seen that one of the transistors of the complementary pair of CMOS gates is always at off condition while the other is at conduction. This situation prevails under static condition when the output is constant. Therefore, in this condition, the power dissipation of a CMOS circuit is extremely small (10 nW) as already mentioned, which is referred to as DC power dissipation. However, at the instant of switching, when output changes state from low to high or high to low, the power dissipation increases. This is due to the fact that during transition both the MOSFETs are in conduction for a small period of time. This leads to spikes in the supply current.

Therefore, during transition, an appreciable amount of drain current flows. Moreover, any output stray capacitance has to be charged before the change in output can take place. The charging of the capacitor requires additional current that is drawn from the power supply resulting in an increase of instantaneous power dissipation.

The average power dissipation of a CMOS device whose output is continuously changing is called the active power dissipation. The active power dissipation increases with the increase in supply voltage as well as the frequency of operation. The power consumption of a CMOS gate is around 10 mW with operating frequency in the 10 MHz region. Thus, at higher frequencies the CMOS circuit losses its advantages.

Propagation delay time. The propagation delay of a standard CMOS gate ranges from 25 ns to 150 ns. This factor depends on the power supply voltage and other factors.

Voltage levels. CMOS circuits can be operated over a voltage range of 3 V to 15 V. A supply voltage of 9 V to 12 V can be used to obtain the overall best performance of a CMOS gate in respect to high speed and noise immunity. When CMOS gates are used in association with TTL gates, the V_{DD} supply voltage is made 5 V, so that the voltage levels of the two families are the same.

Noise immunity. The CMOS family has the highest noise margin among all the logic families. It depends on the operating supply voltage and is typically about 45% of the supply voltage V_{DD}. Noise margin is the same for both high level and low level. For a power supply V_{DD} of 5 V, the noise margin is at least 2.25 V.

Floating inputs. A floating input for a CMOS gate may be considered as either of the logic levels, and noise will be generated at the input terminal. This results in an excessive power dissipation. Therefore, it is recommended that all the unused input terminals must be connected to either V_{DD} or to ground.

Current sourcing and sinking. When the output of a CMOS gate is at low state, it draws current from the driven gate in the order of only 1 µA. This means the sink current of the CMOS gate is 1 µA. Similarly, when output of the gate is at high level, it delivers current to the input of the driven gate. This is the source current and is in order of 1 µA. The worst-case input and output currents of the CMOS devices are as follows.

$$I_{IL}(\text{max}) = -1 \ \mu A, \qquad\qquad I_{IH}(\text{max}) = 1 \mu A$$

$$I_{OL}(\text{max}) = 10 \ \mu A, \qquad\qquad I_{OH}(\text{max}) = -10 \ \mu A$$

Fan out. The fan out of the CMOS gate depends on the type of load being connected. For a CMOS gate driving another CMOS gate the fan out can be calculated from the input and output currents as mentioned above.

Considering the low input state, $I_{OL}(\text{max})/ I_{IL}(\text{max}) = 10 \ \mu A/ 1\mu A = 10.$

Considering the high output state, $I_{OH}(\text{max})/ I_{IH}(\text{max}) = 10 \ \mu A/ 1\mu A = 10.$

Therefore, 10 CMOS gates can be connected at the output of another CMOS gate. Hence the fan out of a CMOS gate is 10.

Several series of the CMOS digital logic family are commercially available in the market. The original design of the CMOS series is recognized from 4000 number designations, which was produced by RCA company. Now a days CMOS devices are available in 74 series numbers and they are pin compatible as well as function compatible with TTL devices having the same number. The CMOS devices of 74 series are indicated by 74C. The performance characteristics of 74C series are about the same as 4000 series. 74HC and 74HCT are the other CMOS series, which are of high speed and pin compatible with the TTL family of 74 series and the CMOS devices of later series are also electrically compatible with TTL family.

At LSI range, CMOS circuits are widely used because of the several advantages already mentioned. RAM, PROM, EPROM, microprocessors, microcontrollers, and several VLSI chips employ CMOS technology.

11.9.4 BiCMOS Logic Circuits

BiCMOS logic circuits are the latest development of digital technology in the silicon fabrication process that combines the speed and driving capabilities of bipolar junction transistors with the density and low power consumption of CMOS devices. BiCMOS technology is employed to develop low-voltage analog circuits, VLSI circuits, Application-specific integrated circuits or ASIC, and high-density gate arrays. The package density of BiCMOS circuits is sacrificed to accommodate the bipolar devices. Because of the low output impedance and increased charging and discharging current of BJTs, the propagation delay of the BiCMOS gates does not increase much as in CMOS gates. Also, it has good compatibility with the voltage levels of ECL and TTL for ready interfacing and minimum loss of switching speed. BiCMOS circuits may be employed in place of CMOS buffers for its high current capability and faster response. In terms of speed, power, and density BiCMOS IC can be compared with ECL.

BiCMOS INVERTER. The circuit of a basic BiCMOS inverter is shown in Figure 11.43. It is the extended version of a CMOS inverter, where two extra bipolar matching transistors T1 and T2 are used with each of the MOS transistors Q1 and Q2 respectively. When input A is low, the *n*-channel MOS Q1 is at cut-off. So, no base current is available for transistor T1 and it is at off state. At the same time the channel MOS Q2 is on. Therefore base current for the transistor T2 is available. So T2 operates at the active region and supplies ($\beta+1$) times the base current at the emitter of T2 to charge the load capacitance C_L. The output Y thus becomes high level. When the logic input is high, it turns on the MOS transistor Q1, and at the same time Q2 becomes off. As Q1 supplies the base current for the transistor T1, a low impedance path is formed for output Y to ground to provide the discharge of load capacitance C_L. Hence the output becomes low.

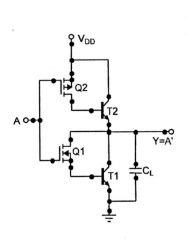

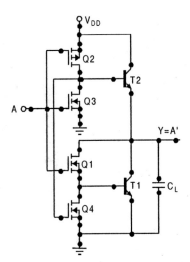

Figure 11.43   **Figure 11.44**

The transfer characteristics and switching speed of the transistors can be improved by providing discharge paths for excess carriers from bases of the transistors T1 and T2. Additional NMOS transistors may be employed for this purpose at the bases of the bipolar transistors. Figure 11.44 shows the inverter circuit with the scheme of provision of the discharge path of excess carriers using additional NMOS transistors Q3 and Q4, which is a more conventional and practical circuit implementation for the BiCMOS inverter.

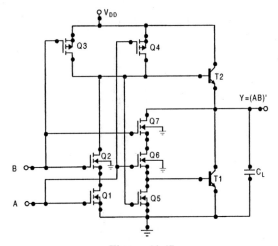

Figure 11.45

BiCMOS NAND. A conventional BiCMOS two-input NAND logic circuit is shown in Figure 11.45. It is the modification of the CMOS NAND circuit in Figure 11.42(b), where bipolar transistors T1 and T2 are provided to form the BiCMOS structure and additional NMOS transistors Q5, Q6, and Q7 are introduced to provide the discharge paths for the excess carriers accumulated at the bases of the bipolar transistors.

BiCMOS NOR. A practical circuit for a BiCMOS two-input NOR logic function is shown in Figure 11.46 with BJTs T1 and T2 providing the BiCMOS characteristics and NMOS transistors Q5, Q6, and Q7 to provide the discharge path for excess carriers accumulated at the bases of bipolar transistors.

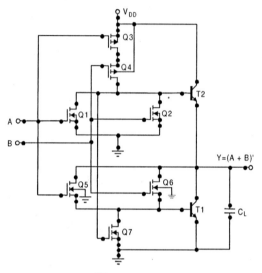

Figure 11.46

11.10 COMPARISON OF DIFFERENT LOGIC FAMILIES

In this chapter, different logic families—their structures and characteristics—have been discussed. Among them, the TTL and the CMOS logic gates are the most widely used logic devices, as they are easily commercially available and economic. I²L and BiCMOS logic devices are gaining popularity and applied at special fields. Each logic family projects different characteristics in terms of static and dynamic performance, size, and cost. A brief comparison of typical performance characteristics of commonly used IC families is tabulated in Figure 11.47.

Logic Families and basic gates	Propagation delay (ns)	Power dissipation (mW)	Noise immunity (V)	Fan out	Logic voltage swing (V)
RTL (NOR)	12	20	0.3	5	2.5
DTL (NAND/NOR)	12	9	0.3	8	4.7
TTL (NAND)	10	10	0.4	10	3.8
STTL (NAND)	3	2	0.5	10	3.8
ECL (NOR/OR)	2	25	0.3	16	3.6
I²L (NAND)	0.7	0.1	0.4	12	3.6
MOS (NAND/NOR)	1	0.1	2.5	10	3.8
CMOS (NAND/NOR)	1	0.002	2.5	10	5

Figure 11.47

11.11 INTERFACING

The output(s) of a circuit or a system should match the input(s) of another circuit or system that has different electrical characteristics. In modular design technique, different circuit modules may be realized with the devices of different logic families. It is always necessary to match the driver circuit module or system with the load circuit in terms of electrical parameters. This is referred to as *compatibility*. Proper interfacing between different logic families is important for compatibility. An interfacing circuit is one that is connected between the driver and the load. The function of an interfacing circuit is to receive the driver output and condition it so that it is compatible with the input requirements of the load.

A circuit designer must take care of the current and voltage characteristics of the two circuits of different logic families while connecting together. For example, an interface circuit is provided, if one of the circuits uses a negative logic system and is connected to a positive logic system. Similarly, an interface circuit is provided to convert an ECL output to a TTL signal if it is connected to a system that employs TTL devices. An interfacing circuit is also needed between a high-speed logic family and a low-speed logic family. Therefore, it is necessary to confer the electrical parameters of the logic families of the driver circuit as well as the load.

The devices of TTL and CMOS families are most commonly used in designing the digital circuits. The worst-case values of electrical parameters of a commercially available series of these families are tabulated in Figure 11.48 for ready reference.

TTL/CMOS series	$V_{IH}(min)$ (V)	$V_{IL}(max)$ (V)	$V_{OH}(min)$ (V)	$V_{OL}(max)$ (V)	$I_{IH}(min)$ (mA)	$I_{IL}(max)$ (mA)	$I_{OH}(min)$ (mA)	$I_{OL}(max)$ (mA)
TTL (74)	2.0	0.8	2.4	0.4	0.04	1.6	0.04	16
TTL (74LS)	2.0	0.8	2.7	0.5	0.02	0.4	0.4	8.0
TTL (74AS)	2.0	0.8	2.7	0.5	0.02	2.0	2.0	20
TTL (74ALS)	2.0	0.8	2.7	0.4	0.02	0.1	0.4	8.0
CMOS (4000B)	3.5	1.5	4.95	0.05	0.001	0.001	1.6	0.4
CMOS (74C)	3.5	1.0	4.9	0.5	0.001	0.001	1.75	1.75
CMOS (74HCT)	2.0	0.8	4.9	0.1	0.001	0.001	4.0	4.0

* The data presented here are the typical values at supply voltage of +5 V. They may differ for gate to gate.

** CMOS driving CMOS gates only.

Figure 11.48

11.11.1 Interfacing CMOS with TTL

The TTL devices operate on 5 V supply whereas the supply voltage for CMOS devices ranges from 3 V to 18 V. As the supply requirement is different, an interfacing circuit is needed.

TTL driving CMOS with the same supply voltage. As the CMOS devices work on a supply voltage ranging from 3V to 18V, which covers the TTL supply voltage of 5 V, an approach to interface the TTL circuit with a CMOS circuit is to use same supply voltage of 5 V. When a TTL load drives a CMOS input, there is no problem for low-level input, as the maximum low-level output for TTL $V_{OL}(max)$ is 0.4 V. This voltage level is always interpreted as low-level input for CMOS, as maximum low-level input voltage for CMOS $V_{OH}(max)$ is

1.5 V. For high-level output from TTL, problems may arise because the minimum high-level output of TTL V_{OH}(min) is 2.4 V, whereas the minimum low-level input for the CMOS family VIH(min) is 3.5 V. Thus the voltage level from 2.4 V to 3.5 V at the input of a CMOS device is indeterminate and will cause malfunction in a CMOS circuit. Therefore, it is recommended to use an external pull-up resistance R_L at the output of the TTL driver to the supply as shown in Figure 11.49. The effect of the pull-up resistance is to raise the output level nearer to 5 V. With the pull-up resistor, if the output of the drive is low, the resistor is grounded and low-level input will be considered by the CMOS load.

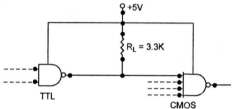

Figure 11.49

The value of the pull-up resistor may be chosen as 3.3K. The current through R_L is 5V/3.3K =1.5 mA approximately, when TTL output is low. The TTL driver will sink this current and must be within the capacity. As for TTL the maximum sink current I_{OL}(max) is 16 mA, so it is well within the range. The gate capacitance of a CMOS load has to be charged through the pull-up resistor R_L. To enhance the switching speed, it is important to decrease the value of R_L. The minimum permissible value of resistor R_L is determined by the maximum sink current of TTL $i.e.$, 16 mA. In this case, if the maximum supply voltage variation is considered as 5.25 V, then the minimum value of the pull-up resistor will be

$$R_L = 5.25V/\ 16\ mA = 328\Omega \equiv 330\Omega\ (approximately).$$

TTL driving CMOS at different supply voltages. For CMOS logic gates, the best performance is achievable when their supply voltage is kept within range of 9 V to 12 V. At lower supply voltage, the performance of CMOS gates deteriorates with the increase of propagation delay and decease in noise immunity. Therefore, it is recommended to use 12 V supply voltage for the CMOS circuits. In this case, if the CMOS circuit is to be driven by a TTL driver, an open-collector TTL gate should be employed, with a pull-up resistor R_L to the supply voltage 12 V, as shown in Figure 11.50.

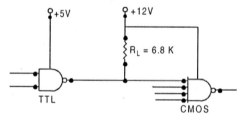

Figure 11.50

The pull-up resistor R_L may be selected as 6.8 K. If output of the TTL driver is low, its maximum output voltage V_{OL}(max) is 0.4 V and it must sink current with the pull-up resistor. The sink current may be calculated as 12V/6.8K = 1.76 mA, which is within limit of the maximum TTL sink current of 16 mA (I_{OL}(max)). When TTL output is at high level, the open collector output of the TTL gate approaches 12 V and this is acceptable as high level for a CMOS gate. So now the TTL outputs are compatible to CMOS input states.

CMOS driving TTL. To interface a CMOS gate with a standard TTL load, the low input of the CMOS output gate $V_{OL}(max)$ must be less than 0.8 V, which is the maximum allowable low-level input $V_{IL}(max)$ of TTL. Similarly, for high-level output from a CMOS driver, the minimum output voltage $V_{OH}(min)$ of CMOS must be higher than 2 V, which is the minimum high input level $V_{IH}(min)$ of TTL. Therefore, as per voltage levels concerned there is no mismatch between CMOS driver and TTL load provided their supply voltages are made the same to 5 V.

But while interfacing, one must not forget the current requirement of the driver gate and the driven loads. The low-level output current of the driver must be greater than the low-level input current to the load, and also the high-level output current of the driver must be greater than high-level input current of the load. The worst-case output currents of CMOS are

$$I_{OL}(max) = 400 \ \mu A, \text{ when output is low,} \qquad \text{and}$$

$$I_{OH}(min) = 40 \ \mu A, \text{ when output is high.}$$

The worst-case input currents for TTL are

$$I_{IL}(max) = 1.6 \ mA, \text{ for a low input,} \qquad \text{and}$$

$$I_{IH}(min) = 400 \ \mu A, \text{ for a high input.}$$

From the above data, we can see that for a high output level, there is no problem of interfacing. However, for a low output from driver, the maximum sink current capacity of CMOS ($I_{OL}(max) = 400 \ \mu A$) is much lower than low-level input current ($I_{IL}(max) = 1.6 \ mA$). So CMOS output is not compatible with standard TTL loads.

Let us consider the case that CMOS is driving the low power Schottky TTL load. Regarding the input/output voltage levels, they are compatible. Regarding the current requirements we observe that:

The worst-case output currents of CMOS are

$$I_{OL}(max) = 400 \ \mu A, \text{ when output is low,} \qquad \text{and}$$

$$I_{OH}(min) = 40 \ \mu A, \text{ when output is high.}$$

The worst-case input currents for TTL are

$$I_{IL}(max) = 400 \ \mu A, \text{ for a low input,} \qquad \text{and}$$

$$I_{IH}(min) = 20 \ \mu A, \text{ for a high input.}$$

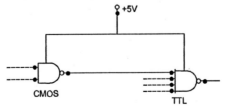

Figure 11.51

It may be noticed from the above data that for a high driver output, the output current from a CMOS driver is sufficient to drive LSTTL (low-power Schottky TTL) gate.

However, for low logic output, maximum low-level output current of CMOS is just equal to the maximum low-level input current of LSTTL. Therefore, with a simple interface circuit configuration as in Figure 11.51, the CMOS driver can drive only one LSTTL load.

The above problem can be eliminated by employing a CMOS buffer, which boosts up the current capability as an interface circuit as shown in Figure 11.52. For example, a CMOS current buffer 74C902 has the worst-case current output of $I_{OL}(max) = 3.6$ mA, for low output, and IOH(min) = 800 µA, for high output. Since $I_{IL}(max) = 1.6$ mA for standard TTL, the CMOS buffer can drive two standard TTL gates.

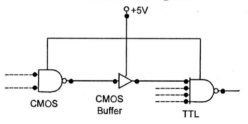

Figure 11.52

11.12 SOME EXAMPLES

Example 11.1. *For the DTL NAND gate as shown in Figure 11.53, $V_{BE}(sat) = 0.8$ V, $V\gamma = 0.5$ V, $V_{CE}(sat) = 0.2$ V, the drop across the connecting diode is 0.7 V and $V\gamma$ (diode) $= 0.6$ V. The inputs of this switch are obtained from the outputs of similar gates.*

(a) *Verify that the circuit functions as a positive NAND and calculate $h_{FE}(min)$.*

(b) *Will the circuit operate properly if D2 is not used?*

(c) *Calculate the noise margin if all the inputs are high.*

(d) *Calculate the noise margin if at least one input is low. Assume for the moment that Q is not loaded by a following stage.*

(e) *Calculate the fan out.*

(f) *Obtain average power.*

    (Assume $h_{FE} = 30$)

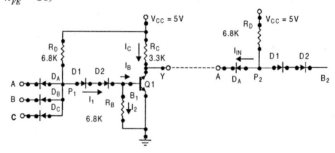

Figure 11.53

    Solution. (a) The logic levels are $V_{CE}(sat) = 0.2$ V for logic 0, and $V_{CC} = +5$ V for logic 1. If any of the inputs A, B, or C is at logic 0 state, then the corresponding diode will conduct and voltage at P_1 is $V_{P1} = 0.2 + 0.7$ V $= 0.9$ V. Condition for diodes D1 and D2 in conduction is that V_{P1} must be greater than the diode drop of D1 plus diode drop of D2 plus V_{BE} drop of the transistor Q1 $= 0.7 + 0.7 + 0.7$ V $= 2.1$ V. Therefore, diodes D1, and D2 are not conducting and also transistor Q1 is not conducting as sufficient voltage is not available at its base. Hence, output Y is 5 V, *i.e.*, logic 1.

If all the inputs are high to 5 V, all the diodes D_A, D_B, and D_C are reverse biased and nonconducting. Hence, diodes D1 and D2 are conducting and transistor Q1 turns on. The voltage at P_1 is

$$V_{P1} = V_{D1} + V_{D1} + V_{BE}(sat) = 0.7 + 0.7 + 0.8 \text{ V} = 2.2 \text{ V}.$$

So the voltage across each diode D_A, D_B, or D_C is 5 V -2.2 V $= 2.8$ V at reverse direction, conforming that they are off. At this time, Q1 is on and so $Y = V_{CE}(sat) = 0.2$ V. Hence is low, thus conforming the NAND operation.

To find $h_{FE}(min)$,

$$I_1 = \frac{V_{CC} - V_{P1}}{R_D} = \frac{5 - 2.2}{6.8 \times 10^3} = 0.412 \text{mA}$$

$$I_2 = \frac{V_{BE}(sat)}{R_B} = \frac{0.8}{6.8 \times 10^3} = 0.112 \text{ mA}$$

$$I_B = I_1 - I_2 = 0.412 - 0.112 = 0.3 \text{ mA}$$

$$I_C = \frac{V_{CC} - V_{CE}(sat)}{R_C} = \frac{5 - 0.2}{3.3 \times 10^3} = 1.45 \text{ mA}.$$

Therefore, $h_{FE}(min) = \dfrac{I_C}{I_B} = \dfrac{1.45 \text{mA}}{0.3 \text{mA}} = 4.83.$

To assure the transistor Q1 for saturation, h_{FE} must be greater than $h_{FE}(min)$.

(*b*) If one diode D2 is not used, then under all inputs low condition, $V_{P1} = 0.2 + 0.7$ V $= 0.9$ V. Voltage at B1 or at the base of the transistor, $V_{B1} = 0.9 - 0.6$ V $= 0.3$ V, where cut-in voltage of the diode is 0.6 V. Since $V_{B1} < V\gamma$, where $V\gamma = 0.5$ V is the cut-in voltage of the transistor, theoretically the transistor Q1 is in cut-off. But a little spike in inputs or noise may turn on the transistor and logic function will be indeterminate.

(*c*) For all inputs are high, the output will be low. $V_{P1} = 2.2$ V. The input diode will start conducting if its cathode terminal voltage falls below $V_{P1} - 0.6$ V $= 2.2 -0.6$ V $= 1.6$ V. This means $V_{IH}(min) = 1.6$ V or the input high voltage must not fall below 1.6 V. The normally available high input voltage is 5 V. Hence noise margin $= 5 - 1.6$ V $= 3.4$ V.

(*d*) If one of the inputs becomes low, then the output is high. Voltage at P_1, $V_{P1} = 0.9$ V and Q1 is off. To drive the transistor just into the active region voltage, P_1 is required to be $V\gamma$ (for D1) $+ V\gamma$ (for D2) $+ V\gamma$ (for Q1) $= 0.6 + 0.6 + 0.5$ V $= 1.7$ V. Hence the noise margin $= 1.7 - 0.9$ V $= 0.8$ V.

(*e*) From part (*a*), we have seen that the maximum collector current of the transistor Q1, unloaded condition, is 1.45 mA, when output becomes low. At this time, if another DTL gate is connected as in Figure 11.53, the driver gate must sink from the driven gate. At low-level condition, the driven gate can deliver the current I_{IN}, which may be calculated as follows.

$$I_{IN} = \frac{V_{CC} - V_{P2}}{R_D} = \frac{5 - 0.9}{6.8 \times 10^3} = 0.60 \text{ mA}$$

$(V_{P2} = 0.9$ V under low input condition. V_{P2} is equivalent to V_{P1}.)

If N number of DTL gates are connected, then the collector current will be,

$$I_C = N \times 0.6 + 1.45 \text{ mA}.$$

Assuming $h_{FE} = 30$,

$$I_C = h_{FE} \times I_B = 30 \times 0.3 \text{ mA} = 9 \text{ mA}.$$

So, $N \times 0.6 + 1.45 = 0.9$.

Or, $N = \dfrac{9 - 1.45}{0.6} = 12.58 = 12.$

Therefore, fan out of the given DTL gate is 12.

(f) When the output is low the power consumption P_{LOW} is given by

$$P_{LOW} = V_{CC} (I_1 + I_C) = 5(0.412 + 1.45) = 5.31 \text{ mW}.$$

When the output is high, at least one input diode conducts. $I_C = 0$, as the transistor is at cut-off. Power is drawn by R_D only. Current through R_D is

$$I_1 = \frac{V_{CC} - V_{P1}}{R_D} = \frac{5 - 0.9}{6.8 \times 10^3} = 0.6.$$

Power dissipation in the gate at high output state, P_{HIGH} is

$$P_{HIGH} = V_{CC} \times I_1 = 5 \times 0.6 = 3 \text{mW}.$$

Therefore, average power dissipation is

$$P_{AVERAGE} = \frac{P_{LOW} + P_{HIGH}}{2} = \frac{5.31 + 3}{2} = 6.155 \text{mW}.$$

Example 11.2. *(a) Determine the fan out of a 74LS00 NAND gate. (b) How many 74LS00 inputs can a 7400 output drive?*

Solution. (a) From the data sheet of 74S00, under the worst-case condition, we get,

$$I_{OH}(\text{min}) = 0.4 \text{ mA}, \quad I_{OL}(\text{max}) = 8 \text{ mA}$$

$$I_{IH}(\text{min}) = 0.02 \text{ mA}, \quad I_{IL}(\text{max}) = 0.4 \text{ mA}.$$

Fan out (at high state) $= \dfrac{I_{OH}(\text{min})}{I_{IH}(\text{min})} = \dfrac{0.4}{0.02} = 20.$

Fan out (at low state) $= \dfrac{I_{OL}(\text{max})}{I_{IL}(\text{max})} = \dfrac{8}{0.4} = 20.$

Therefore, fan out of 74LS00 gate is 20.

(b) As per the 74 series data sheet, the currents at worst-case conditions are,

$$I_{OH}(\text{min}) = 0.4 \text{ mA}, \quad I_{OL}(\text{max}) = 16 \text{ mA}$$

$$I_{IH}(\text{min}) = 0.4 \text{ mA}, \quad I_{IL}(\text{max}) = 1.6 \text{ mA}.$$

At high state, the number of 74LS00 gates that can be driven by 7400,

$$= \frac{I_{OH}(\text{min}) \text{ of } 7400}{I_{IH}(\text{min}) \text{ of } 74LS00} = \frac{0.4}{0.02} = 20.$$

At low state, the number of 74LS00 gates that can be driven by 7400,

$$= \frac{I_{OL}(max) of 7400}{I_{IL}(max) of 74LS00} = \frac{16}{0.4} = 40.$$

Therefore, the number of 74LS00 gates can be driven by a 7400 gate is 20.

Example 11.3. *Draw the interconnection of I²L gates to form a 2 to 4 decoder.*

Solution. We know, for a 2 to 4 decoder there are two inputs and four outputs. If the inputs are designated as A and B, then the outputs will be A'B', A'B, AB', and AB. The circuit diagram with I²L gates to realize above functions is in Figure 11.54.

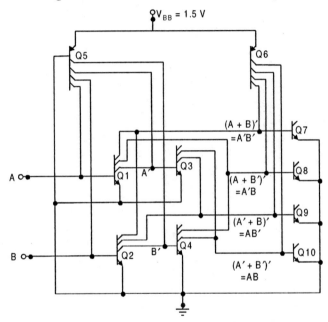

Figure 11.54

Q1, Q2, Q3, and Q4 are the multiple-collectors *npn* transistors. Q5 and Q6 are the *pnp* transistors serving as the active pull-ups. Q7, Q8, Q9, and Q10 are the output transistors to show the completeness of the circuit.

REVIEW QUESTIONS

11.1 Define logic family?

11.2 Explain the governing parameters of the logic families.

11.3 Describe the difference between the bipolar integrated circuits and MOS integrated circuits.

11.4 What is noise immunity? What is propagation delay?

11.5 Describe the advantages and disadvantages of totem pole output configuration.

11.6 Explain current sourcing and current sinking.

11.7 Draw and explain the circuit diagram of a 3-input I^2L NOR gate.

11.8 What is the fastest logic family? Explain.

11.9 Explain why an open TTL gate acts as logic high.

11.10 What is the advantage of ECL logic family?

11.11 What are the Schottky diode and Schottky transitor?

11.12 What is a multiple emitter transistor?

11.13 Discuss why wired logic should not be used for active pull-up outputs.

11.14 Describe the characteristics of MOS logic.

11.15 Describe enhance type MOS and depletion type MOS with constructional details.

11.16 Why is CMOS faster than PMOS/NMOS?

11.17 Write a note on interfacing of CMOS with TTL and vice versa.

11.18 What is the necessity of an interfacing circuit?

11.19 Explain the purpose of totem pole at the TTL output configuration.

11.20 What is the function of the diode at output stage of a totem pole output configuration

11.21 Draw a 4-input CMOS NAND gate. Repeat for a 4-input NOR gate with CMOS.

11.22 Draw the basic BiCMOS inverter and explain.

11.23 What are the advantages of a BiCMos logic circuit?

11.24 What is tri-state logic. What is its application.

11.25 Compare the characteristics of different logic families.

11.26 (a) Determine the high-level output voltage of the RTL gate for a fan out of 5.

   (b) Determine the minimum input voltage required to drive an RTL transistor to saturation, when $H_{FE} = 20$.

   (c) From the results from (a) and (b), determine the noise margin of the RTL gate when the input is high and fan out is 5.

11.27 Show that the output transistor of the DTL gate of Figure 11.9 goes to saturation when all the inputs are high. Assume $h_{FE} = 25$.

11.28 Connect the output Y of the DTL gate shown in Figure 11.9 to N inputs of other DTL gates. Assume that the output transistor is saturated and its base current is 0.44 mA. Its h_{FE} is 20.

   (a) Calculate the current through a 2 K resistor.

   (b) Determine the current coming from each input connected to the gate.

   (c) Determine the total collector current in the output transistor as a function of N.

   (d) Find the value of N that will keep the transistor in saturation.

   (e) What is the fan out of the gate?

11.29 Let all the inputs of the open-collector TTL gate of Figure 11.13 be at high state of 3 V.

   (a) Calculate the voltages in the base, collector, and emitter of all the transistors.

   (b) Calculate h_{FE}(minimum) of Q2 that ensures the saturation of the transistor.

   (c) Assuming the minimum h_{FE} of Q3 is 6.18, determine the maximum current that can be tolerated in the collector to ensure saturation.

   (d) What is the minimum value of R_L while ensuring the saturation of Q3?

11.30 (a) Show with the truth table that when two open collector true outputs are wired connected with an external resistor to V_{CC}, the output will produce an AND function.

(b) Prove with the truth table that when two open collector complemented outputs are wired connected with an external resistor to V_{CC}, the output will produce a NOR function.

11.31 Why should the totem pole outputs not be tied together to form wired logic? Explain with the circuit diagram. Show that the load current (base current plus collector current of Q4) in Figure 11.17 is about 32 mA. Compare this value with the recommended load current in the high sate of 0.4 mA.

11.32 Realize the logic diagram for the function F= ABCDEFGH using open collector two-input AND gates. Use IC 7409, which is a quad two-input AND gate IC. How many such ICs are required to implement the above function?

11.33 Calculate the emitter current I_E across R_5 in the ECL gate of Figure 11.35. When

(a) At least one input is high at –0.8 V.

(b) All inputs are low at –1.8 V.

Calculate the voltage drop across the collector resistor R_9 in each case and show it is about 1 V as required. Assume $I_C = I_E$.

11.34 Show that when NOR outputs of ECL gates are wired connected, the output produces an OR function.

11.35 Calculate the noise margin of an ECL gate.

11.36 The MOS transistors are bilateral, *i.e.*, current may flow from source to drain or from drain to source. Using this property, realize a circuit that implements the following Boolean functions—

$$F = (AB + CD + AED + BEC)'.$$

11.37 (a) Draw the circuit diagram of a 4-input NAND gate using CMOS transistors.

(b) Repeat for a 4-input NOR function.

Appendix 1

ALTERNATE GATE SYMBOLS

Different types of logic gates and their symbols have been discussed in this book with their applications. However, readers may find alternate gate symbols while referring to other books. The alternate gate symbols are with the circles or bubbles at the inputs. The circles or the bubbles imply that the inputs are complemented first followed by the normal operation of the gates. The following table describes the graphic symbols of these gates and their functions.

Graphic Symbol of logic gates	Functions	Boolean Expressions
A –◁○▷– Y	INVERTER	$Y = A'$
A, B gate – Y	AND	$Y = AB$
A, B gate – Y	NAND	$Y = (AB)'$
A, B gate – Y	OR	$Y = A+B$
A, B gate – Y	NOR	$Y = (A+B)'$

❑ ❑ ❑

431

74 SERIES INTEGRATED CIRCUITS

This section presents the available TTL 54/74 series of ICs for ready reference for logic designers. The IC numbers given here are with the reference of their functions. However, the actual IC numbers as specified by the manufacturers also indicates the exact section of the logic family they belong, though their logic functions are the same. As an example, 74LS00 indicates quad 2-input NAND gates of the low-power Schottky family, whereas 74H00 indicates quad 2-input NAND gates of high-speed TTL. Both 54 and 74 series ICs have the same functions, but the 54 series is used for military application, for the operating temperature is −55°C to + 125°C, whereas 74 series ICs are used in industrial and commercial application with a temperature range of 0°C to +85°C. The pin configuration of both series is similar.

IC Number	Function
7400	Quad 2-input NAND gates
7401	Quad 2-input NAND gates (open collector)
7402	Quad 2-input NOR gates
7403	Quad 2-input NAND gates (open collector)
7404	Hex Inverters
7405	Hex Inverters (open collector)
7406	Hex Inverter Buffers/Drivers (open collector)
7407	Hex Buffers/Drivers (open collector)
7408	Quad 2-input AND gates
7409	Quad 2-input AND gates (open collector)
7410	Triple 3-input NAND gates
7411	Triple 3-input AND gates
7412	Triple 3-input NAND gates (open collector)
7413	Dual 4-input NAND Schmitt Triggers
7414	Hex Schmitt Trigger Inverters
7415	Triple 3-input AND gates (open collector)

7416	Hex Inverter Buffers/Drivers (open collector)
7417	Hex Buffers/Drivers (open collector)
7420	14-input NAND gates
7421	Dual 4-input AND gates
7422	Dual 4-input NAND gates (open collector)
7423	Expandable Dual 4-input NOR gates with Strobe
7425	Dual 4-input NOR gates with strobe
7426	Quad 2-input NAND Buffers (open collector)
7427	Triple 3-input NOR gates
7428	Quad 2-input NOR Buffers
7430	8-input NAND gates
7432	Quad 2-input OR gates
7433	Quad 2-input NOR Buffers (open collector)
7437	Quad 2-input NAND Buffers
7438	Quad 2-input NAND Buffers (open collector)
7439	Quad 2-input NAND Buffers
7440	Dual 4-input NAND Buffers
7441	BCD-to-Decimal Decoder/Driver (open collector)
7442	BCD-to-decimal Decoder
7443	Excess-3-to-Decimal Decoder
7444	Excess-3-Gray-to-BCD Decoder
7445	BCD-to-Decimal Decoder/Driver (open collector)
7446	BCD-to-Seven Segments Decoder/Driver (open collector)
7447	BCD-to-Seven Segments Decoder/Driver (open collector)
7448	BCD-to-Seven Segments Decoder/Driver (open collector)
7449	BCD-to-Seven Segments Decoder/Driver (open collector)
7450	Expandable Dual 2-wide 2-input AND-OR-INVERT gates
7451	Dual 2-wide 2-input AND-OR-INVERT gates
7452	Expandable 4-wide AND-OR gates
7453	Expandable 4-wide 2-input AND-OR-INVERT gates
7454	4-wide 2-input AND-OR-INVERT gates
7455	Expandable 2-wide 4-input AND-OR-INVERT gates
7460	Dual 4-input Expanders
7461	Triple 3-input Expanders
7462	4-wide AND-OR Expander
7464	4-wide AND-OR-INVERT gates
7465	4-wide AND-OR-INVERT gates (open collector)

7470	AND-gated J-K Flip-flop
7471	AND-gated R-S Flip-flop
7472	AND-gated J-K Flip-flop
7473	Dual J-K Flip-flops
7474	Dual D Flip-flops
7475	Dual 2-bit Transparent Latch
7476	Dual J-K Filp-flops
7477	Dual 2-bit Transparent Latch
7478	Dual JK Flip-flops
7480	Gated Full Adder
7482	2-bit Binary Adder
7483	4-bit Binary Adder
7485	4-bit Magnitude Comparator
7486	Quad XOR gates
7487	4-bit True/Complement, Zero/One Element
7489	16 × 4 RAM (open collector)
7490	BCD Counter
7491	8-bit Serial-in Serial-out Shift Register
7492	Divide-by-12 Counter
7493	4-bit Binary Counter
7494	4-bit Serial/Parallel-in Serial-out Shift Register
7495	4-bit Serial/Parallel-in Parallel-out Shift Register
7496	5-bit Serial/Parallel-in Parallel out, Serial-in Serial-out Shift Register
7499	4-bit Bidirectional Universal Shift Register
74100	Dual 4-bit Transparent Latch
74101	AND-OR-gated J-K Flip-flop
74102	Dual J-K Flip-flops
74103	Dual J-K Flip-flops
74106	Dual J-K Flip-flops
74107	Dual J-K Flip-flops
74108	Dual J-K Flip-flops
74109	Dual J-K Flip-flops
74110	AND-OR-gated J-K Flip-flop
74111	Dual J-K Flip-flops
74112	Dual J-K Flip-flops
74113	Dual J-K Flip-flops
74114	Dual J-K Flip-flops

74116	Dual 4-bit Transparent Latches
74121	Monostable Multivibrator
74122	Retriggerable Monostable Multivibrator
74123	Dual Retriggerable Monostable Multivibrators
74124	Dual Voltage-controlled Oscillators
74125	Tri-state Quad Buffers
74126	Tri-state Quad Buffers
74128	Quad 2-input NOR Buffers
74132	Quad 2-input NAND Schmitt Triggers
74133	13-input NAND gate
74134	Tri-state 12-input NAND gate
74135	Quad XOR/XNOR gates
74136	Quad XOR gates (open collector)
74138	1:8 Demultiplexer
74139	Dual 1:4 Demultiplexer
74141	BCD-to-Decimal Decoder/Driver (open collector)
74145	BCD-to-Decimal Decoder/Driver (open collector)
74147	Priority Encoder (Decimal to Binary)
74148	Priority Encoder (Octal to Binary)
74150	16:1 Multiplexer
74151	8:1 Multiplexer
74152	8:1 Multiplexer
74153	Dual 4:1 Multiplexers
74154	1:16 Demultiplexer
74155	Dual 1:4 Demultiplexers
74156	Dual 1:4 Demultiplexers (open collector)
74157	Quad 2:1 Multiplexers
74158	Quad 2:1 Multiplexers
74159	1:16 Demultiplexer
74160	Decade Up-counter
74161	4-bit Binary Up-counter
74162	Decade Up-counter
74163	4-bit Binary Up-counter
74164	8-bit Serial-in Parallel-out Shift Register
74165	8-bit Serial-in/Parallel-in Serial-out Shift Register
74166	8-bit Serial-in/Parallel-in Serial-out Shift Register
74168	Decade Up-down Counter

74169	4-bit Binary Up-down Counter
74174	Hex D Flip-flops
74175	Quad D Flip-flops
74176	Presettable BCD Counter
74177	Presettable 4-bit Binary Counter
74178	4-bit Universal Shift Register
74179	4-bit Universal Shift Register
74180	8-bit Parity Generator/Checker
74181	Arithmetic Logic Unit (ALU)
74182	Look-ahead Carry Generator
74184	BCD-to-Binary Code Converter
74185A	6-bit Binary-to-BCD converter
74190	Decade Up-down Counter
74191	4-bit Binary Up-down Counter
74192	Decade Up-down Counter
74193	4-bit Binary Up-down Counter
74194	4-bit Bidirectional Universal Shift Register
74195	4-bit Serial/Parallel-in Parallel-out Shift Register
74196	Presettable BCD Counter
74197	Presettable 4-bit Binary Counter
74198	8-bit Bidirectional Universal Shift Register
74199	8-bit Serial/Parallel-in Parallel-out Shift Register
74206	256-bit RAM (open collector)
74221	Dual Monostable Multivibrator
74240	Tri-state Octal Inverter Buffers
74241	Tri-state Octal Buffers
74244	Tri-state Octal Buffers
74246	BCD-to-Seven Segment Decoder/Driver (open collector)
74247	BCD-to-Seven Segment Decoder/Driver (open collector)
74248	BCD-to-Seven Segment Decoder/Driver
74249	BCD-to-Seven Segment Decoder/Driver (open collector)
74256	Dual 4-bit Addressable Latches
74260	Dual 5-input NOR gates
74266	Quad XNOR gates (open collector)
74273	Octal D Flip-flops
74276	Quad J-K Flip-flops
74279	Quad S-R Flip-flops

74280	9-bit Parity Generator/Checker
74283	4-bit Binary Adder with Fast Carry
74289	16×4 RAM (open collector)
74290	BCD counter
74293	4-bit Binary Counter
74352	Dual 4:1 Multiplexers
74365	Tri-state Hex Buffers
74366	Tri-state Hex Buffers
74367	Tri-state Hex Buffers
74368	Tri-state Hex Buffers
74375	Dual 2-bit Transparent Latches
74376	Quad J-K Flip-flops
74381	Arithmetic Logic Unit/Function Generator
74386	Quad XOR gates
74390	Dual Decade Counters
74393	Dual 4-bit Binary Counters
74425	Tri-state Quad Buffers
74426	Tri-state Quad Buffers
74445	BCD to Decimal Decoder/Driver (open collector)
74490	Dual BCD Counters

❑ ❑ ❑

PIN CONFIGURATION OF 74 SERIES INTEGRATED CIRCUITS

This section presents the pinout configuration of some commonly used integrated circuits of 74 series for ready reference. These may be helpful for logic designers. Open collector ICs of TTL are marked with a (*). Similar to the open collector TTL devices, CMOS devices with open drain output configuration are also marked with a (*).

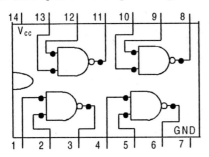

7400/7403*/7426*/7437/7438*/74132

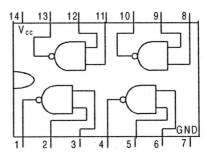

7401*/7439

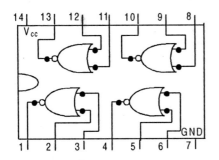

7402/7428/7433*/74128

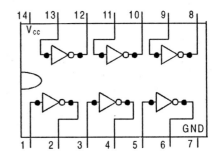

7404/7405*/7406*/7416*

439

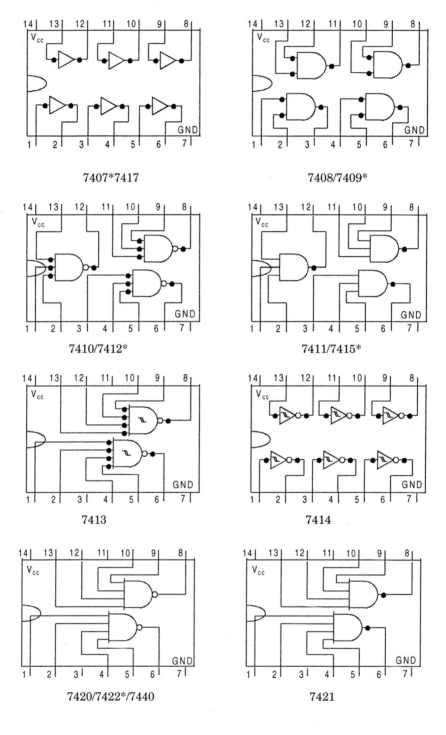

7407*7417

7408/7409*

7410/7412*

7411/7415*

7413

7414

7420/7422*/7440

7421

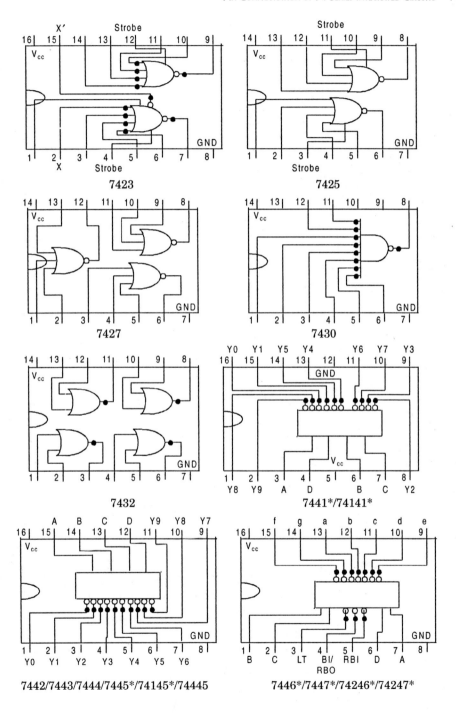

7423

7425

7427

7430

7432

7441*/74141*

7442/7443/7444/7445*/74145*/74445

7446*/7447*/74246*/74247*

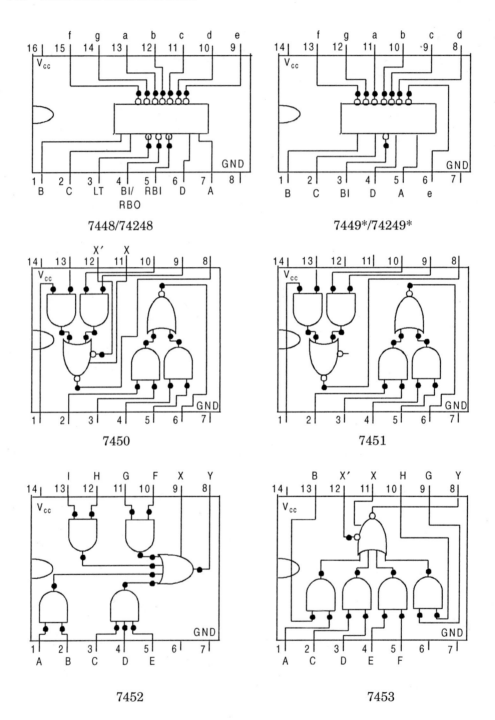

7448/74248

7449*/74249*

7450

7451

7452

7453

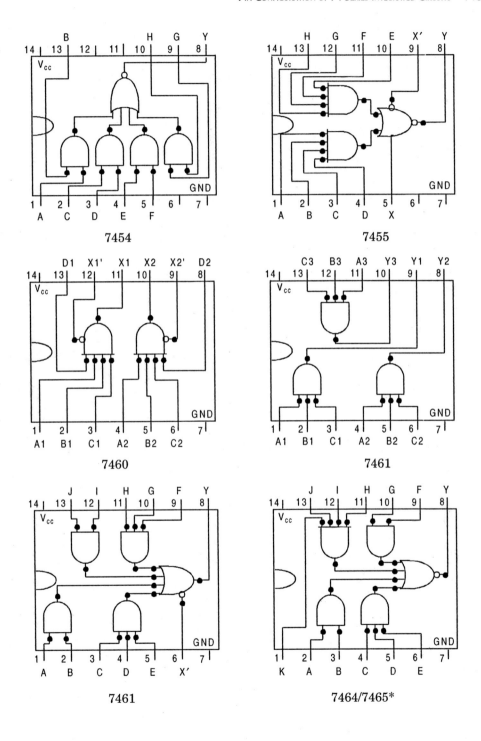

7454

7455

7460

7461

7461

7464/7465*

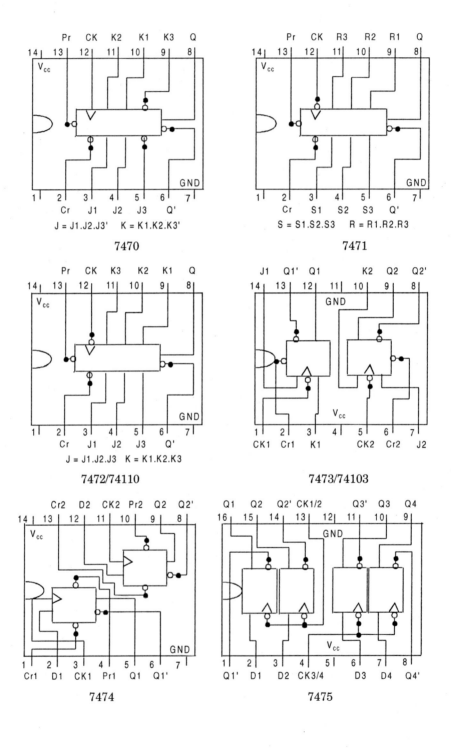

7470

J = J1.J2.J3'   K = K1.K2.K3'

7471

S = S1.S2.S3   R = R1.R2.R3

7472/74110

J = J1.J2.J3   K = K1.K2.K3

7473/74103

7474

7475

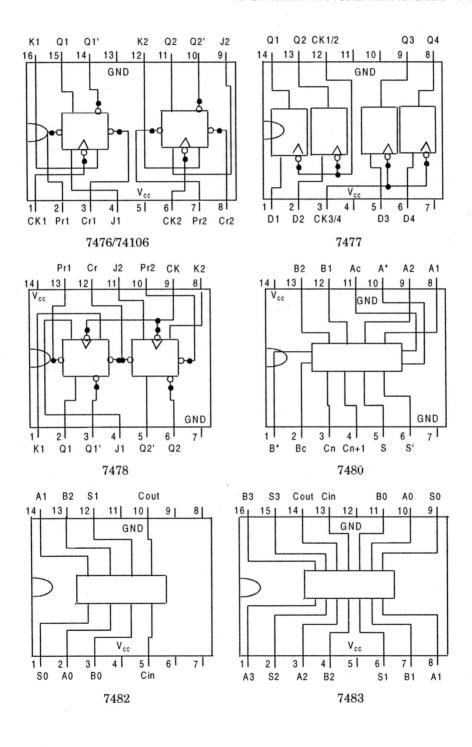

7476/74106

7477

7478

7480

7482

7483

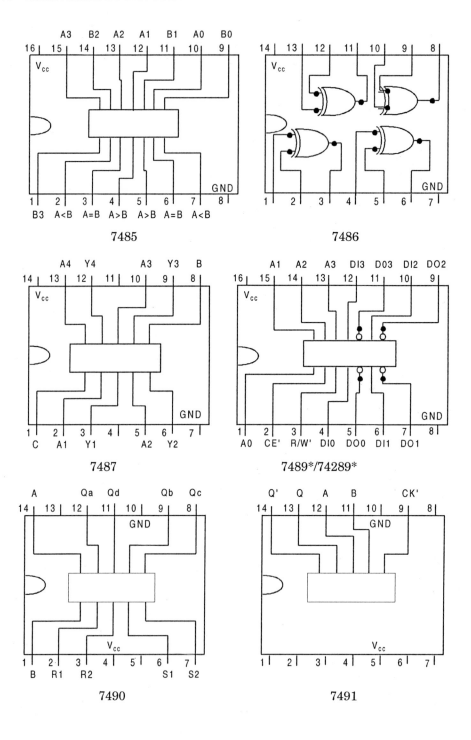

7485

7486

7487

7489*/74289*

7490

7491

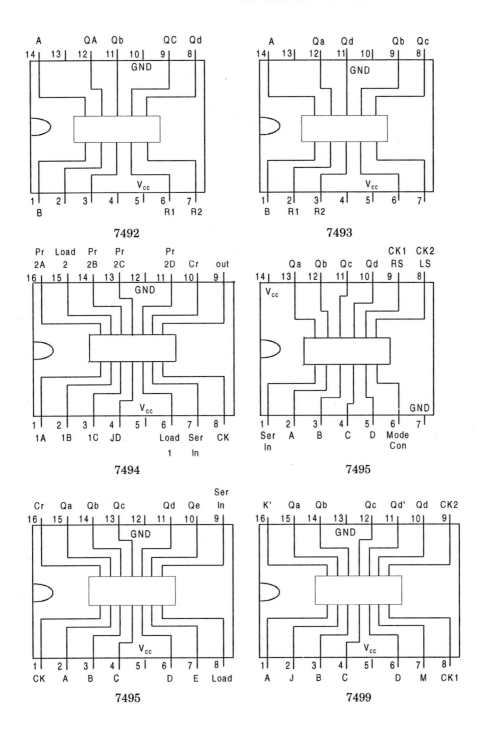

7492

7493

7494

7495

7495

7499

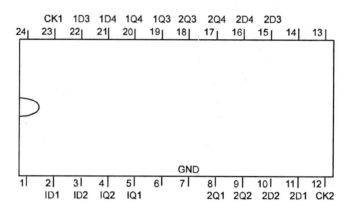

74100

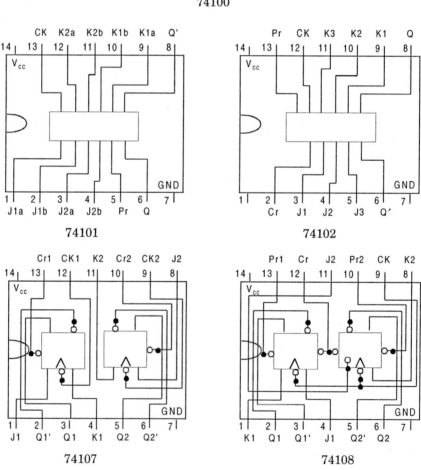

74101

74102

74107

74108

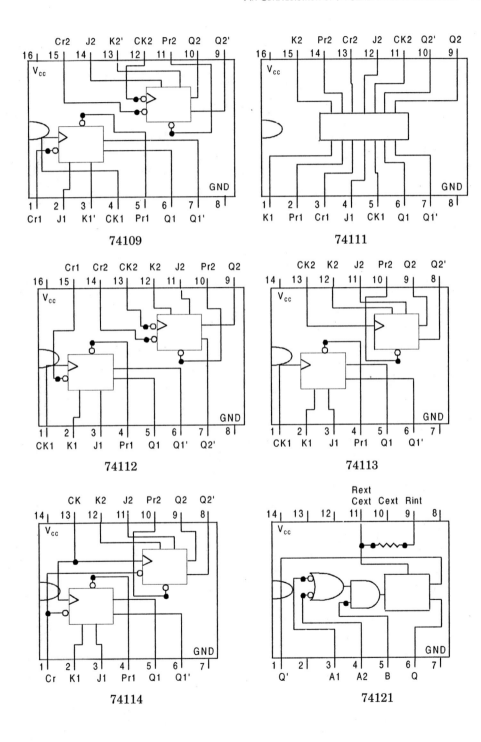

74109

74111

74112

74113

74114

74121

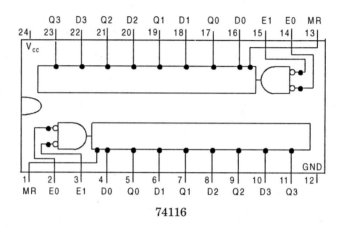

74116

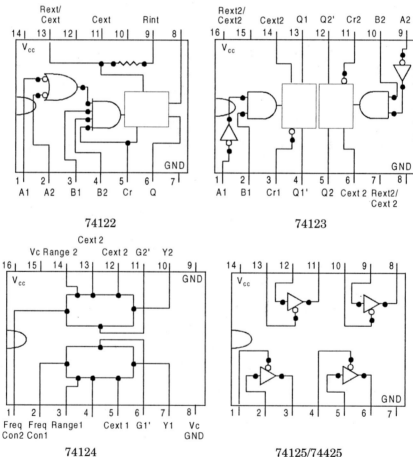

74122

74123

74124

74125/74425

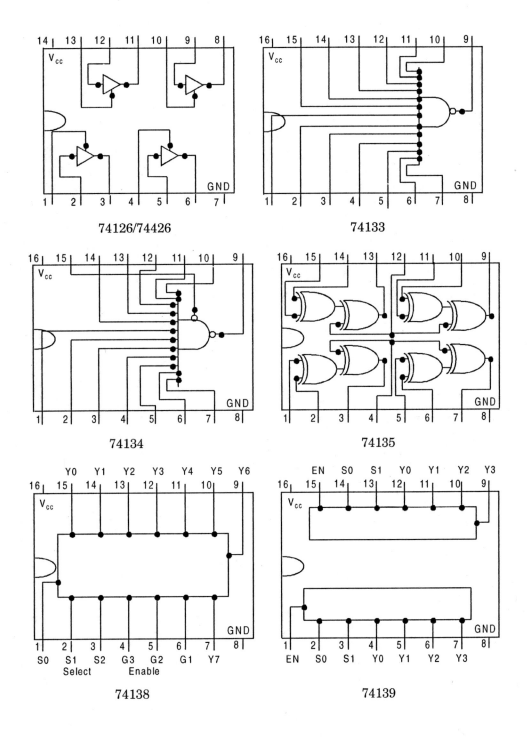

74126/74426

74133

74134

74135

74138

74139

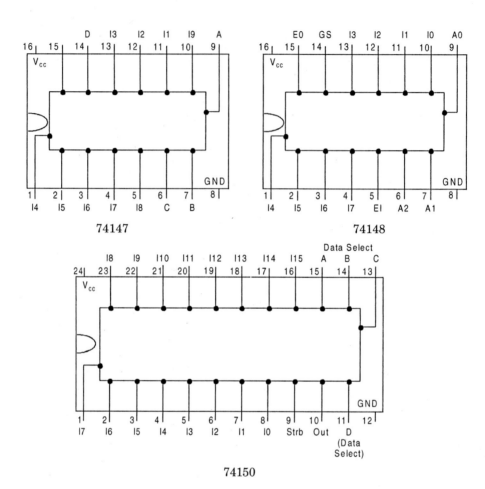

74147

74148

74150

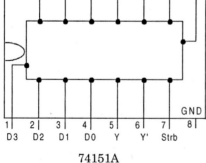

74151A

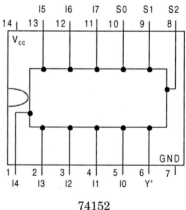

74152

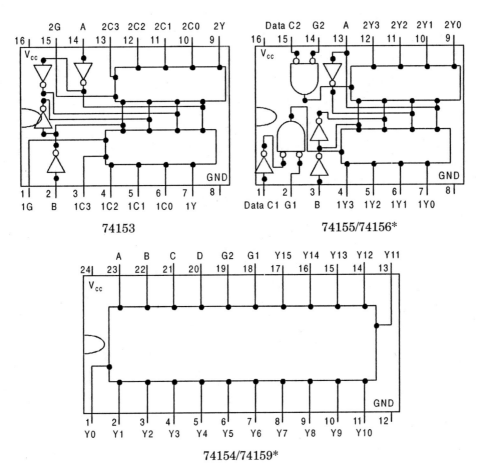

74153

74155/74156*

74154/74159*

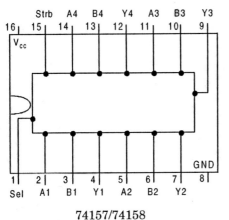

74157/74158

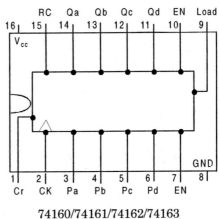

74160/74161/74162/74163

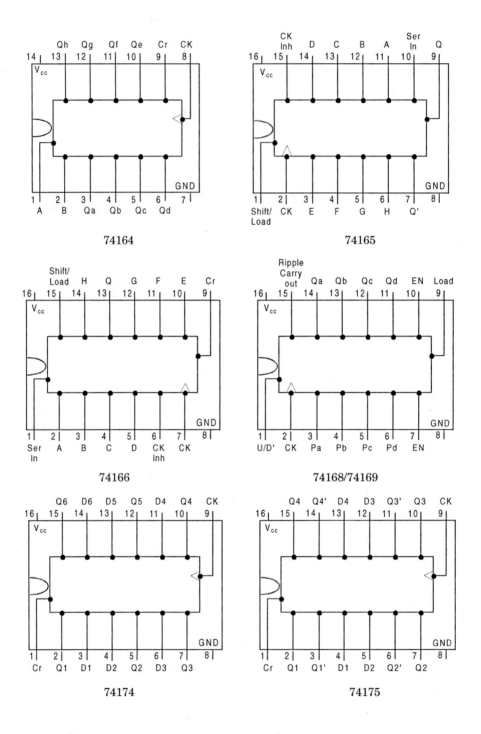

74164

74165

74166

74168/74169

74174

74175

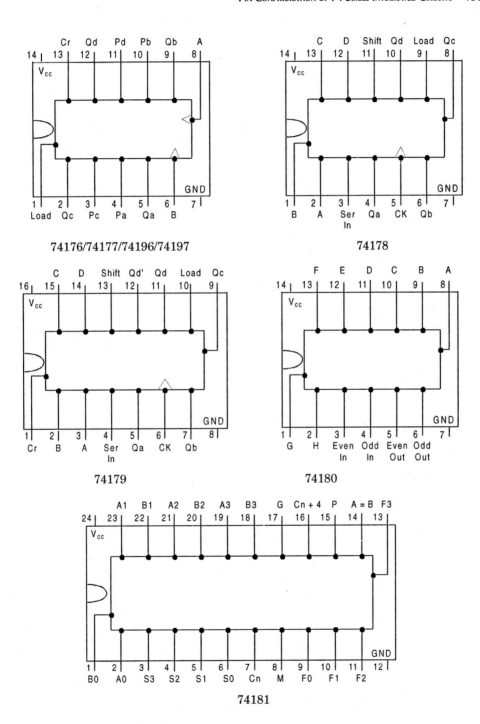

74176/74177/74196/74197

74178

74179

74180

74181

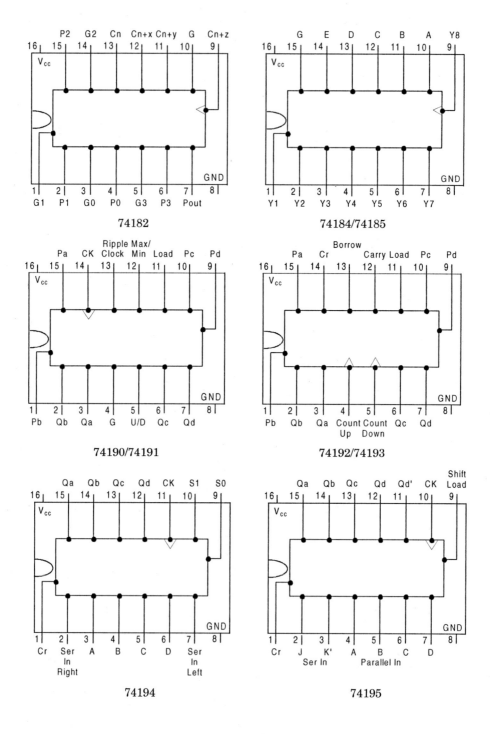

74182

74184/74185

74190/74191

74192/74193

74194

74195

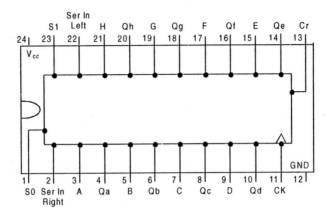

74198

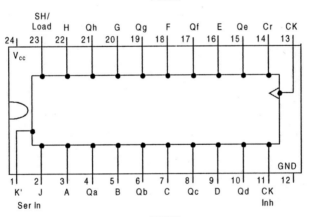

74199

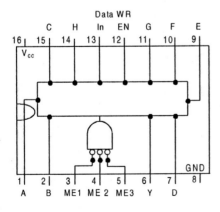

74206*

74221

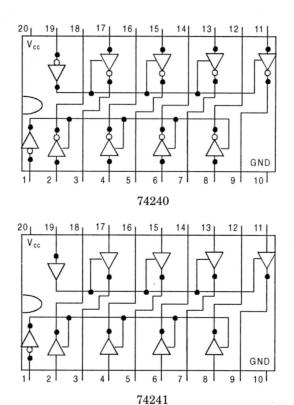

74240

74241

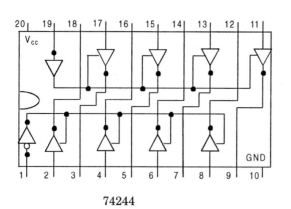

74244

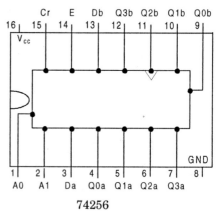

74256

Appendix 4

4000 Series
Integrated Circuits

The 4000 series of integrated circuits represents the CMOS IC logic family. This section presents the available 4000 series of ICs for ready reference for logic designers. The IC numbers given here are with the reference of their functions. However, the actual IC numbers as specified by the manufacturers also indicate the electrical and physical characteristics.

IC Number	Function
4000	Dual 3-input NOR gates, one inverter
4001	Quad 2-input NOR gates
4002	Dual 4-input NOR gates
4006	Shift register with variable length (maximum 18 bits)
4007	Dual Complementary pair MOS, one inverter
4008	4-bit Full-Adder
4009	Hex Inverters/Buffers
4010	Hex Buffers, noninverting
4011	Quad 2-input NAND gates
4012	Dual 4-input NAND gates
4013	Dual D Flip-flops with Preset and Clear
4014	8-bit Parallel Input Shift Register
4015	Dual 4-bit Serial-in Parallel-out Shift Register
4016	Quad Bilateral Analog/Digital Switches
4017	Synchronous Decimal Up-counter with Decimal Decoder
4018	5-bit Programmable Divider/Counter
4019	Quad 2:1 Common Addressable Multiplexers
4020	Asynchronous 14-bit Binary Up-counter
4021	8-bit Parallel Input Shift Register
4022	Synchronous Octal Counter with Decoded Outputs

4023	Triple 3-input NAND gates
4024	Asynchronous 7-bit Binary Up-counter
4025	Triple 3-input NOR gates
4026	Decimal Up-counter with Seven Segment Decoder
4027	Dual Positive Edge-triggered J-K Flip-flop with Preset and Clear
4028	BCD-to-Decimal Decoder
4029	Synchronous 4-bit Binary Up-down Counter with Preset
4030	Quad 2-input XOR gates
4031	64-bit Serial Shift Register
4032	Triple Serial Adder
4033	Decimal Up-counter with Seven Segment Decoder and Ripple Blanking
4034	8-bit Bi-directional Bus Register
4035	4-bit Parallel Input/Parallel Output Shift Register with Clear
4036	4 × 8-Bit RAM
4037	Triple AND/OR Combination Gates
4038	Triple Serial Adder (Negative Logic)
4039	4 × 8-Bit RAM
4040	Asynchronous Binary 12-bit Counter
4041	Quad True/Complement Buffers
4042	Quad D Flip-flops with Common Clock
4043	Quad NOR-gated R-S Flip-flops
4044	Quad NAND-gated R-S Flip-flops
4045	Asynchronous Binary 21-bit Up-counter
4046	Phased Locked Loop (PLL)
4047	Monostable/Astable Multivibrator
4048	8-input Multifunction Gate
4049	Hex Inverters/Buffers
4050	Hex Buffers/TTL Drivers
4051	8:1 Analog/Digital Multiplexer
4052	Dual 4:1 Analog/Digital Multiplexers
4053	Triple 2:1 Analog/Digital Multiplexers
4054	4-segment LCD Driver
4055	BCD-to-Seven Segment Decoder/LCD Driver
4056	BCD-to-Seven Segment Decoder/LCD Driver with Memory
4059	16-bit Programmable Divider
4060	Asynchronous Binary 14-bit Up-counter with Internal Oscillator
4062	200-bit Dynamic Serial Shift Register

4063	4-bit Comparator
4066	Quad Bilateral Analog/Digital Switches
4067	16-channel Analog/Digital Multiplexer/Demultiplexer
4069	Hex Inverters
4070	Quad 2-input XOR gates
4071	Quad 2-input OR gates
4072	Dual 4-input OR gates
4073	Triple 3-input AND gates
4075	Triple 3-input OR gates
4076	4-bit D-Latches, noninverting
4077	Quad 2-input XNOR gates
4078	8-input OR/NOR gate
4081	Quad 2-input AND gates
4082	Dual 4-input AND gates
4085	Dual 2-input AND/NOR Combination gates
4086	Dual 2 × 2-input Expandable AND/NOR Combination gates
4089	Binary Rate Multiplier
4093	Quad 2-input NAND Schmitt Trigger
4094	8-bit Shift Register with Output Latches
4095	Positive Edge-triggered J-K Flip-flop AND Input with Preset and Clear
4096	Positive Edge-triggered J-K Flip-flop AND Input with Preset and Clear
4097	Dual 8-channel Analog/Digital Input Multiplexers/Demultiplexers
4098	Dual Monostable with Schmitt Trigger Input and Clear
4099	8-bit Addressable D Latch
40014	Hex Schmitt Trigger
40085	4-bit Comparator
40097	Hex Buffers/Drivers
40098	Hex Inverters
40100	32-bit Left/Right Serial Shift Register
40101	9-bit Parity Generator/Parity Checker
40102	Synchronous 2-decade Down-counter with Preset
40103	Asynchronous 8-bit Down-counter with Preset
40104	4-bit Parallel-in/Parallel-out Left/Right Shift Register
40105	16 × 4-bit FIFO
40106	Hex Schmitt Trigger Inverters
40107	Dual 2-input NAND Drivers
40108	4 × 4 Multiport Register

40109	Quad Level Changer, noninverting
40110	Decimal Up-down Counter with Register, Seven Segment Decoder Driver
40117	Dual 4-bit Data Switch
40147	BCD Priority Encoder
40160	Synchronous Decimal Up-down Counter with Preset and Clear
40161	Synchronous 4-bit Binary Up-down Counter with Preset and Clear
40162	Synchronous Decimal Up-down Counter with Preset and Clear
40174	6-bit D Register with Clear, noninverting
40175	4-bit D Register with Clear, noninverting
40181	4-bit ALU
40182	Carry Unit for 40181
40192	Synchronous Decimal Up-down Counter with Preset
40194	4-bit Parallel-in/Parallel-out with Right/Left Shift
40195	4-bit Universal Shift Register
40198	Synchronous 4-bit Binary Up-down Counter with Preset
40208	4 × 4-bit Multiport Register
40240	8-bit Bus Line Driver with 2 Enable Inputs, inverting type
40244	2 × 4-bit Bus Driver with Separate Enable Input, noninverting
40245	Octal Bus Transceiver, noninverting
40257	Quad 2:1 Multiplexers
40373	8-bit D Latch with Enable, noninverting
40374	8-bit D Latch with Enable, noninverting
40511	BCD to Seven Segment Decoder/Memory/Driver (Hexadecimal Code)
4104	Quad TTL/CMOS Level Changer
4402	Dual 4-input NOR gate with Transistor Output
4412	Dual 4-input NAND gate with Transistor Output
4415	Quad Precision Timers/Drivers
4426	Decimal Up-counter with Seven Segment Decoder
4428	Binary to Octal Decoder
4433	Decimal Up-counter with Seven Segment Decoder and Ripple Blanking
4441	Quad True/Complement Buffers
4449	Hex True/Complement Buffers
4490	Hex Contact Bounce Eliminator
4500	1-bit Processing Unit
4501	Dual 4-input NAND gate, 2-input NOR/OR gate
4502	Hex Inverters/Buffers with Inhibit and Enable
4503	Hex Buffers/Drivers

4504	Hex TTL/CMOS or CMOS/CMOS Level Shifter
4505	64 × 1-bit RAM
4506	Dual Expandable AND/OR/INVERTER Combination gates
4507	Quad 2-input XOR gates
4508	Dual 4-bit R-S Flip-flops
4510	Synchronous Decimal Up-down Counter with Preset
4511	BCD-to-Seven Segment Decoder/Memory/Driver
4512	8:1 Data Selector/Multiplexer with Enable
4513	BCD to Seven Segment Decoder/Memory/Driver with Ripple Blanking
4514	4-bit Binary Decoder/Demultiplexer with Input Latch
4515	4-bit Binary Decoder/Demultiplexer with Input Latch
4516	Synchronous 4-bit Binary Up-down Counter with Preset
4517	Dual 64-bit Serial Shift Register
4518	Dual Synchronous Decimal Up-counter
4519	Quad Common Addressable 2:1 Multiplexers
4520	Dual Synchronous Binary Up-counter
4521	24-bit Binary Counter/Divider
4522	BCD Down-counter with Preset
4524	256 × 4-bit ROM
4526	4-bit Binary Down-counter with Preset
4527	Decimal Rate Multiplier
4528	Dual Post Triggerable Monostable with Clear
4529	Dual 4-channel Analog Multiplexers
4530	Dual 5-input Majority Logic Gate
4531	12-bit Parity Unit
4532	3-bit Priority Encoder
4534	Pental Count Decades with Multiplexed Outputs
4536	Programmable Timer
4537	256 × 1-bit RAM
4538	Dual Post-Triggerable Precision Monostable
4539	Dual 4:1 Data Selector/Multiplexer
4541	Programmable Timer with RC Oscillator
4543	BCD-to-Seven Segment Decoder/Memory/-Driver
4544	BCD-to-Seven Segment Decoder/Memory/Driver with Ripple Blanking
4547	BCD-to-Seven Segment Decoder/Memory/Driver
4548	Dual Post-Triggerable Precision Monostable
4549	8-bit Register for Successive Approximation in A/D Changer
4551	Quad 2-channel Analog Multiplexers
4552	64 × 4-bit RAM

4553	3-place Decimal Up-counter
4554	Dual 2-bit Parallel Binary Multiplier
4555	Dual 4-channel Demultiplexers
4556	Dual 4-channel Demultiplexers
4557	64-bit Presettable Serial Shift Register
4558	BCD-to-Seven Segment Decoder
4559	8-bit Register for Successive Approximation in A/D Changer
4560	4-bit BCD Adder
4561	9's Complementer
4562	128-bit Shift Register with Parallel Outputs
4566	Universal Time Base Generator
4568	Phase Comparator and Counter with Preset
4569	Dual Fast 4-bit Down-counter with Preset
4572	Quad INVERTERs, 2-input NOR gate and 2-input NAND gate
4573	Quad Programmable OPAMP
4574	Quad Programmable Comparators
4575	Dual Programmable OPAMP and Comparator
4578	Comparator with Voltage Follower
4580	4 × 4-bit Multiport Register
4581	4-bit ALU
4582	Carry Generator
4583	Dual Schmitt Trigger with Presettable Hysterisis
4584	Hex Schmitt Trigger Inverters
4585	4-bit Comparator
4597	8-bit D Latch with Address Counter Bus Compatible
4599	8-bit Addressable D Latch
4720	256 × 1-bit RAM
4723	Dual 4-bit Addressable D Latch with Clear
4724	8-bit Addressable D Latch
4731	Quad 64-bit Serial Shift Register
4737	4½ Decade Counter
4738	IEC/IEEE Bus Interface
4750	Frequency Syntheszer
4751	Universal Programmable 5-stage Frequency Divider
4752	AC Motor Control Circuit
4753	Universal Time Base Generator
4754	18 Elements Bar Graph LCD Driver
4755	Transceiver Serial Data Communication

❑ ❑ ❑

PIN CONFIGURATION
OF *4000* SERIES
INTEGRATED CIRCUITS

T his section presents the pinout configuration of some commonly used integrated circuits of 4000 series for ready reference. These may be helpful for logic designers. The pin configurations given here are for DIP packages only. Designers must refer to the data books of the manufacturers for other types of packages like SMD (Surface Mount Devices), etc.

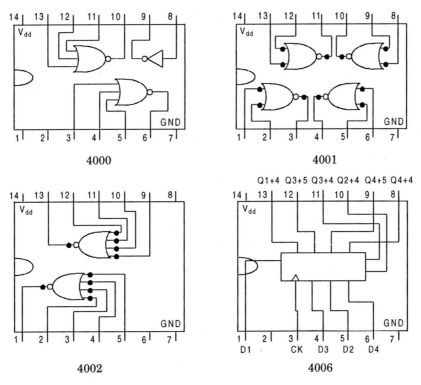

4000

4001

4002

4006

465

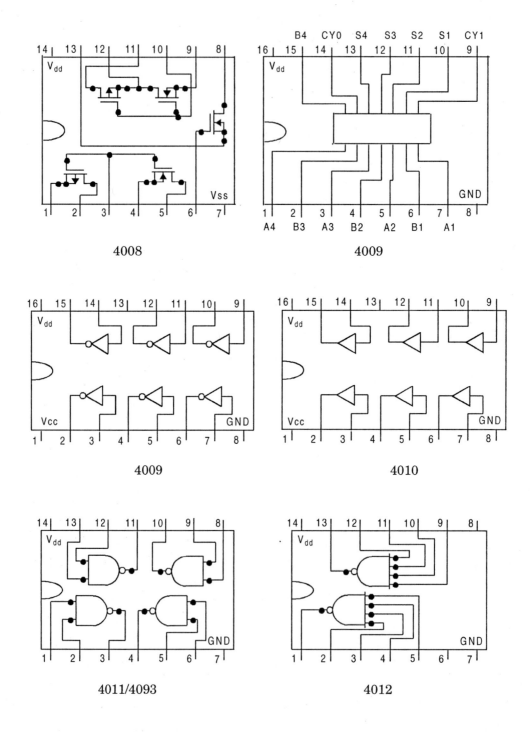

4008

4009

4009

4010

4011/4093

4012

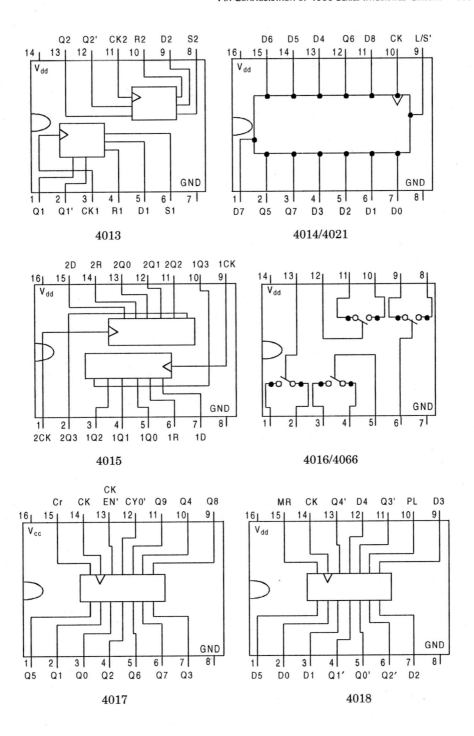

4013

4014/4021

4015

4016/4066

4017

4018

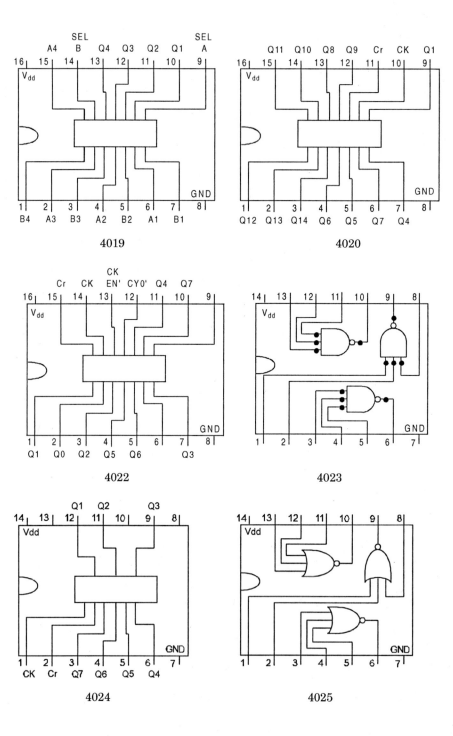

4019

4020

4022

4023

4024

4025

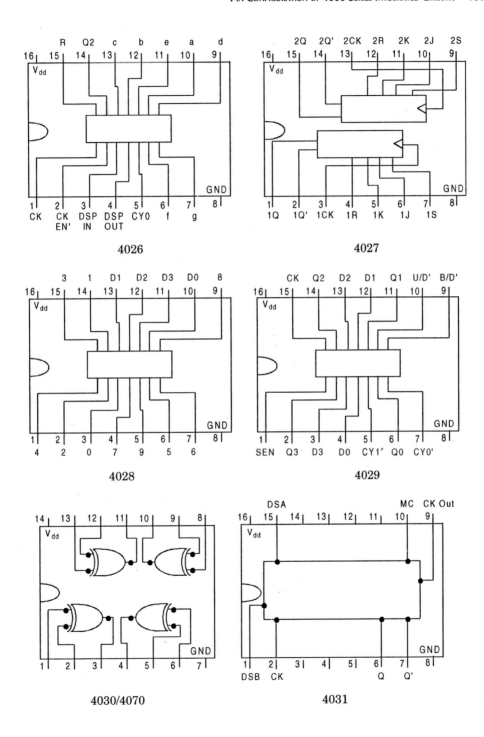

4026

4027

4028

4029

4030/4070

4031

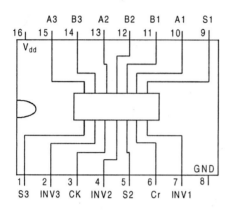

4032/4038

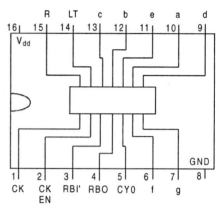

4033

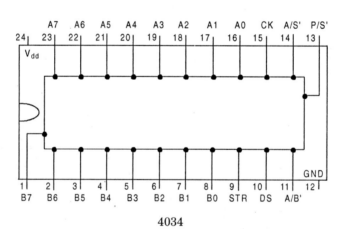

4034

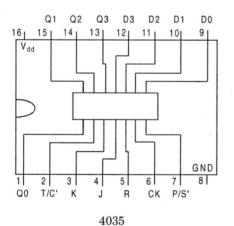

4035

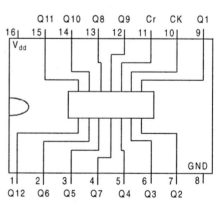

4040

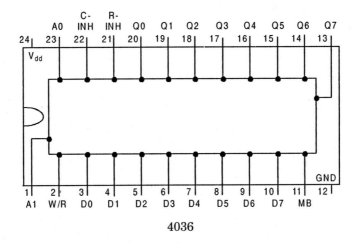

4036

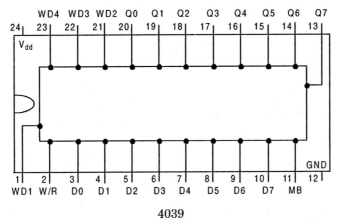

4039

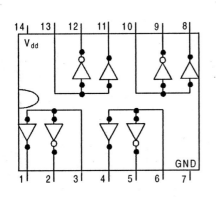

4041

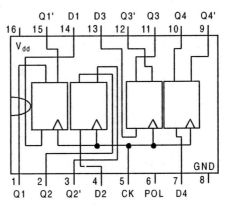

4042

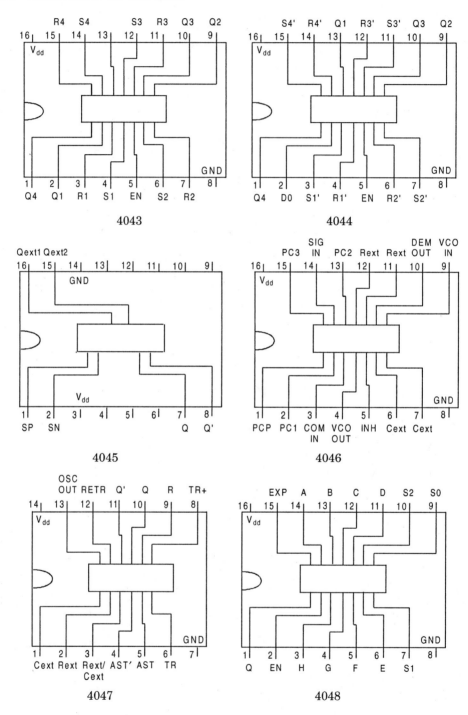

4043

4044

4045

4046

4047

4048

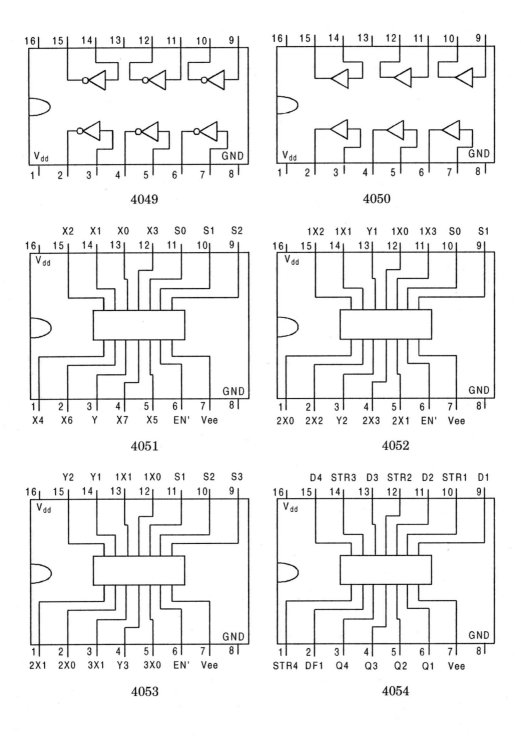

4049

4050

4051

4052

4053

4054

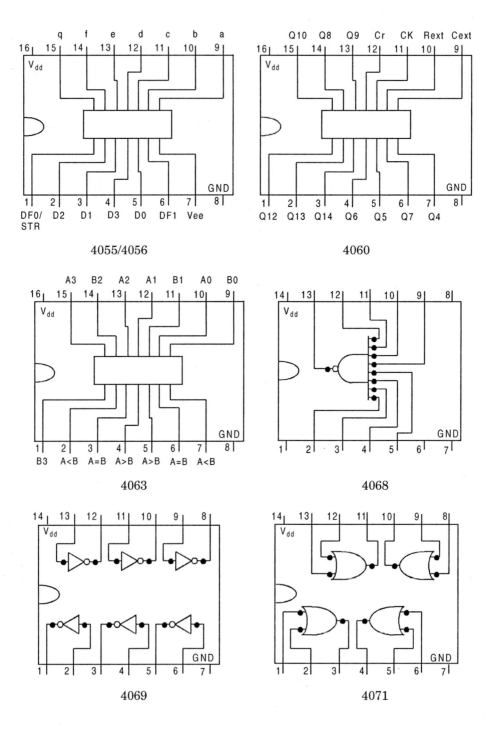

4055/4056

4060

4063

4068

4069

4071

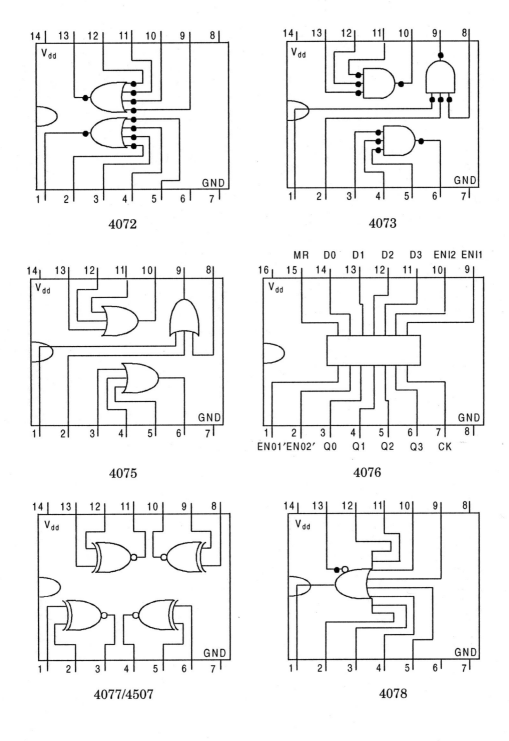

4072

4073

4075

4076

4077/4507

4078

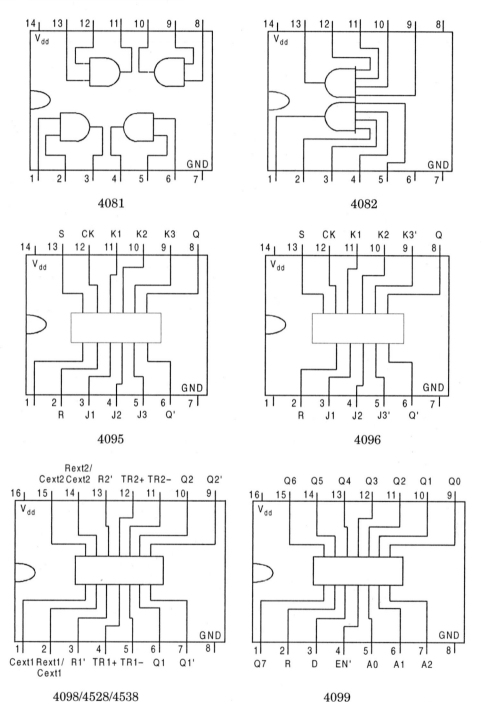

4081

4082

4095

4096

4098/4528/4538

4099

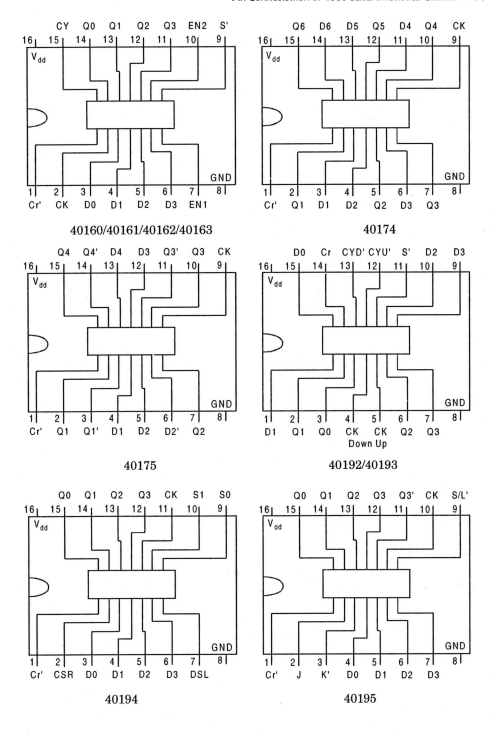

40160/40161/40162/40163

40174

40175

40192/40193

40194

40195

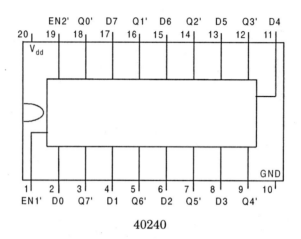

40240

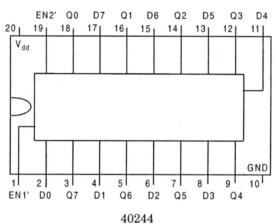

40244

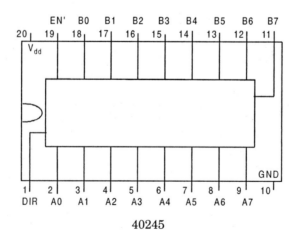

40245

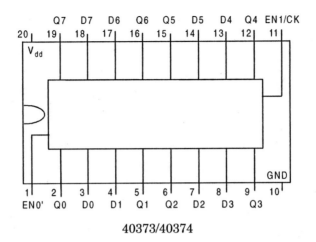

40373/40374

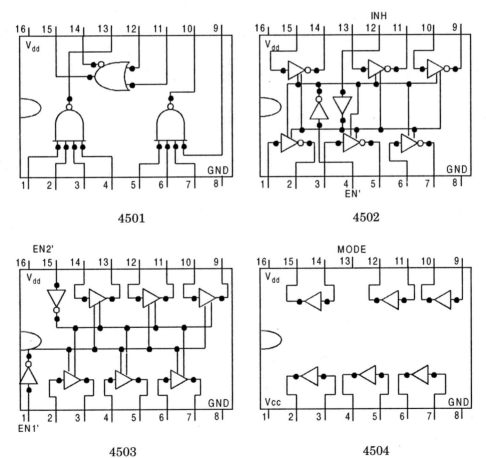

4501

4502

4503

4504

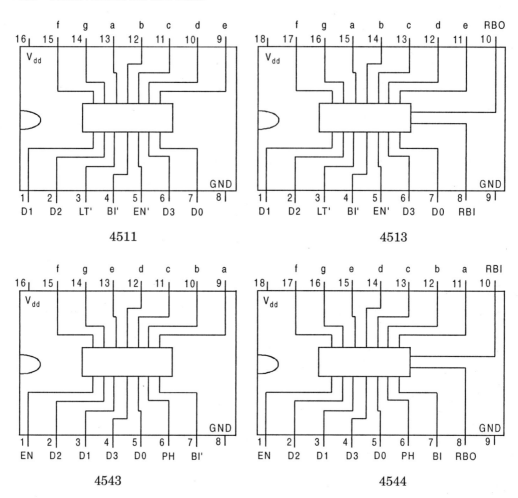

4511

4513

4543

4544

Appendix 6 ABOUT THE CD-ROM

- Included on the CD-ROM are simulations, figures from the text, third party software, and other files related to topics in digital principles and logic design.
- See the "README" files for any specific information/system requirements related to each file folder, but most files will run on Windows 2000 or higher and Linux.

GLOSSARY

Access Time: Time between the memory receiving a new input address and the output data's becoming available in a read operation.

Active Load: A transistor that acts as a load for another transistor.

Active Low: The low state is the one that causes the circuit to become active rather than the high state.

Active Power Dissipation: The power dissipation of a device under switching conditions. It differs from static power dissipation because of the large current spikes during output transitions.

ADC: Analog-to-digital converter.

Addend: Number to be added to another.

Alphanumeric Codes: Codes that present numbers, letters, punctuation marks, and special characters.

Arithmetic-Logic Unit (ALU): Digital circuit used in computers to perform various arithmetic and logic operations.

Assert: To activate. If an input line has a bubble on it, the input can be asserted by making it low.

Astable Multivibrator: Digital circuit that oscillates between two unstable output states.

Asynchronous Transfer: Data transfer performed without the aid of a clock.

Augend: Number to which the addend is added.

Bilateral Switch: CMOS switch which acts like a single-pole, single-through switch (SPST) controlled by an input logic level.

Binary Numbers: A number code that uses only the digits 0 and 1 to represent quantities.

Bipolar: Having two types of charge carriers; free electrons and holes.

Bistable Multivibrator: Name that is sometimes used to describe a flip-flop.

Bit: An abbreviation for binary digit. It combines the first letter of *binary* and the last two letters of *digit*.

Buffer/Driver: Circuit designed to have a greater output current and/or voltage capability than an ordinary circuit.

Buffer Register: Register that holds digital data temporarily.

Byte: A binary number with 8 bits.

Checksum: Special data word stored in the last ROM location. It is derived from the addition of all other data words in the ROM, and is used for error checking purposes.

Compatibility: Ability of the output of one device to drive the input of another device.

Contact Bounce: Opening and closing of a set of contacts as a result of the mechanical bounce that occurs when the device is switched.

DAC: Digital-to-analog converter.

Data Selector: A synonym for multiplexer.

Decoder: A circuit that is similar to demultiplexer, except there is no data input. The control input bits produce one active output line.

Demultiplexer: A circuit with one input and many outputs.

Differential Linearity: A measure of the variation in size of the input voltage to an A/D converter which causes the converter to change from one state to the next.

DIP: Dual-in-line package. This is the most common type of IC package.

Don't-Care Condition: An input-output condition that never occurs during normal operation. Since the condition never occurs, one can use an X on the Karnaugh map. This X can be a 0 or 1, whichever is preferable.

Edge Detector: Circuit that produces a narrow positive spike that occurs coincident with the active transition of a clock input pulse.

Encoder: Digital circuit that produces an output code depending on which of the inputs is activated.

EPROM: An erasable programmable read-only memory. With this device, the user can erase the stored content with ultraviolet light and electrically store new data.

Even Parity: A binary number with an even number of 1s.

Fall Time: The time required for a signal to transition from 90 percent of its maximum value down to 10 percent of its maximum.

Fan out: Maximum number of standard logic inputs that the output of a digital circuit can drive reliably.

Full-Adder: A logic circuit with three inputs and two outputs. The circuit adds 3 bits at a time, giving a sum and carry output.

Glitch: Very narrow positive or negative pulse that appears as an unwanted signal.

Half-Adder: A logic circuit with two inputs and two outputs. The circuit adds 2 bits at a time, giving a sum and carry output.

Hold Time: The minimum amount of time that data must be present after the clock trigger arrives.

Inhibit Circuits: Logic circuits that control the passage of an input signal through to the output.

Karnaugh Map: A drawing that shows all the fundamental products and the corresponding output values of a truth table.

LED: A light-emitting diode.

Logic Circuit: A digital circuit, a switching circuit, or any kind of two-state circuit that duplicates mental processes and behaves according to a set of logic rules.

Low-Power Schottky TTL (LS-TTL): TTL subfamily that uses the identical Schottky TTL circuit but with larger resistor values.

Low-Power TTL: TTL subfamily that uses basic TTL standard circuit except that all resistor values are increased.

LSB: Least significant bit.

Millman's Theorem: A theorem from network analysis which states that the voltage at any node in a resistive network is equal to the sum of the currents entering the node divide by the sum of the conductances connected to the node, all determined by assuming the voltage at the node is zero.

Modulus: Defines the number of states through which a counter can progress.

MSB: Most significant bit.

Multiplexer: A circuit with many inputs and one output.

Natural Count: The maximum number of states through which a counter can progress. It is given by 2^n, where n is the number of flip-flops in the counter.

Nibble: A binary number with 4 bits.

Octet: Eight adjacent 1s in a 2x4 shape on a Karnaugh map.

Odd Parity: A binary number with odd number of 1s.

Offset Error: Deviation from the ideal zero volts at the output of a digital-to-analog converter when the input is all 0s. In reality, there is a very small output voltage for this situation.

Overflow: An unwanted carry that produces an answer outside the valid range of the numbers being represented.

Overlapping Groups: Using the same 1 more than once when looping the 1s of a Karnaugh map.

Pair: Two horizontally or vertically adjacent 1s on a Karnaugh map.

Percentage Resolution: Ratio of the step size to the full-scale value of a digital-to-analog converter. Percentage resolution can also be defined as the reciprocal of the maximum number of steps of a digital-to-analog converter.

Priority Encoder: Special type of encoder that senses when two or more inputs are activated simultaneously and then generates a code corresponding to the highest-numbered input.

Product-of-Sums Equation: A Boolean equation that is the logical product of logical sums. This type of equation applies to an OR-AND circuit.

Quad: Four horizontal, vertical, or rectangular 1s on a Karnaugh map.

Quantization Error: The error inherent in any digital system due to the size of the LSB.

Redundant Group: A group of 1s on a Karnaugh map that are all part of other groups. Redundant groups may be eliminated.

Rise Time: The time required for a signal to transition from 10 percent of its maximum value up to 90 percent of its maximum.

SAR: Sequential approximation register, used in a sequential ADC.

Setup Time: The minimum amount of time required for data inputs to be present before the clock arrives.

Static Power Dissipation: The product of DC voltage and current.

Strobe: An input that disables or enables a circuit.

Sum-of-Products Equation: A Boolean equation that is the logical sum of logical products. This type of equation applies to an AND-OR circuit.

Timing Diagram: A picture that shows the input-output waveforms of a logical circuit.

Truth Table: Logic table that depicts a circuit's output response to the various combinations of the logic levels at its inputs.

Wired-AND: Term used to describe the logic function created when open-collector outputs are tied together.

BIBLIOGRAPHY

1. Taub. H., and D. Schilling, *Digital Integrated Electronics*, New York; McGraw-Hill Book Co., 1977.

2. Roth Charles H., Jr., *Fundamentals of Logic Design*, Mumbai, Jaico Publishing House, 2003.

3. Brown Stephen, and Vranesic Zvonko, *Fundamentals of Digital Logic With Verilog Design*, New Delhi, Tata McGraw-Hill Publishing Co. Ltd., 2002.

4. Malvino Albert Paul, and Leach Donald P., *Digital Principles and Applications*, New Delhi, Tata McGraw-Hill Publishing Co. Ltd., 1996.

5. Salivahanan S, and Arivazhagan S, *Digital Circuits and Design*, New Delhi, Vikash Publishing House Pvt. Ltd., 2005.

6. Mano. M. Morris, *Digital Logic and Computer Design*, New Delhi, Prentice Hall of India, 2001.

7. Tocci Ronald J., *Digital Systems Principles and Applications*, New Delhi, Prentice Hall of India, 2001.

8. Jain R P, *Modern Digital Electronics*, New Delhi, Tata McGraw-Hill Publishing Co. Ltd., 2001.

9. Lee, Samuel C., *Digital Circuits and Logic design*, New Delhi, Prentice Hall of India, 2001.

10. Peatman, J. P., *The Design of Digital Systems*, New York; McGraw-Hill Book Co., 1972.

11. *The TTL Data Book for Design Engineers*. Dallas, Texas: Texas Instruments Inc., 1976.

12. *MECL Integrated Circuits Data Book*. Phoenix, Ariz.: Motorola Semiconductor Products, Inc., 1972.

13. *RCA Solid State Data Book Series: COS/MOS Digital Integrated Circuits*. Somerville, NJ: RCA Solid State Div., 1974.

14. Givone, Donald D., *Digital Principles and Design*, New Delhi, Tata McGraw-Hill Publishing Co. Ltd., 2002.

❑ ❑ ❑

INDEX